BIORESOURCES FOR FOOD SECURITY AND RURAL LIVELIHOOD

BIORESOURCES FOR FOOD SECURITY AND RURAL LIVELIHOOD

Edited by :

G.K. Kulkarni
Department of Zoology,
Dr. B.A. Marathwada University,
AURANGABAD-431004 (M.S.)

and

B.N. Pandey
Department of Zoology,
Maghadh University
BODH GAYA-824231

NARENDRA PUBLISHING HOUSE
DELHI-110006 (INDIA)

First Published in 2010

ISBN: 978-93-80428-04-8

Published by :

NARENDRA PUBLISHING HOUSE
Publisher and Distributor
1417, Kishan Dutt Street, Maliwara,
DELHI-110006 (India)
Phones: 91-011-23268470, 91-011-23259412
E-mail: info@nph-india.com
Website: www.nph-india.com

Printed in India

Laser Typeset by Amrit Graphics, Shahdara, Delhi-110032

Contents

Preface

Over one billion people experience the hardship that hunger imposes, a figure which continues to rise even amidst the riches of the 21st century. Engulfed within a vortex of population growth, economic instability and climate change, food security has become the most intractable challenge for national and global governance. The feeble outcome of the 2009 World Summit on Food Security suggests that the richer countries have exhausted their energies and resources in rescuing the discredited banking sector

The UN Food and Agriculture Organization (FAO) released a report on 5 November 2009 entitled "Food Security and Agricultural Mitigation in Developing Countries: Options for Capturing Synergies," at the Barcelona Climate Change Talks.

The report calls for considering food security, agricultural mitigation, adaptation and development in global agendas and national strategies for climate change and food security, as well as up-front financing.

Biological diversity comprises countless plants that feed and heal people, many crop varieties and aquatic species with specific nutritional characteristics, livestock species adapted to harsh environments, insects that pollinate fields and micro-organisms that regenerate agricultural soils. Biodiversity, essential for agriculture and food production, is threatened by urbanization, deforestation, pollution and the conversion of wetlands. Due to agricultural modernization, changes in diets and population density, humankind increasingly depends on a reduced amount of agricultural biological diversity for its food supplies. A dozen species of animals provide 90% of the animal protein consumed globally and just four crop species provide half of plant-based calories in the human diet. FAO estimates that of 6 300 animal breeds, 1 350 are endangered or already extinct. This rapidly diminishing gene pool is cause for concern. Reduction of biodiversity entails a reduction of options for ensuring more diverse nutrition, enhancing food production, raising incomes, coping with environmental constraints and managing ecosystems. Recognizing, safeguarding and using the potential and diversity of nature is critical for food security and sustainable livelihood.

Since its founding more than 60 years ago, national and international action and advice on policies, programmes and research have been the driving forces of FAO's goal to achieve a food-secured world, free from hunger and malnutrition for present and future generations. The concept of 'food security' has developed over the past three decades. Concerns about food security up to the end of the 1970s were directed more at the national and international level, and concerned the ability of countries to secure adequate food supplies. The very first corporate strategy put forward in the Strategic Framework for FAO 2000-2015, "Contributing to the eradication of food insecurity and rural poverty" states FAO's aim of ensuring "sustainable rural livelihoods and more equitable access to resources". It is recognized that for people, countries and the world to be truly food secure at all times, our livelihood systems must be economically, socially and environmentally sustainable. FAO has developed and promoted a broad food security framework that identifies four key dimensions – availability, access, utilization and stability. A livelihoods approach applied to FAO's food security work has functioned well in both development and emergency contexts and has added value to the food security paradigm.

Biodiversity is the basis for human livelihoods and survival. We have lived as gatherers, hunters and fishing people during the largest part of our history. Later, agriculture and livestock keeping was based on a selection of the - for us -most useful species. Biodiversity is absolutely crucial for a sustainable food production to feed the growing population of the world. Livestock play an important part in securing the livelihoods of poor farmers around the world. Integrating the production of animals and plants can enhance yields of both and improve overall food security and income for rural livelihood.

We are deeply indebted to Dr. G. Madhavan Nair, General President, 97th Indian Science Congress held at Kerala University, Thiruvananthapuram from Jan. 3 to 7, 2010 for encouraging to hold a symposium with focal theme "Bioresources as a tool for food security and rural livelihood" in the section on Animal, Veterinary and Fishery Sciences. We are sincerely thankful for all authors who have contributed their articles for publication in this proceeding.

20.11.2009

G.K. Kulkarni

B.N. Pandey

List of Contributors

A. Moitra
Department of Zoology,
T.M. Bhagalpur University,
BHAGALPUR-812007, (Bihar)

A.A. Phadatare
Department of Animal Science and
Dairy Science, Post Graduate Institute,
Mahatma Phule Krishi Vidyapeeth,
RAHURI - 413 722, (Maharastra)

A.K. Jain
Department of Animal Genetic and Breeding,
Guru Angad Dev Veterinary and
Animal Science University,
LUDHIANA (Punjab)

A.K. Mathur
Department of Zoology,
Embroy Transfer on Frieswal Cow
E.T. Lab, Project Direction on
Cattle Grass Farm Road,
MEERUT-250001 (U.P.)

A.K. Srivastva
Department of Zoology,
National Dairy Research Institute,
KARNAL-132001 (Haryana)

A.K. Tripathi
Department of Zoology,
Magadh University,
BODH GAYA - 824234 (Bihar)

A.K. Tyagi
Department of Zoology,
National Diary Research Institute
KARNAL-132001 (Haryana)

Agniswar Sarkar
Central Inland Fisheries Research Institute,
BARRACKPORE-700120
Kolkata (W.B.)

Anu Yamuna Joseph
Molecular Biology Laboratory,
SCMS Institute of Bioscience and
Biotechnology Research and Development,
South Kalamassery,
COCHIN-682033, (Kerla)

Asheesh Shivam Mishra
Aquatic Biodiversity Unit,
Department of Zoology,
H.N.B. Garhwal University,
SRINAGAR-246174 (U. Khand)

Ashok K. Saxena
Department of Zoology,
D.A.V. College,
KANPUR-208001 (U.P.)

B. Indira
College of Fisheries,
Kerala Agricultural University,
PANANGAD-682506 (Cochin)

B. Bandhophadhyay
Oriental Institute of Science & Technology,
Vidyasagar University,
MIDNAPORE (W.B)

B.K. Behera
Central Inland Fisheries Research Institute
BARRACKPORE-700120
Kolkata (W.B)

B.K. Joshi
National Bureau of Animal Genetic Resources,
G.T. Road, bypass
KARNAL-132001 (Haryana)

B.N. Pandey
Department of Zoology,
Magadh University
BODH GAYA-824231

B.P. Mishra
National Bureau of Animal Genetic Resources,
G.T. Road bypass,
KARNAL-132001 (Haryana)

C. Kethirvelan
Department of Zoology,
Veterinary College and Research Institute,
NAMAKKAL (T.N.)

C. Kethirvelan
Department of Zoology,
Veterinary College and Research Institute,
NAMAKKAL (T.N.)

C. Mohankumar
Molecular Biology Laboratory
SCMS Institute of Bioscience and
Biotechnology Research and Development,
South Kalamassery,
Cochin-682033 (Kerla)

C. Suresh
Department of Biochemistry,
National Institute of Nutrition,
Tarnaka,
HYDERABAD-500 007 (A.P.)

Chittaranjan Baruah
Department of Zoology
Gauhati University,
GUWAHATI-781 014 (Assam)

D.K. Belsare
Member, Expert Committee,
Central Institute of Freshwater Aquaculture
Center
POWARKHEDA-461110 (M.P.)

D.K. Sharma
Department of Zoology,
Gauhati University,
GUWAHATI-781 014 (Assam)

D.M. Choudhari
Department of Animal Science and Dairy Science,
Mahatma Phule Krishi Vidyapeeth,
RAHURI-413 722 (Maharastra)

Debabrata Das
Central Inland Fisheries Research Institute,
BARRACKPORE-700120
Kolkata (W.B.)

G.K. Kulkarni
Department of Zoology,
Dr. B.A. Marathwada University,
AURANGABAD-431004 (Maharastra)

G.R. Parihar
Department of Zoology,
Jai Narain Vyas University,
JODHPUR-342001 (Raj.)

J.N. Khedkar
Department of Animal Science and Dairy
Science, Post Graduate Institute,
Mahatma Phule Krishi Vidyapeeth,
RAHURI-413 722 (Maharastra)

J.P. Shukla
Department of Zoology
S.H. Kisan P.G. College,
BASTI-272001 (U.P.)

Jyoti Verma
Aquatic Biodiversity Unit,
Department of Zoology,
H.N.B. Garhwal University,
SRINAGAR-246174 (U. Khand)

Jyotid D. Vora
Department of Biochemistry and Food Science
and Quality Control,
Ramnaraiu Ruia College, Matunga,
MUMBAI-400 019

K. Sivakumar
Department of Zoology,
Centre of Advanced Study in Marine Biology,
Annamalai University
PARANGIPETTAI-608502

K.V. Jayachandran
College of Fisheries,
Kerala Agricultural University,
PANANGAD-682506, Cochin (Kerala)

L. Kannan
Department of Zoology,
Centre of Advanced Study in Marine Biology,
Annamalai University
PARANGIPETTAI-608502

M. Bhaskar
Division of Animal Biotechnology,
Department of Zoology,
Sri Venkateswara University,
TIRUPATI-517502. (A.P.)

M. Murlidhar
Department of Zoology
Chanakya Mahavidyalaya,
BHOPAL-462026

M. Pandian
Department of Zoology,
S.S & L.S. Patkar College of Arts & Science
S.V. Road, Goregaon (West),
MUMBAI-400062

M.L. Mehra
Department of Animal Genetic and Breeding,
Guru Angad Dev Veterinary and
Animal Science University,
LUDHIANA-141001 (Punjab)

M.N. Saha
Ramel Biofertilizer and Pesticide Division,
Barasat,
Kolkata-700126

Mitalli Dhimani
Central Inland Fisheries Research Institute,
BARRACKPORE-700120 (Kolkata)

Mousumi Saha
Ramel Biofertilizer & Pesticide Division,
Barasat,
KOLKATA-700126 (W.B.)

Nilima Prabhu
Department of Zoology,
S.S & L.S. Patkar College of Arts & Science
S.V. Road, Goregaon (West),
MUMBAI-400062

P. Kathiravan
National Bureau of Animal Genetic Resources,
G.T. Road, bypass
KARNAL-132001 (Haryana)

P. Perumal
Centre of Advanced Study in Marine Biology,
Annamalai University,
PARANGIPETTAI-608502 (T.N.)

P. Santhanam
Department of Marine Science,
School of Marine Sciences,
Bharathidasan University,
TIRUCHIRAPPALLI-620 024 (T.N.)

P.D. Prasada Rao
Environmental Biotechnology Division,
National Environmental Engineering
Research Institute,
NAGPUR-440020 (Maharastra)

P.D. Raut
Department of Environmental Science, Shivaji
University,
KOLHAPUR-416012 (Maharastra)

P.K. Dubey
National Bureau of Animal Genetic Resources,
G.T. Road bypass
KARNAL-132001 (Haryana)

P.L. Paul
Department of Zoology,
Hislop College,
NAGPUR-440001 (Maharastra)

Prabhat K. Bajpai
Department of Zoology,
D.A.V. College,
KANPUR-208001 (U.P.)

Prahlad Dube
Biodiversity Research Laboratory
Department of Zoology,
Government College,
KOTA-324001 (Raj.)

Pramanick Kousik
P.G. Department of Zoology,
Asutosh College 92,
S.P. Mukherjee Road
KOLKATA-700026

R. Tahseen
Department of Zoology,
T.M. Bhagalpur University,
BHAGALPUR-812007 (Bihar)

R.S. Kataria
National Bureau of Animal Genetic Resources,
G.T. Road, bypass
KARNAL-132001 (Haryana)

Rachna Nautiyal
Aquatic Biodiversity Unit,
Department of Zoology,
H.N.B. Garhwal University,
SRINAGAR-246174 (U. Khand)

Ranjit Singh
Department of Zoology,
Magadh University
BODH GAYA-824231

S. Anisha
Molecular Biology Laboratory
SCMS Institute of Bioscience and Biotechnology
Research and Development,
South Kalamassery,
COCHIN-682033 (Kerla)

S.D. Belsare
Member, Expert Committee,
Central Institute of Freshwater Aquaculture
Center,
POWARKHEDA-461110 (M.P.)

S.K. Maurya
Department of Zoology,
Embroy Transfer on Frieswal Cows
E.T. Lab, Project Direction on
Cattle Grass Farm Road,
MEERUT-250001 (U.P.)

S.S. Lal
Department of Zoology,
Ch. Charan Singh University,
MEERUT-250005 (U.P.)

S.S. Mishra
Central Inland Fisheries Research Institute,
BARRACKPORE-700120
Kolkata

S.S. Pawar
Department of Zoology,
S.S & L.S. Patkar College of Arts & Science
S.V. Road,
Goregaon (West),
MUMBAI-400062

Sadhana Deshpande
Department of Zoology,
VP's ASC College
Vidya Nagari,
BARAMATI-413133

Sanjive Shukla
P.G. Department of Zoology,
B.S.N.V.P.G. College,
LUCKNOW-226001 (U.P.)

T. Thangarandjou
Department of Zoology,
Centre of Advanced Study in Marine Biology,
Annamalai University
PARANGIPETTAI-608502

T.K. Ghosh
Environmental Biotechnology Division,
National Environmental Engineering Research
Institute,
NAGPUR-440 020 (Maharastra)

U.D. Sharma
Prawns Research Centre,
Department of Zoology,
University of Lucknow,
LUCKNOW-226007 (U.P.)

V. P. Semwal
Aquatic Biodiversity Unit,
Department of Zoology,
H.N.B. Garhwal University,
Srinagar-246174 (U. Khand)

V.S. Somvanshi
A-1/701, Ridhi Gardens,
Film City Road,
Goregaon (E)
MUMBAI-400097

Bioresources for Food Security and Rural Livelihood, Pages 1–18
Edited by : G.K. Kulkarni and B.N. Pandey

CRUSTACEANS AS A SOURCE OF FOOD SECURITY FOR RURAL LIVELIHOOD

G.K. Kulkarni

ABSTRACT

India has witnessed two successful revolutions as far as Food Security and Rural Development are concerned. They are :

1. Green Revolution : Concerned with Agriculture Production.
2. White Revolution : Concerned with Milk Production.
3. Now India is embarking upon third revolution i.e. blue revolution which is concern with Aquaculture Production of
4. Shellfishes like Crustaceans [Prawns, Shrimps, Crabs, Lobsters etc.] and Mollusks [Bivalves, Oysters etc].
5. Fin fishes like major carps, etc.

As the Green Revolution was acclaimed as the means to end world hunger, the Blue Revolution often, is hailed as a way to increase incomes and the available supply of affordable food among the poor globally. India with an annual production of 70,000 mt of shrimp, occupied the 5th ranking in shrimp production in the world in 1998-1999.

Crustaceans are major constituents to aquatic ecosystems that provide a variety of ecological and economic services. Individual crustacean species are adept at occupying diverse niches and their success, in part, stems from neuroendocrine signaling cascades that regulate physiology in response to environmental and internal cues. Peptide hormones are major signal transducers in crustaceans. Neurohormones in crustaceans was first demonstrated by Koller (1925,1927) and Perkins (1928). Crustaceans possess both classical epithelial type endocrine glands (Y-organ, Androgenic gland, Ovary and Mandibular organs) and also those that are of neural origin (Eyestalks, Brain, Subesophageal ganglion and Thoracic ganglia). A variety of physiological processes are regulated by neurohormones including gonad maturation and gametogenesis. Reproduction is under the control of 2 antagonistic hormones e.g. gonad maturation inhibiting hormone from the eyestalk, and gonad maturation stimulating hormone secreted by the brain and thoracic ganglion. In crustaceans it was established that a substance from the neurosecretory system is a neurohormone by demonstrating its :

1. Presence in a neurosecretory system;
2. Release into the circulation;
3. Effectors that respond to the substance at its circulation concentration;
4. Sufficient stability in the hemolymph so that the substance is not degraded or inactivated before it reaches its target.

A few biogenic amines like 5-HT, NE, DA etc function mainly as neurotransmitters and neuroregulators stimulating the release of neurohormones. 5-HT stimulates the release of Gonad Stimulating Hormone from brain and thoracic ganglion. Among the vertebrate steroid hormones , estrogen, progesterone and testosterone were found to be present in gonads and stimulate gametogenesis and spawning in prawns and crabs. These findings could be of immense importance to boost aquaculture practices to meet challenges of 21st century for food security and rural livelihood.

Keywords : Crustaceans, Aquaculture, neurohormones, biogenic amines, steroids, genes

INTRODUCTION

Aquaculture refers to the culture of aquatic organisms including fish, molluscs, crustaceans and aquatic plants. Globally there is an increasing demand for commercially valuable aquatic products. The oceans cover over 70% of the earth's surface but, even so, fourteen of the sixteen major fishing areas in the world are over-fished. Traditional approaches, such as finfish, shellfish and seaweed farming, are helping to satisfy the demand and reduce the pressure on the wild stocks but these are restricted by natural maximum yields and the availability of suitable sites for aquaculture. One goal of fisheries management is to identify existing populations within a species, and control the harvest so that all populations, and the individual genetic characteristics they possess, are conserved. The differences among populations reflect the degree to which different spawning groups are isolated from each other and genetic diversity within a population is critical for the long-term survival of the fisheries they support.

According to FAO (2001) global production of cultured fish and shellfish more than doubled in weight and value between 1987 and 1999, from 15.5 million tons (mt) to 36 million mt and from US $ 24.4 billion to 50.4 billion $. While yields from capture fisheries reached a plateau around 1989. Yields from aquaculture have steadily increased both in absolute amounts as well as in the percentage of total food available. It has been pointed out that, India, after introducing the new liberal policy in late 1980s and early 1990s, launched its shrimp monoculture officially in the country in several states through the Coastal Aquaculture Projects of 1986 in its 6th 5 year plan (which was implemented during 1988-93). Thus, shrimp monoculture, and species specific commercial culture of brackish water shrimp gained momentum in India in the mid 1990s and the total potential area for brackish water shrimp culture was estimated to be 11,90,800 ha among which 13,816 ha was under cultivation in 2000. The contribution of shrimp to India's export reveals why the Government has taken such a promotional role to develop the industry. By 1997-98 India had registered the export turnover of Rs. 4,120 crores of which shrimp alone contributed 2,700 crores.

Currently aquaculture crops account for more than 25% of the total fish and shellfish consumed directly by humans. The high and rising trends of exports of prawns/shrimps has been visualized since late 1970s and early 1980s. These crustaceans are being produced from the sea, estuaries, brackish and fresh waters by both capture and culture methods. As many as 70% of the foreign exchange derived from marine exports has been obtained from exports of prawns/shrimps from India (Dogra, 1992a). It is further reported by Dogra (1992b) that the world shrimp farming produced of 6,90,000 mt comes mainly from the developing countries especially China (1,45,000 mt), Indonesia (4,40,000 mt), Thailand (1,10,000 mt), Ecuador (1,00,000 mt), India (35,000 mt) and Bangladesh (25,000 mt).

Contribution of aquaculture to human food is expected to expand significantly in future in order to meet the growing demand and anticipated declines in capture fisheries. The FAO predicts that aquaculture crops may increase to more than 50% of the total value of the global food fish catch by 2010 (Stonich & Bailey, 2000).

Shrimp farming in India is one of the success stories of modern aquaculture. In the eighties and nineties, industrialized shrimp aquaculture has witnessed unprecedented growth, and it has become one of the most important food production industries of the country. Total farmed shrimp production has increased from 30,000 mt in 1990 to 1,30,000 mt in 2005 (Anonymous, 2005). The area under shrimp farming during the same period expanded from 65,000 ha to about 1,60,000 ha. The export revenue of farmed shrimp crossed Rs. 4,000 crore mark during 2004-05 (Anonymous, 2005). These figures clearly signify the relevance of inputs of improved and new technologies in aquaculture. Indeed the predominant culture systems employed at a global level are semi-intensive, dealing with single species such as *Penaeus monodon*, *Litopenaeus* (= *Penaeus*) *vannamei* and *Fenneropenaeus* (= *Penaeus*) *indicus* As such there is need for more information on the production mechanisms especially the ecological, physiological and biochemical (including genetical), aspects concerning the different sp.

Shrimps are important components in capture and aquaculture. Demand for shrimp in the international markets has increased greatly in the past, leading to over-exploitation of the natural shrimp stocks in many parts of the world. Decline in capture fisheries triggered increased focus on aquaculture production as an alternative source for sustaining and expanding the seafood industry. Aquaculture has developed rapidly in the past two decades, with a present annual global production of 7,00,000 metric tons. The shrimp fishery in India contributes to about 13 % of the total marine fish landings, earning foreign exchange of Rs. 45000 million annually. In recent times, due to the shortfall in shrimp production from the seas, development of coastal aquaculture of shrimp gained importance

According to MPEDA (2005) India had 43,433 hectares of prawn ponds that produced 42,820 tons of prawns, and 1,40,000 hectares of Shrimp ponds that produced 1,43,000 mt of shrimps. According to Kutty (2009) the annual increase in farmed fresh water prawn production in the world during the decade ending 2001 has been estimated as 29% and during 1999-2001 as 48%, the production in 2001 being about 3,00,000 mt (all *Macrobrachium* sp). Between 1999-2003 the annual increase in farmed *M. rosenbergii* production in India was about 80%, production reached to 30,450 mt in 2002-2003.

Aquaculture technology for penaeid shrimps has evolved remarkably and at present the shrimps are cultured on a large scale. One of the most important aspects of any culture operations is controlled breeding to obtain seeds. Many shrimp hatcheries have been using the technique of eyestalk ablation for controlled breeding, leading to acceleration of gonad maturation, breeding and spawning. Neuroendocrine factors are known to control the process of gonad maturation.

In crustaceans, neurohormones were first demonstrated by Koller (1925,1927) and Perkins (1928) in different centers of nervous system (Fig. 1). The list of putative neurohormones in Crustacea is long, and the experimental evidence supporting their roles

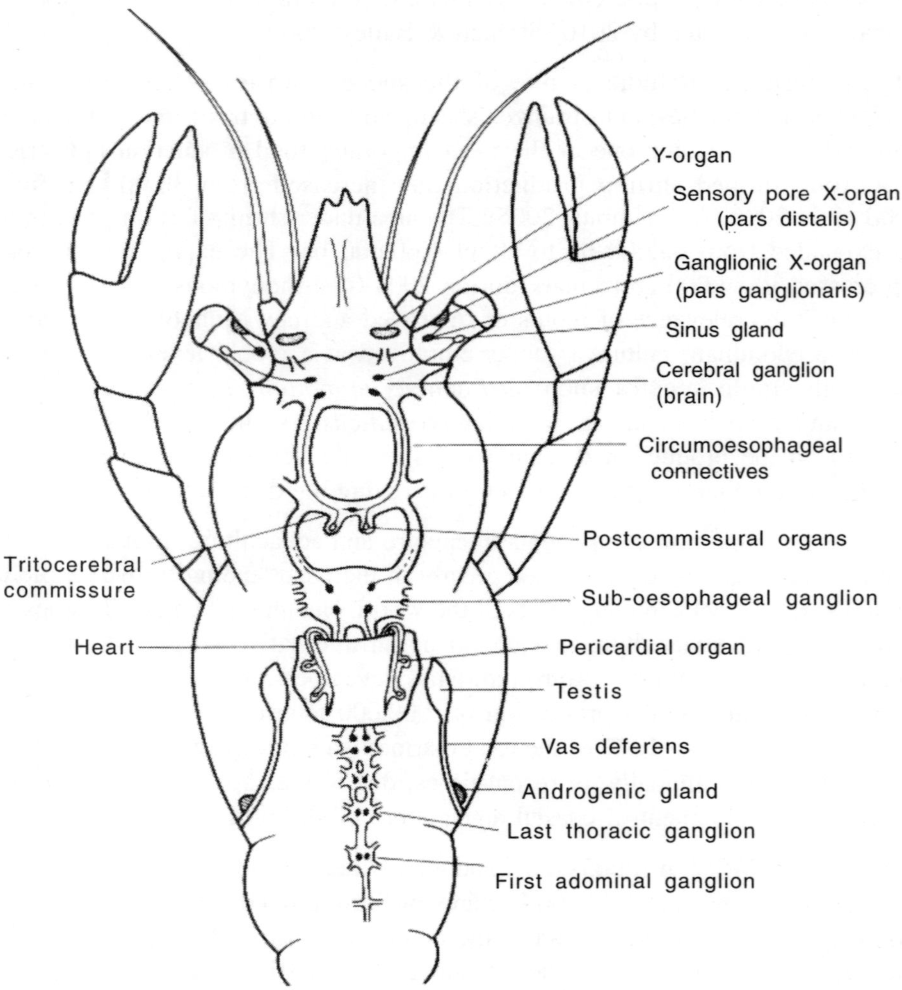

Fig. 1 : Neurosecretory centers in a typical crustacean (Prawn/shrimp)

as neurohormones varies considerably. To date, more than 20 candidate neurohormones have been identified in crustaceans. Crustacean endocrine glands secreting different hormones regulate a wide array of physiological activities. The neurohaemal tissue sinus gland located in the eyestalk is the most fully characterized in terms of its anatomy, physiology, ultra structure, and the chemistry of its peptide contents. The time-course and frequency of action potentials serve to regulate the amount of hormone release (Cooke, 1977). Ultra structural aspects of neurohormone release from the sinus gland terminals shows a process of exocytosis of granule contents .. In the crabs, *Paratelphusa jacquemontii* (Nirmal, 1964) and *Scylla serrata* (Deshmukh and Rangnekar, 1965) four types of neurosecretory neurons in the central nervous system are described.

NEUROENDOCRINE GLANDS

1. **Eyestalks** : These are paired structures attached to each optic lobes of the brain in many crustaceans. Hanstrom (1931) provided the first description of such a structure in decapod crustaceans, naming it the "X-Organ". It is a compact structure, consisting of a cluster of several closely packed lobules. The presence of sensory type cells and secretory granules suggested that this organ may have a dual function. This organ has a role as photoreceptor, chemoreceptor and baroreceptor. Durand (1956) reported 4 types of neurosecretory cells (NSCs) in the eyestalk of a crayfish, *Cambarus*, whereas 3 types (A, B, & C) of NSCs have been reported in eyestalks of the crab, *Barytelphus cunicularis* (Diwan and Nagabhushanam, 1975) . It is a sense organ secreting physiologically active hormones.
2. **Brain** : This is a fused bilobed structure located in the rostral region. It is connected to the eyestalks through optic nerve on the dorsolateral sides and to the thoracic ganglion through circumesophageal commissures. It bears different types of neurosecretory neurons. Enami (1951) was the first to report the occurrence of 3 types of NSCs in the brains of a crab, *Sesarma* and the crayfish, *Cambarus* whereas Durand (1956) observed 4 types of NSCs in the brain of the crayfish, *Cambarus*. 3 types of NSCs in the brain of the crab, *Barytelphusa cunicularis* (Diwan and Nagabhushanam, 1975) and 4 types of NSCs in the brain of the prawn, *Macrobrachium lamerri* (Sarojini *et al.*, 1990) have also been reported.
3. **Thoracic Ganglion** : It is a large mass formed by the fusion of all the ganglia from the mandibular to the first abdominal ganglion in prawns, crayfish etc, and to the last abdominal segment in case of crabs. These are distinguished by the presence of neuropiles in it. The secretory neurons are normally found outside the neuropiles. 3 types of NSCs in the thoracic ganglia of crabs, *Barytelphusa cunicularis* (Diwan and Nagabhushanam, 1975) and *Scylla serrata* (Deshmukh and Rangnekar, 1965) have been reported.
4. **The Neurosecretory Neurons** : The neurosecretory cells (NSCs) are the modified neurons, in which substances that may act as hormones can be identified by selective histochemical stains. The secretory material which is formed in the cell body may be

transported by axoplasmic flow either to be released into the circulatory system or to be stored in modified axonal terminals called neurohaemal organs (Carlisle and Knowles, 1953).

NON-NEURAL ENDOCRINE GLANDS

1. **Y-organs** : These are paired ectodermal glands, first described by Gabe (1953). In shrimps, they are attached to the epidermis in the maxillary segment of the head. In crabs, they are ovoid, compact structures lying in the anterior end of the gill chamber. These secrete molting hormone (MH) ecdysterone (Hopkins *et. al.*, 1979), steroidal in nature, and contain non innervated single type of cells. The cells change in size during the molting cycle. They are smallest during inter molt (stage C), and largest (almost double) in late pre molt (D3-4). They begin to increase early in pre molt (D0-D1) and to decrease at post molt (A-B) (Highnam and Hill, 1977).
2. **Androgenic Glands** : Cronin (1947) discovered this gland in male crab, *Callinectes* as a "ductless accessory gland", attached to the posterior vas deferens. Cells arranged in intertwining cords separated by connective tissues and blood sinuses. Charniaux-Cotton (1955) named it as 'androgenic gland' in an amphipod, *Orchestia gammarella.* Particularly in malacostracans (shrimps and prawns) it controls the sexual differentiation of males, including the testes. The androgenic gland of *Procambarus clarkii* contains 2 types of cells. Chemical nature of androgenic gland hormone is reported to be both steroidal and proteinaceous, Khlaila *et al* (2002) reported direct inhibitory action of sinus gland hormones on the androgenic gland in the crayfish, *Cherax quadricarinatus*, and suggested an endocrine axis-like relationship between the sinus gland, the androgenic gland, and the male reproductive system in decapod crustaceans. Okuno *et al*. (1999) and Ohira *et al*. (2003) reported that male sexual characteristics inducing androgenic gland hormone (AGH) was characterized from the terrestrial isopod, Armadillidium vulgare and its cDNA cloned to elucidate the structure of AGH. The molecular cloning of cDNAs encoding AGH precursors from two other isopods, *Porcellio scaber* and *Porcellio dilatatus* were also done. It is observed that AGH transcripts were detected only in cDNA synthesized using total RNA from the testis carrying the androgenic glands, but not in that from testis only, seminal vesicle, vas deferens or hepatopancreas.
3. **Ovary** : The crustacean ovary is a lobed structure, and shows changes in coloration as it matures. The ovaries of amphipods and isopods have been known for several years to be the source of a hormone that induces the formation of the female secondary sexual characteristics. In decapods, Nagamine and Knight (1965) showed that implantation of ovarian tissue into male prawns, *Macrobrachium. rosenbergii* whose androgenic glands had been removed, resulted in the appearance of female secondary sexual characters like ovigerous and ovipositing setae and a brood chamber. Chemically ovarian hormone is steroidal in nature. Estrogen and Progesterone like substances are reported.

4. **Mandibular Organs (MO)** : These ductless organs first described by Le Roux (1968) are located on the posterior side and at the base of the posterior mandibular adductor muscle, above the maxillary muscles. In crabs they are oval, spherical or crescent shaped depending upon the species. Each gland receives one or two branches from the sternal artery. The cells are 20-30 ìm with large oval or polymorphic nuclei, up to 18 ìm long, with one or two large nucleoli. Methyl farnesoate (MF) and Farnesoic acid are secreted by crab MOs and chemically are terpenoids. At least in females MO have a normal role of stimulating ovarian development and promoting vitellogenesis (Laufer *et al.* 1986). Laufer *et al.* (1998) also found stimulation of ovarian maturation in the crayfish *Procambarus clarkii* by exogenously injected MF. MF levels in the hemolymph of treated females undergoing vitellogenesis were found to be twice as great as those of the controls. Kalavathy *et al.* (1999) found stimulatory action of MF on testicular growth in the fresh water crab, *Oziotelphusa senex senex* evidenced by increased testicular weight, testicular index and testicular follicle diameter. These results strongly supported the concept that MF acts as a gonadotropin in crustaceans.

BIOTECHNOLOGICAL ADVANCEMENTS TO BOOST AQUACULTURE POTENTIAL

Aquatic Biotechnology is defined as the application of science and engineering in the direct or indirect use of aquatic organisms or parts or products of living aquatic organisms in their natural or modified forms. Biotechnology has the potential to help aquaculture to enhance a cultured organism's growth rate, reproductive potential, disease resistance, and ability to tolerate adverse environmental conditions. Aquatic biotechnology includes 3 major components :

- **Aquaculture biotechnology** (animal health and brood stock optimization);
- **Aquatic bioprocessing** (obtaining valuable compounds from aquatic organisms);
- **Aquatic bioremediation** (use of microorganisms to degrade toxic chemicals in the aquatic environment).

Implications of Neuroendocrine Manipulations (Ablation and Reinjection/Reimplantation Therapies) (Model – 1)

The eyestalks of decapod crustaceans contain Gonad Inhibiting Hormone (GIH) that inhibits vitellogenesis in the ovaries (Charniaux-Cotton, 1955). After breeding season its activity is highest, and in the early part of the breeding season its activity is absent (Klek-Kawinska and Bomirski, 1975; Kulkarni *et al.*, 1991). GIH from eyestalks of *Cancer magister, Panulirus argus* and *Uca pugilator* has MW between 2000 to 5000 Daltons. Inhibitory action of eyestalk hormones on oogenesis has been confirmed in *Uca, Rhithrophoropeus, Crangon*, and *Macrobrachium lamerrii*, *Procambarus clarkia*

(Kulkarni *et al.*, 1991) and *Parapenaeopsis hardwickii* (Kulkarni *et al.*, 1979, Nagabhushanam *et al.*, 1980).

Vitellogenesis Inhibiting Hormone (VIH) of Crustacea is in three forms . (1) VIH from eyestalks of male crab *Cancer magister* tested upon female shrimp *Crangon crangon* has M.W. 2000 Dalton. (2) VIH from eyestalks of lobster *Panulirus argus* bioassayed in the female crab *Uca pugilator* has M.W.- 5,000 Daltons. (3) VIH from sinus glands of *Homorus gammarus* tested on the female prawn, *Palaemonetes varians* has MW 7000 to 8000 Dalton.

Aguilar *et al.* (1992) purified a neuropeptide with VIH activity from sinus gland (SG) of the crayfish, *Procambaurs bouviert* and the yield was 5.4 ng/SG with a mol. Wt. of 8388 daltons.

In contrast the brain and thoracic ganglion of several species appear to contain gonad-stimulating hormone (GSH) (Otsu, 1964; Kulkarni and Nagabhushanam, 1980; Kulkarni and Nagabhushanam, 1981; Kulkarni *et al.*, 1984). Ngernsoungnern *et al* (2008) demonstrated the existence of GnRH-like peptides in the central nervous system (CNS) and ovary of giant freshwater prawn, *Macrobrachium rosenbergii* using immunocytochemistry. The immunoreactivity (ir) of lamprey (L) GnRH-III was detected in the soma of neuronal cluster number 11 in the middle part of supraesophageal ganglion, whereas ir-octopus (oct) GnRH was observed in the soma of neurons in thoracic ganglia. In addition, ir-L-GnRH-I was observed in the cytoplasm of late previtellogenic oocyte and early vitellogenic

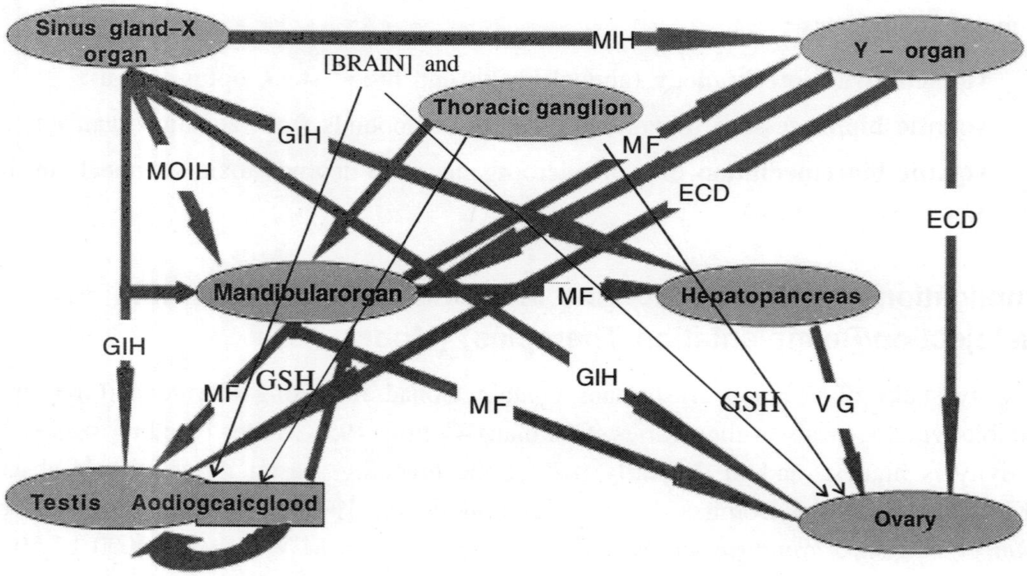

Model 1 : Blue arrows are stimulatory and red are inhibitory. (See colour vew also).

oocyte. These data suggest that *M. rosenbergii* contain at least three isoforms of GnRH: two GnRH isoforms closely related to L-GnRH-III and oct-GnRH in the CNS, whereas another isoform, closely related to L-GnRH-I, was localized in the ovary, supporting the results that ir-GnRH-like peptides exist in the CNS of decapod crustaceans. Liu *et al.* (2009) evidenced the presence of a putative egg-laying hormone, which has been previously described in the shrimp, *Penaeus monodon*, in four additional crustaceans like a crayfish *Cherax destructor*, rock lobster *Jasus edwardsii*, snow crab, *Chionoecetes opilio* and the blue swimmer crab, *Portunus pelagicus*. Female cerebral ganglia, ventral nerve cord and gonads were studied for the distribution of the immunoreactive hormone. Immunoreactivity was found in all tissues, and distribution patterns showed similarity within the four species. It is concluded that putative egg-laying hormone like neuropeptide is wide spread in crustaceans.

Methods to remove eyestalk : Generally eyestalks are cut at the base and the wound is sealed instantly by ice cube (cold cautery) or with electrocautery (hot cautery) to stop loss of blood. Preferably electrocautery seals the cut more quickly, and assures cent per cent survival.

Several species of prawns mature after eyestalk ablation and repeatedly spawn viable eggs. In one intermolt period of 20-30 days, *Penaeus monodon* spawn upto 6 times, at intervals of 3-5 days without appreciable decline in the number of eggs spawned; but hatching rate declined in later spawnings.

Penaeus kerathurus spawns upto 8 times after eyestalk ablation at intervals of less than 10 days with no reduction in the number of eggs even after repeated spawning (Lumare, 1979).

Preference of unilateral eyestalk ablation to induce maturation : The lack of success with bilateral eyestalk ablation may be attributed to the fact that crustacean eyestalk produces, apart from the GIH, a number of neurosecretory hormones which :

1. Regulate lipid metabolism and protein synthesis in the hepatopancreas;
2. Induce hyperglycemia in the blood to combat stress;
3. Regulate calcium metabolism during cuticle formation in molting;
4. Affect the water balance during ecdysis;
5. Inhibit the production of molting hormones (MIH) by the Y-organ;
6. Influence the movements of pigments in the chromatophores.

Since so many physiological functions are affected by bilateral eyestalk ablation, it is bound to have repercussions in the maturation of the gonads also. Therefore, unilateral eyestalk ablation was resorted to by lateral workers, since the presence of one eyestalk seems to ensure the normal functioning of all the metabolic process.

Implications of Pharmacological Agents in Gonad Maturation (Model- 2)

Well established neurotransmitters in crustacean central nervous system are acetylcholine (Ach), gamma amino butyric acid (GABA), Glutamate, Octopamine (OA), Dopamine (DA), 5-Hydroxytryptamine (5-HT), Epinephrine (E) and Norepinephrine (NE) which are playing important role in various physiological activities as neuroregulators. Among the candidate neurotransmitters tested for possible roles in crustacean reproduction, 5-HT, DA and OA were found promising. 5-HT immunoreactive neurons were found in 3 (ME, MI and MT) out of 4 optic ganglia in eyestalks . The brain, thoracic ganglion and every ganglion in the ventral nerve cord displayed at least a few immunoreactive cell bodies. In addition to identification and localization of 5-HT immunocytochemically, precise quantitative measurement of 5-HT using RP-HPLC in various parts of nervous system has been reported (Kulkarni and Fingerman, 1992 a,b). Stimulation of the release of GSH is reported in crabs, *Uca pugilator* (Kulkarni *et al.*, 1992; Sarojini *et al.*, 1993), *Barytelphusa cunicularis* (Parvathi and Kulkarni, 2001) and the crayfish, *Procambarus clarkii* (Kulkarni *et al.*, 1992).

A series of experiments revealed that 5-HT stimulates ovarian and testicular development into marine and freshwater crabs, and crayfishes. An increased and dose-dependent gonadal maturation has been reported. Supporting evidence for this hypothesis was obtained by determining the effects of two 5-HT agonist (fenfluramine-5-HT releaser; fluoxetine-5-HT potentiator) and antagonist (5,6-DHT) on the gonadal maturation and development of crab and crayfish (Kulkarni and Fingerman, 1986; Sarojini *et al.*, 1993). 5-HT alone and 5-HT in combination with fenfluramine and fluoxetine stimulated gonadal maturation and development, whereas 5,6-DHT alone and even in the presence of 5-HT has inhibited the gonadal maturation. It has been proposed that 5-HT does not act directly on the gonads, instead stimulates the release of GSH from brain and thoracic ganglion. The crayfish given DA plus 5-HT had significantly smaller OI than the crayfish that received 5-HT alone, but comparatively large one than the crayfish that received DA alone. It is hypothesized that 5-HT exerts its effect on gonads by releasing GSH. Sarojini *et al.* (1994) observed that the inhibitory action of DA on ovarian maturation could have been due to I) Inhibition of GSH release; II) Stimulating release of GIH from eyestalk and III) both I and II . DA when injected alone into female crayfish, *P. clarkii* inhibited ovarian maturation with smaller ovarian index. Chen *et al.* (2003) observed in the fresh water prawn, *Macrobrachium rosenbergii* that DA depressed vitellogenin synthesis while 5-HT enhanced the process. DA was able to inhibit vitellogenin synthesis in eyestalk-ablated prawns also as in intact prawns. The inhibitory action of DA is at the thoracic ganglia through inhibition of vitellogenesis-stimulating hormone release, but not at the eyestalk level through stimulation of VIH release from the X-organ-Sinus gland complex.

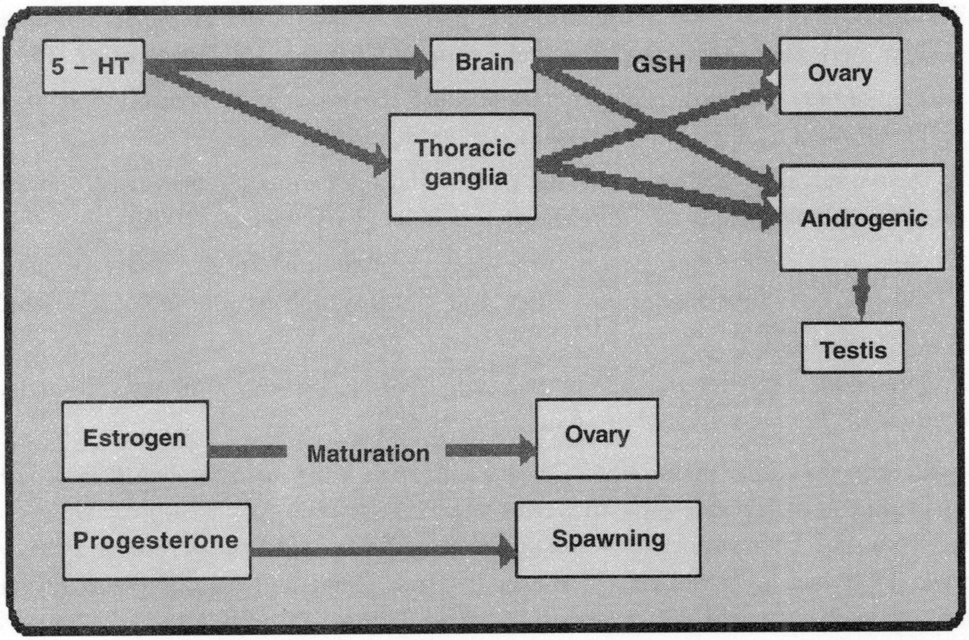

OA has been detected particularly in the contraction of ovarian wall liberating the oocytes (Fingerman & Nagabhushanam 1997). RPCH, in addition to its role in regulating pigmentation, appears to stimulate Methyl Farnesoate (MF) synthesis in the MO of *P. clarkii*. MF is involved in the regulation of crustacean reproduction. Mancillas *et al* (1981) have found immunoreactivity for opioids like Leu-Enk and Met-Enk in retinular cells, follicle cells and avitellogenic and early vitellogenic oocytes (but not in fully ripe oocytes), testis, spermatogonia, spermatocytes and spermatids (but not in mature spermatozoa) of spiny lobster, *Panulirus interruptus* and crayfish *P. clarkii*. Injection of synthetic Met-Enk into female fiddler crab, *U. pugilator* significantly slowed ovarian maturation, which was dose-dependent. In contrast injection of opioid antagonist, nalaxone, produced dose-dependent ovarian maturation. It was hypothesized that the opioids (1) stimulate GIH release; (2) inhibit GSH release or (3) do both 1 and 2. (Fingerman et.al. 1994)

Vertebrate Steroids Hormones and their Implications in Gonad Maturation (Model – 3)

(A) Occurrence

- α-estradiol and estradiol-17β in ovaries of *H. americanus* (Lisk, 1961).
- Progesterone, testosterone and estrone in whole body extract of *Euphausia superba* (Nikitina *et al.* 1977).
- Estrone and Estradiol-17β in ovaries of shrimp *Parapenaeus fissurus* (Jeng *et al.* 1978).

- Testosterone in the testis and androgenic glands of the lobster *Homarus americanus* (Gilgan and Idler, 1967) and the crab, *Carcinus maenas* (Blanchet *et al.* 1972).
- Pregnenolone, 17α-hydroxypregnenolone, and testosterone in hemo-lymph of both sexes of *Astacus leptodactylus* (Ollevier *et al.* 1986).
- Progesterone and estradiol-17β in the mandibular organ, green gland, hepatopancreas, ovary and hemolymph of *Homarus americanus* (Couch *et al.* 1978, 1987).
- 5α-dihydroxytestosterone, testosterone and pregnenolone in the ovary of shrimp, *Nephrops noregicus*, whereas eggs and hemolymph contained 17β-estradiol (Fair *et al.* 1989).
- Pregnenolone, dehydroepiandrosterone, progesterone, 17β-estradiol and estrone in the ovaries of *Penaeus monodon* (Fair *et al.* 1990).

Induction of ovarian maturation in *Parapenaeopsis hardwickii* (Kulkarni *et al.* 1979) and spawning in *Parapenaeopsis stylifera* (Nagabhushanam *et al.,* 1980) by injection of progesterone has been reported. Testosterone propionate and testosterone acetate were observed to induce hyperactivity in the androgenic gland and resulting into testicular maturation in the panaeid prawn, *Parapenaeopsis hardwickii* (Table – 1: Kulkarni and Nagabhushanam, 1981).

(B) Effects of progesterone on induction of ovarian maturation and spawning of shrimp, *Metapenaeus ensis*.

Treatment	Body length (cm)	Body weight (g)	Number of females showing various stages of ovarian development at each sample time				Shrimp Mortality (%)
			Day 0A	Day 14	Day 21	Day 28	
			I II III IV VB	I II III IV V	I II III IV V	I II III IV V	
Progesterone (0.1 μg/g body weight)	11.1 – 12.1	19.8 – 26.5	15 0 0 0 0 0	5 3 3 0 0	2 3 3 2 0	2 2 4 0 2C	33.3
Control (0.1 μl pure ethanol/g body weight)	11.5 – 12.2	21.3 – 26.4	15 0 0 0 0	7 6 0 0 0	2 9 0 0 0	4 7 0 0 0	26.7

A - Day after injection

B - Stages of ovarian development

C - Spawning occurred nocturnally 30 and 31 days after injection.

No. of shrimps: 15; No. of injection : 01

(C) Effects of 17 α-hydroxy-progesterone on Vg secretion in kuruma prawn, Penaeus japonicus

Treatment	Body length (cm)	Body weight (g)	Vg concentration in sera (μ g/m l)		Rate of Vg increase in sera x ± SD (Va/g)
			Initial (Vi) x ± SD	48 h afte injection (Va) x ± SD	
Control I (no injection)	14.1 – 15.5	34.0 – 44.3	1.74 ± 0.87	5.72 ± 3.87	3.04 ± 1.00
Control II (0.1 μl pure ethanol)	14.1 – 15.7	34.4 – 49.2	5.52 ± 2.42	12.82 ± 7.95	2.42 ± 1.16

Table 1 : Changes in the endrogenic gland of *P. hardwickii* after different treatments

	Control		TSP		TSA	
	10 days	20 days	10 days	20 days	10 days	20 days
Cell size (μ m) ± S.D.	1.5	1.8	3.4 ±	4.9	3.1	4.6
	0.2	0.4	0.7	0.6	0.6	0.8
Cytoplasm	H	H	GR	GR	GR	GR
Nucleus size (μ m) S.D.	0.6	0.7	1.8	2.3	1.6	2.1
	0.02	0.04	0.1	0.3	0.5	0.4

H = homogenous; GR = granular; S.D. = standard deviation

Genetic Engineering (Model – 4)

Lutz (2003) stated that spermatophore microinjection appears to be a viable means of introducing genetic constructs into *Macrobrachium* sp. Gene transfer may therefore complement the promise of "Conventional" selection and hybridization as a means of improving performance in fresh water prawn culture. According to Lutz (2003) the most important traits that might be improved are, in addition to growth rate, cold tolerance, the duration of larval development and the tolerance of reduced salinity during the larval cycle. Baghel & Saxena (2001) proposed a number of biotechnological tools to enhance freshwater prawn production. These include chromosomal engineering, molecular genetics, genetic engineering and tissue culture engineering. Noting that fresh water prawns were less susceptible to the white spot syndrome virus than marine shrimps, these authors believe that gene transfer should be used to produce strains having a growth hormone gene construct.

Dr. Tom Chen (Personal communication) , Director of the University of Connecticut Biotechnology Center, has transplanted a foreign gene of fish Growth Hormone into a crayfish for the first time in history! The gene inserted was a "reporter" gene - a gene that is easy to detect and confirm in the animal's cells. The gene was directly inserted into the immature gonads, using a small, sharp knife. A small hole was made in the exoskeleton of an immature crayfish near its thorax, where the testes and ovaries are located. Then delivered a new gene, in the form of a "pantropic retroviral vector" - a small segment of the desired DNA encased in the shell of a virus. The new gene then appears in the eggs or sperm and is perpetuated in subsequent generations.

While feeding animal hormones can accomplish the goals of faster growth and enhanced reproduction, a growth hormone gene successfully transferred to an animal allows the animal itself to produce the hormone, and the trait may be carried on to its offspring. Chen explains that transplanting genes in crustaceans is difficult because it requires access to unfertilized eggs, and the eggs laid by wild females are usually fertilized by the time they are collected.

Chen explains that transplanting genes in crustaceans is difficult because it requires access to unfertilized eggs, and the eggs laid by wild females are usually fertilized by the time they are collected.

EPILOGUE

The shrimp farming experiences have made the aqua farmer and other stakeholders aware that while aquaculture has a high potential in ensuring food and nutritional security and livelihood, besides profit to the farmer and socio-economic benefits to local communities, it is likely to slip from sustainable development pathways if adequate understanding, improved management practices, governmental policies and plans for implementation and regulation to ensure sustainability are not in place.

Shrimp farming has played a big part in uplifting rural living standards and reversing the trend to urban migration from rural area. Rural livelihood is at the mercy of economic realities in the international market places, which has become increasingly competitive in recent years. Hopefully a viable domestic market can be developed in India in the future, as purchasing power rises among more people. This will give us another sales outlet, which should mean better prices and living for farmers.

Aquaculture as a major production sector is still in it's infancy in India, and it has to be guided and supported to grow and play a useful role to ensure food security for rural livelihood. The fisheries play a significant role in meeting the people's nutritional requirements by augmenting food supply, earning foreign exchange and generating employment.

- 75% of edible crustacean population today in India is from capture fisheries. There is stagnation in production from this practice due to over exploitation of near-shore resources.
- 25% of crustacean population is today contributed from culture fisheries, which is quite inadequate to meet growing demands for protein-rich diet.
- Harnessing the biological wealth from the aquatic environment by way of biotechnological means will not only help to increase the production but it will be on a sustainable basis.

One of the major problems in the field of aquaculture technology is to obtain gametes, because most of the commercially imp. sp are incapable of spontaneous maturation under artificial conditions.

Presently most employed method of stimulating maturation is the eyestalk ablation, which occasionally results in high mortality of precious species. It is high time that an alternative technologies be evolved for the acceleration of gonadal maturation by simply injecting/feeding micro quantities of purified hormones/ or stimulatory neurotransmitters.

The concept of neurohormones and neuroregulators has shown promising hopes in acceleration of gonadal maturation in crustaceans. These hormonal occurrences and their implications in regulating reproductive activities in crustaceans appear to be rivaling the vertebrate endocrine system in complexity. Much work is required to know about their roles other than the known one, because known actions are tip of the iceberg.

Neuroendocrine manipulations, injections of a few biogenic amines (5-HT, dopamine, naloxon) , vertebrate type hormones (estrogen, progesterone & testosterone) and gene therapy have proved experimentally to accelerate gonadal maturation and spawning in crustaceans.

These can be employed as an alternative tools to boost aquaculture potential of crustaceans for sustainable rural development and to ensure food security for mankind for which there is a need to develop effective delivery technique to be employed in a large scale aquaculture practice.

ACKNOWLEDGEMENTS

I am grateful to Prof. R. Nagabhushanam and Dr. Mrs. R. Sarojini (Aurangabad), Prof. T. Subramoniam (Chennai), Prof. S.N.Dwivedi (Ex-Director, CIFE, Mumbai), Prof. P.V. Dehadrai (Ex-DDG, ICAR, New Delhi), Late Prof. K.G. Adiyodi (Kochi), Prof. Milton Fingerman (New Orleans, USA), Prof. Gerald Bakus (Los Angeles, USA) Prof. Tom Chen (USA), Prof. D. Borst (USA), and Prof. Amir Sagi and Dr. Tom Moshe (Israel), for providing guidance and necessary academic support. I am thankful to Dr. Satish

Mokashe (Lecturer in my dept), Dr. G. R. Bharaswadkar, Mr. Atul R.Chourpagar, Ms. Rumana Shaikh, Mr. Praveen Nalawade, Mr. Pappu Kudnar, Mr. Anant Harkal and Mr. Gajanan Sontakke (all research students) for their untired technical support.

REFERENCES

Aguilar M. B., Quackenbush L.S., Hund, D. T., Shabanowitz, J. and Huberman, A. Comp. Biochem. Physiol. 102B **(3):** 491-498,1992.

Baghel and Saxena 2001: CF. News of Scientific and Research Importance 2000-2003.

Blanchet M. F., Junera H. and Meusy J.J. Experientia, **31:** 865-867. 1975.

Carlisle D. B and F. G. W. Knowles Nature, **171:** 404, 1953

Chen Y. N., Fan H. F., Hsieh S. L. And Kuo C. M. Aquaculture, 228: 383-395, 2003

Cooke I. M. In: Peptides in Neurobiology, (H. Gainer, ed; Plenum Press, New York). pp 345-374, 1977

Couch E. F., Adejuwon C. A. and Koide S. S. Am. Zool. **18:** 610. 1978

Couch E. F., Hagino N. and Lee J. W. Comp. Biochem. Physiol., **87A:** 765-770. 1987

Cronin L. E. J. Morphol. **81:** 209-239, 1947.

Deshmukh P. B. and P. V. Rangnekar J .Biol. Sci. **8:** 16-25, 1965..

Diwan A. D and R. Nagabhushanam, Rev. di Biol. **68:** 79-99, 1975.

Dogra 1972a. CF. S. Pattanaik Inst. Economic Growth, Univ. Delhi. Working paper Series No. E/274/2006.

Dogra 1972b. CF. S. Pattanaik Inst. Economic Growth, Univ. Delhi. Working paper Series No. E/274/2006.

Durand J. B. Biol. Bull. **11:** 62-96. 1956

Fairs N. J., Evershed R. P., Quinlan P. T. and Goad L. J.; **74:** 199-208. 1989

Fairs N. J., Zuinlan P.T., and Goad L. J. Aquaculture, **89:** 83-99. 1990

FAO 2001 CF. S. Pattanaik Inst. Economic Growth, Univ. Delhi. Working paper Series No. E/274/2006.

Fingerman M., R. Sarojini and R. Nagabhushanam, Neuroendocrinology, **60:** Suppl. 1, 74. 1994

Gabe M. C. R. Acad. Sci. Paris. **237:** 1111-1113, 1953.

Hanstrom B. I. Z. Morphol. Okol. Tiere. **23:** 80-236. 1931

Highnam K. C. and Hill L. (eds) The comparative endocrinology of the invertebrates. Arnold Publ. London. 1977

Hopkins P. M, D. E. Bliss, S. W. Shechan and J. R. Boyer Gen. Comp. Endocrinol. **39:** 192-207.1979.

Jeng S. S., Wan W.C.-M., and Chang C. F.. Gen. Comp. Endocr. **36:** 211-214, 1978

Kalavathy Y., Mamatha P. and Reddy P. S. Naturwissenschaften, **86(8):** 394-395. 1999

Khalaila I., Manor R., Weil S., Granot Y., Keller R. And Sagi A. Gen. Comp. Endocr. **127(2):** 147-156. 2002

Klek-Kawinska E. and A. Bomirski Gen. Comp. Endocrinol. **25:** 9-13.1975

Koller G. Verh. Stsch. Sool. Ges. **30:** 120-132. 1925.

Koller G. Z. Vergl. Physiol. **5:** 191-246, 1927.

Kulkarni G.K., L. Glade and M. Fingerman J. Crust. Biol. **11:** 513-522. 1991.

Kulkarni G.K.and R. Nagabhushanam Aquaculture, **19:** 13-30. 1980.

Kulkarni G.K., R. Nagabhushanam and P.K. Joshi Hydrbiologia, **108:** 281-289. 1984

Kulkarni G. K., Nagabhushanam R. and P. K. Joshi Indian J. exp. Biol. **17:** 986-987. 1979.

Kulkarni G.K. and R. Nagabhushanam Aquaculture, **23:** 19-27. 1981.

Kulkarni G. K. and M. Fingerman Biol. Bull. **182:** 341-347. 1992.

Kulkarni G. K. and M. Fingerman Comp. Biochem. Physiol. **101C:** 419-423. 1992

Kulkarni G.K., R. Nagabhushnam, G. Amaldoss, R.B. Jaiswal and M. Fingerman Invert. Reprod. Develop. **21(3):** 231-240. 1992.

Kulkarni G. K. and M. Fingerman Trends in Life Sci. (India) **1(2):** 111-119. 1986.

Kutty M. N. Aquacult. Res. **36(3):** 255-263, (2009)

Laufer H., M. Landau, D. Borst and E. Homola In : Advances in Invertebrate Reproduction, 1986, p 135-143 (M. Porchet, J. C. Andries and A. Dhainaut eds, Elsevier Sci. Publ. North Holland).

Le Roux A. C. R. hebd Acad. Sci. Ser. D Sci. Nat. **266:** 1414-1417, 1968.

Liu Zh., Donald J., Hanna P., Nuurai P. and Sobhon P. Invertebrate Neurosci., **9(1):** 21-28. 2009

Lutz 2003 CF. News of Scientific and Research Importance 2000-2003.

Mancillas J. R., J. F. McGinty, A. I. Selverston, H. Karten and F. E. Bloom Nature, **293:** 576-578. 1981

MPEDA 2005 CF: The Hindu daily 14 June 2006

Nikitina S. M., Savchenko O. N., Kogan M.E. and Ezhkova N. S.. Zh. Evol. Biokhim. Fiziol., **13:** 443-447. 1977

Nirmal H.B. J. Biol. Sci. **7:** 15-19, 1964.

Ngernsoungnern A., Ngernsoungnern, P., Kavanaugh, S., Sower, S. A., Sobhon P., and Sretarugsa, P. Invert. Neurosci. **8(1):** 49-57, 2008.

Ohira T., Hasegawa Y., Tominaga, S., Okuno A. And Nagasawa, H. Zoological Sci. **20(1):** 75-81. 2003

Okuno A., Hasegawa, Y., Ohira T., Katakur Y. And Nagasawa, H. Biochem. Biophys. Res. Communi. **264(2):** 419-423. 1999

Ollievier F., DeClerck D., Diederik H. and De Loof A. Gen. Comp. Endocr. **61:** 214-228. 1986

Otsu T. Gumna Symp. On Endocrinol. **1:** 25-28. 1964.

Parvathi K .and G.K. Kulkarni J. Aqua. Biol. **18 (1):** 215-226..2001.

Perkins E. M. J. exp. Zool. **50:** 71-195, 1928.

Sarojini R.*et al.*: Comp. Biochem. Physiol. **108C:** 33-38 1994

Sarojini R., R. Nagabhushanam and M. Fingerman Comp. Biochem. Physiol. **106C:** 321-325. 1993

Sarojini R., R. Nagabhushanam and A Mary. J. Aqua. Trop. **5:** 1-7, 1990.

Stonich and Bailey, 2000 CF. S. Pattanaik Inst. Economic Growth, Univ. Delhi. Working paper Series No. E/274/2006.

Bioresources for Food Security and Rural Livelihood, Pages 19–26
Edited by : G.K. Kulkarni and B.N. Pandey

ENERGY BUDGET IN INDIAN FRESH WATER TELEOST AND ITS RELEVANCE IN AQUACULTURE

B.N. Pandey and Ranjit Singh

ABSTRACT

Bioenergetic modelling provides a powerful technique for estimating the growth and rate of food consumption in natural population. Bionergetic model assumes that growth (P) is the differences between the energy income from the food (C) and the energy losses in the form of faeces (F), excretory losses (U) and metabolism (R). The surplus of energy of the food is actually utilized for growth of somatic and reproductive tissues. The basic model is written as :

$$C = F + U + R + P.$$

In the present article an attempt has been made to study bioenergetics and metabolic costs of living of Indian fresh water air breathing teleosts. Its comparison has been made with other Indian teleostean fishes.

Keywords : Energy Budget, fresh water teleosts

INTRODUCTION

Studies on bioenergetic modelling in fishes provide a powerful technique for estimating the growth and rate of food consumption in natural population. Workers involved in the study of fish growth from the stand point of either management or aquaculture, have favoured the use of bioenergetic models. An energy budget provides a powerful framework for identifying the most important aspects of the life of any fish. Cho *et al*. (1982), Kitchell (1983) and Cui and Wootton (1988) have made extensive studies on the bioenergetics and metabolic costs of European species of fishes but a perusal of literature indicates that our informations in this regards on Indian species of fishes is scanty (Alam *et al*. 1994; Alam, 1995; Rahmani, 2009) as such the present work has been taken into account. The present work provides informations on daily food intake, estimation of energy value of ration, faeces and excretory material, energy value of respiratory metabolism, maintenance energy expenditure and energy value of growth in an air breathing murrel, *Channa punctatus* (Bloch). Its comparison has been made with other teleostean species.

Live specimens of *Channa punctatus* of almost the same body weight (50-60g) were procured from local fish dealers at Gaya. The fishes were transported to the laboratory in plastic bucket where they were kept in glass aquaria for proper acclimatization in the laboratory condition for about a week. Unhealthy and diseased fishes were discarded. After seven days the fishes were kept in three different glass aquaria (3 replicas) each containing six fishes. The fishes were fed once daily (between 11. am to 2.0 pm) with weighed chopped goat liver *ad libitum.* The uneaten food if found was removed, dried and weighed. This experiment was kept continued for ten days and mean values for different parameters (as recorded in Table 1) were obtained. The energy content of the food and faecal matter were obtained by Bomb calorimeter as given by Alam *et al.* (1994). All the energy values were converted to Joules (4.2 Joules/cal). Details of the methods employed in the collection of faeces were those of Ray and Patra (1987). The details of the methods employed in the determination of routine oxygen uptake were those of Ojha *et al.* (1978). The basic models used for studying energy budget were as follows :

$$C = F + U + R + \wedge P + MA$$

Where

C = total energy value of food material ingested, U and F are the energy lost in urine and faecal material respectively.

R = energy lost in metabolism or respiration,

^ P = production of new tissues or growth and

MA = energy lost in miscellaneous activities.

The data showing the average values for daily food intake, energy values of ration, total faecal discharge, total urea discharge, total ammonia discharge and total excretory products are summarized in Table 1.

The data for energy value of food, energy values of faeces, absorption efficiency, energy value in excretory product (urine), energy value in metabolism and energy values in growth in different weight groups of *Channa punctatuts* (Bloch) have been presented in Table 2.

The important findings are as follows :

1. Daily food intake (g/fish/day) ranged from 0.65 – 7.07 g/fish/day within the weight range 16.3 to 104.0 g ; the daily food intake ranged from 4.0 – 6.8% within the weight group.
2. The energy value of ration (J/fish/day) ranged from 2440 – 26,710 Joules (J) within the weight range 16.3 – 104.0g
3. The energy value of faecal production (g/day; dry weight) ranged from, 0.11 to 1.64 g/day within this weight range.

Table 1 : Showing daily food intake and energy value of ration, total faecal discharge and total excretion in different weight groups of Channa punctatus (Bloeh) at 31.5 ± 1°C

S.N.	Body weight (g)	Daily food intake g/ fish/ day	Daily food intake % of body wt.	Energy value of ration in joules	Total faecal discharge (g / day) dry wt.	Total urea discharge mg/day)	Total Ammonia discharge (mg / day)	Total excretory Product mg/day
1.	16.3	0.65	4.0	2440	0.11	3.9	3.12	7.03
2.	21.4	0.94	4.4	3570	0.18	5.14	4.11	9.25
3	34.5	1.59	4.6	6000	0.32	8.28	6.62	14.90
4.	42.3	2.03	4.8	7690	0.44	10.15	8.12	18.27
5.	56.5	3.05	5.4	11150	0.65	13.56	10.85	24.41
6.	64.5	3.61	5.6	13650	0.82	15.48	12.38	27.86
7.	72.2	4.33	6.0	16380	1.0	17.33	13.86	31.19
8.	82.3	5.18	6.3	19570	1.23	19.75	15.78	35.53
9,	95.5	6.30	6.6	23810	1.45	22.92	18.34	41.26
10.	104.0	7.07	6.8	26710	1.64	24.96	19.97	44.93

Table 2 : Showing energy values in food, faeces, absorption efficiency, energy value in urine (calculated) energy value in metabolism in different weight groups of Channa punctatus (Bloch) at 31.5 ± 1.0°C.

Body weight (g)	Total energy value of food	Energy values of faeces in Joules in *Joules*	Energy values of faeces % of C	Absorption efficiency (%)	Energy Value of urine (J) in Joules	Energy Value in Metabolism in Joules	Energy Value in Metabolism in%	Energy Value in Growth in Joules	Energy Value in Growth in%
16.3	2440	348.9	14.3	85.7	73.2	1100.4	45.0	917.5	37.7
21.4	3570	538.8	15.1	84.9	107.1	1588.6	44.5	1335.5	37.4
34.5	6000	966.0	16.1	83.9 -	180.0	2652.0	44.2	2202.0	36.7
42.3	7690	1284.0	16.2	83.8	230.7	3522.0	43.8	2853.0	37.1
56.5	11510	2014.2	17.5	82.5	345.3	4903.3	42.6	4247.2	36.9
64.5	13650	2470.6	18.1	81.9	*409.5*	5705,7	41.8	5064.2	37.1
72.2	16380	3145.0	19.2	80.8	491.4	6781.3	41.4	5962.3	36.4
82.3	19570	3894.4	19.9	80.1	587.1	7992.7	40.8	7095.8	36.3
95.5	23810	4801.0	20.2	79.8	714.3	9595.4	40,6	8700.3	36.2
104.0	26710	5609.1	21.0	79.0	801.3	10684.0	40.0	9615.6	36.0
			Av.17.76	Av.82,2%			AV.42.47		AV.36.78

4. The energy value of faeces (F) fluctuated between 348.9 to 5609.1 (J) within the above noted weight range.
5. The energy values of F ranged from 14.3-21.0% (Av 17.76%) of the total energy values of C.
6. The absorption efficiency (%) ranged from 85.9 to 79% (Av 82.2%) in this fish.
7. The total excretory product (U) ranged from 7.03 to 44.93 mg/day/fish within the above noted weight range.
8. Energy value of excretory product (U) in this fish was calculated on the principle of 3% of C as proposed by Winberg, (1956) and modified by Nielsen and Jurgensen, (1982) to fluctuate between 73.2-801.3 Joules within the weight range 16.3 – 104.0 g.
9. The energy value of total respiratory metabolism (R) ranged from 1100.4-10,684.0 J/day/fish within the above noted weight range.
10. The percentage of total energy of food (C) used in metabolism (R) ranged from 40 – 45% with a clear trend of decrease in R value with increasing body weight.
11. Overall maintenance energy expenditure in this fish ranged from 62.3% or 1522. J in 16.3 g fish) to a value of 64% (or 17,094.4 J in 104.0 g fish)
12. The energy value utilized for growth/production (P) ranged from 37.7% (or 917.5 J in 16.3 g fish) to a value of 36.0% (or 9615.6 J in 104 .0 g fish.)
13. The relative energy budgets in different weight group of *Channa punctatus* are different but the average value of different components or bioenergertic models are as follows:

$$C = F + U + R + P$$

i.e.

$$C = 17.76 + 3 + 42.47 + 36.76$$

In nut shell, we can say that energy utilized for F + U, R and ^P were 20.76, 42.47 and 36.76% respectively in this fish.

It is well established fact that the nutritional requirements of fishes must be known to obtain a high rate of growth, for a limited time period. Bioenergetic modelling provides a powerful technique for estimating the growth and rate of food consumption in natural population. The bioenergetic model assumes that growth (P of ^B) is the difference between the energy income from food (C) and the energy losses in the form of faecal (F) and excretory losses (U) plus metabolism (R). The surplus energy of the food is actually utilized for growth of the somatic and reproductive tissues. The main energy input to the fish, the income to the budget is the food consumed.

Of about 20000 fish species recognized (Nelson, 1976), Love (1980) has meticulously summarized the diets of about 600 species. A detailed analysis of Love's summary shows that about 85% of species are carnivorous, 6% herbivorous, 4% omnivorous, 3% detritivorous and 2% scavengers and parasites.

Temperature has an important effect on the food intake of fishes belonging to different climatic zones [Jobling (1983), Cui and Wootton (1988)]. The feeding rate of temperate fishes ranged from 1.8 to 17.3% body weight / day (Av 5.9%) and that of tropical fishes from 4.1 to 36.00% (Mean 16.7%; Pandian and Vivekanandan. (1986). Our data on food intake in *Channa punctatus* are well within the range (17.76%) reported by above investigators. It also corroborates the findings of Alam (1995). Singh (2000) and Rahmani (2009).

Winberg (1956) estimated the values of faecal discharge (F) at 15-17% and U at 3 – 5% which is slightly low as compared to the value reported by Nielsen and Jurgnsen (1982) in eels. Our value for F in the present work is well the range reported by Winberg (1956), Pandey *et al.* (2006) and Rahmani (2009).

The starting point for any discussion of growth processes lies in the simple statement : that what remains in the body must equal that what goes in minus that what comes out, and that which remains is growth. At levels of food intake above those required for maintenance, energy is available for growth/production. Growth/production can be viewed as the accretion of new body tissue, dependent upon biochemical synthetic processes analogous to those involved in the repair and renewal of body tissues during maintenance. Our data on growth value (P) in the present study are very similar to these reported by Colman (1970). Brown (1970). Pandey *et al.* (2006) and Rahmani (2009) in different species of fishes.

The relative energy budgets for various fish species showing average percentage losses of energy via different routes are as follows :

Name of the fish	**Energy Equations** **C = F + U + R + P + MA**	**References**
Theoretical	100 = 16 + 4 + 60 + 20	
Young Carnivores	100 = 20 + 7 + 44 + 29	Winberg (1956)
Young Herbivores	100 = 41 + 2 + 37 + 20	Brett and Groves (1979)
Young eel (fresh water)	100 = 11 + 5 + 35 + 46	Brett and Groves (1979)
Channa punctatus	100 = 17.76 + 3 + 42.47 + 36.78	Present study
Channa sacbua	100 = 15.6 + 7 + 40.2 + 33.2 + 4	Pandey *et al.* (2006)
Heteropneustes fossilis	100 = 18.5 + 3 + 42.6 + 35.9	Prabhakar and Pandey (1997)
Anabos testudineus	100 = 12.8 + 7 + 39.5 + 36.5 + 4.28	Singh (2000)
Rita rita (Ham)	100 = 15.2 + 7 + 39.26 + 33.73 + 4.8	Rahmani (2009)

An energy budget is a balance sheet of energy income set against energy expenditure. Thus, an energy budget provides a powerful framework for identifying the important aspect of the life if any fish. Bioenergetic model may provide a powerful technique for estimating the growth and rates of food consumption of fish in natural population

(Kitchell, 1983). The primary aim of the aquaculture is to find the minimum amount of food (at the lowest price) required to produce the largest yield of marketable flesh at the shortest period of time. This is of particular benefit in making rapid predictions of effects on growth of a certain diet formulation or feed regime or an acute culture stress.

REFERENCES

Alam, N., B.N. Pandey, A.Q. Siddiqui and Bachchan Gupta (1994). Studies on some aspects of bioenergetics in an air breathing fish, *Channa punctatus* (BL.) I. Food intake and faecal production. Him. *J. Env. Zool.* **8**: 150-153.

Alam, Naushad (1995). Bioenergetics and metaboloic costs of living of an air breathing fish, *Channa punctalus* (Bloch, Ph.D. thesis, Magadh university, Bodh Gaya.

Besra, S (1993). Rate of ammonia excretion in relation to seasonal temperature and body size of an air breathing perch, *Anabas testudineus. J. Freshwater Biol.* **5(2):** 169-175.

Brett, J.R and C.A. Zala (1979). Daily pattern of nitrogen excretion and oxygen consumption of sockeye salmon under controlled condition. *J. fish. Res. Bd Canada.* **32:** 2479.

Brett, J. Rand T.D.D. Groves (1979). Physiological energetics. In W.S. Hoar, D.J. Randall, and J.R. Brett (eds). In fish physiology vol VII: 279-352 N.Y.A.P.

Brown, M.E (1970). The growth of brown trout fry *J. Exp. Biol.* **22:** 118-129.

Cho, C.Y.; S.J. Slingerand H.S. Bayley (1982). Bioenergetics of salmonid fishes; energy intake, expenditure and productivity. *Compo Biochem. Physiol.* **73B:** 25-41.

Cui Y and R J Wotton (1988). Bioenergetics of growth of a Cyprinid, *Phoxinus phoxinus* the effect of ration, temperature and body size on food consumption, faecal and nitrogen excretion. *J. Fish Boil.* **33:** 431-443.

Jobling, M (1981). A short review and critique of methodology used in fish growth and nutrition studies. *J. Fish Bioi.* **23:** 685-703.

Kitchell, J.F (1983). Energetics In Fish Biomechanism (eds. P. B. Weib and D. Weihs) : 312-338. New York, Praeger.

Love, R.M. (1980). The chemical biology of fishes, P. 947, London A.P.

Nelson, J.S (1976). Fishes of the world. pp. 416. Wiley, New York.

OJha, J; O.P. Dandotia, A.K. Patra and J.S.D. Munshi (1978). Bimodal oxygen uptake in relation to body weight in a freshwater murrel, *Channa gachua, Z. Tier-stysiol.* **40:** 57-66.

Pandey, B.N. and Abhiranjan Kumar (2006). Influence of some hormones on food consumption, food efficiency, growth and bioenergetioes in a freshwater snake headed murrel, *Channa gachua* (Ham). *proc. Zool. Soc. India.* **5(2):** 29-34.

Pandian, T.J. and E. Vivekanandan (1986). Effect of feeding on growth and swimming activity in an obligatory air breathing fish. *Hydrobiologia.* **49:** 33-39.

Prabhakar, A.K. and B.N. Pandey (1999). Energy budget in relation to varies dietary conditions in a siluroid fish, *Heteropneustus fossilis* (BL). *Him J Env. Zool.* **10:** 7-10.

Rahmani, N.A. (2009). Study on energy budget in a freshwater teleostean fish, *Rua rua* (Ham). Ph.D thesis, Magadh University, Bodh Gaya.

Ray A.K and B.C. Patra (1987). A method for collecting fish faeces for studying the digestibility of feeds. *J. lnland fish Soc India.* **19(1):** 71-74.

Singh, R.K (2000). Bioenergetic studies in an air breathing fish, *Anabas testudineus* (Bloch). Ph.D thesis, Magadh University, Bodh Gaya.

Winberg, G.G. (1956). Rate of metabolism and food requirements of fish, *J. Fish. Res. Bd. Can.* Translation series no. **194:** 1960 pp. 1-253.

Bioresources for Food Security and Rural Livelihood, Pages 27–34
Edited by : G.K. Kulkarni and B.N. Pandey

FISHERIES AND AQUACULTURE: THE POTENTIAL FOOD SECURITY SECTOR

P. Perumal and P. Santhanam

ABSTRACT

Indian fisheries sector has contributed significantly since independence and greatly helped to the food security of the country with an annual production rate of over 7 million tones of finfish and shellfish from both capture and culture fisheries in 2007. India is the second largest producer of fish and is playing a significant role in global fisheries. India has 8,118 km coastline with an exclusive economic zone of 2.02 million km^2. In India, the fishes used for direct human consumption per capita is 4.8 kg/year and it is expected to be increased to a level of 11 kg per annum for fish eating population for which production and distribution has to be scaled up appropriately. With a total fishermen population of about 14.5 million (Livestock census, 2003) and rich marine and inland water resources, fisheries and aquaculture forms an important sector with regard to employment, livelihood and food security. Fish products also form a significant commodity for overseas trade. During the past decades the Indian fisheries and aquaculture has witnessed improvements in craft, gears and farming methods. Indian fisheries industry has made great strides with the annual production increasing from 0.75 million tones in 1950 to about 7.3 million tones in 2007.India's estimated marine resources potential is 3.93 million tones as per the latest update of 2000. During 2007, the marine fish catch was 7.3 million tones. There are 3827 fishing villages with 1914 traditional fish landing centers. There are 79% of full-time fishers and 72.3% of the part-time fishers who are from the coastal states and union territories working towards food security. The country's freshwater resources consist of 195210 km of rivers and canals, 2.9 million hectares of minor and major reservoirs, 2.4 million hectares of ponds and lakes and about 0.8 million hectares of flood plain lakes besides derelict water bodies. During the ten-year period of 1995-2004, inland capture production grew from 60, 0000 to 80, 0000 tones and at present contributes to 13% of total fish production of the country. The estimated area of brackishwater available for aquaculture is about 1.2-1.4 million hectares, of which, about 14% area has been brought under farming. The brackishwater shrimp farm practices yields from 300 to 1000 kg/ha/year through extensive and semi intensive culture. About 81% of the fish catch is marketed as fresh or chilled and forms staple food along the coastal and inland landing centers. About 6% of the catch goes for drying and curing. About 5% production accounts for frozen, while 4.7 % goes for reduction to fish meal, 0.7 % for offal reduction and 0.53 % for other miscellaneous purpose. The utilization by fish canning industry is only 0.6 % of the total catch. Value added products of different descriptions are slowly becoming popular as convenience food. The fish processing industry is well developed in the country as there are about 625 registered exporters. Commercial production is mostly directed towards export. The total exports of fish and fishery products were 163,000 tones in 2004. The country exports twelve major commodity groups to over 40 countries. Shrimp products accounts for 65-70% of the export earnings. Fisheries play an important role in the national economy, providing full-time or part-time employment to 14.66 million people. The contribution of fisheries to GDP at the current prices (2003-04) is 1.07%. India's future fisheries development plans are aimed at making substantial contributions to doubling of food production, improving the welfare of fishers, promoting exports and providing food and livelihood security to rural population. Aquaculture is recognized as an important source for meeting future demands for protein food in the country. A number of schemes have been instituted by state and central sectors to augment production from brackishwater and freshwater aquaculture sectors. Considering the growing global demand for seafood, developing the export production with due care for safety and product competitiveness has been embarked ahead.

Keywords: Finfish, Shellfish, Brackishwater, Freshwater, Aquaculture

INTRODUCTION

India's estimated fish potential is 7,308,230 metric tones as per the latest update of 2007. During 2006, the fish production (capture & culture) was 7,014,140mt. With a total fishermen population of about 14.5 million (Livestock census, 2003) and rich marine and inland water resources, fisheries and aquaculture forms an important sector with regard to employment, livelihood and food security. Food products also form a significant commodity for overseas trade. During the past decades the Indian fisheries and aquaculture has witnessed improvements in craft, gears and farming methods. India is the second largest producer of fish and it is playing an important role in global fisheries. During the last five decades, Indian fisheries has made great strides, with the annual production increasing from 0.75 million tones of fish in 1950 to about 7.3 million tones in the year 2007, indicating over nine-fold increase during the period. The inland fisheries contribution is in appreciable level and the marine fishery also been contributing significantly in recent years. The country has also emerged as one of the giants in export front, recording a peak during the year 2004, with earnings of US$ 1,365,495,000. Creation of required harvest and post-harvest infrastructure has been receiving due attention of the central and state governments.

MARINE FISHERIES

The sea is the largest food security system through income generation, besides employment generation. The fish protein is recognized as a valuable ingredient in a balanced diet with chief protein with all essential amino acids and fatty acids. The Indian marine fisheries has risen from a subsistence oriented traditional avocation to the status of a developing industrial sector through successive stages of technological development that has taken place during different periods of time, first with a change of natural fibres to synthetic fiber in net fabrication and concurrent introduction of mechanized trawlers in the fifties, second with the introduction of mass harvesting gear in the seventies followed by motorization and proliferation of innovative gears in the eighties. India has 8,118 kms long coastline, 0.53 million sq. km shelf area and 2.02 million sq. km of exclusive economic zone. The marine fisheries development was put on a sound footing through its researches on biology of commercially important species and monitoring their stocks for proper management, sustainable utilization and conservation. India's estimated marine resources potential of 3.10 million t. during the year 2007-2008. Off these about 60% was taken from the west coast and the rest from the east coast. The fishing units consist of 208000 traditional crafts, 55000 traditional motorized crafts, 1250 mechanized boats and about 100 deep-sea fishing vessels. There are 3827 fishing villages and 1914 traditional fish landing centers are our potential force for Indian fishery sector. There are about 79% of the full time fishers and about 72% of the part time fishers are from the coastal states and union territories are working towards for our food security. Different kinds of fishing gears, including trawls, seines, lines, bag nets, stake nets and lift nets are deployed for fish catching.

The marine fish landings of India during the year 2007 has been estimated as 2.88 million tones showing an increase of about 1.7 lakh tones (6.3%) . Among these, the pelagic finfishes constituted 57%, demersal fishes 25%, crustaceans 14% and mollusks 4% (of the total landings). The sector- wise contributions during the same year were: 68% by mechanized, 28% by motorized and the 4% by artisanal crafts. The estimated landings have showed an appreciable increase in the groups such as oil sardine with 4,97,264 t (26% increase compared to that of 2006), penaeid prawn, 2,02,053 t (17.2%), Indian mackerel, 1,78,734 t (25.9%), croakers, 1,68,031 t (40.7%), lesser sardines, 94,827 t (6.5%), silverbellies, 69,856 t (8.1%) and other clupeid, 67,750 (62.1%).But marked decreasing landings were reported by the groups such as non-penaeid prawn, 1,39,052 t (18.6%), ribbon fishes, 1,32,388 t (43.7%), bombay duck, 1,12,721 t (4.9%), threadfin breams, 94,221 t (15.4%) and cephalopod, 94,804 t (30.3%). The estimated region-wise production showed that the north-east region, comprising West Bengal and Orissa coasts contributed 13% to the total production. South-east region consisting of Andhra Pradesh, Tamil Nadu and Puducherry coasts contributed 22%. In the west coast, the north-west region comprising Maharastra and Gujarat coasts contributed 30% of the total, whereas, the south-west region comprising Kerala, Karnataka and Goa coasts contributed to the maximum of 35%. The stock assessment of deep sea fishes in the continental slope of Indian EEZ were studied by CMFRI (Kochi) based on trawl survey at nine stations by FORV Sagar Sampada (Cruise No. 252) during 2007 and the survey reported that a total of 63 species of finfishes belonging to 53 genera, 42 families and 19 orders were recorded at the shelf edge off North Andaman.

INLAND FISHERIES

India has huge inland water resources consisting of 195210 rivers and canals, 0.3 million ha of estuaries, 0.19 million ha of backwaters and lagoons, 3.15 million ha of reservoirs, 0.2 million ha of floodplain wetlands and 0.72 million ha of upland lakes. During the ten year period of 1995-2004 inland capture production grew from 6,00,000 to 8,00,000 tones and at present contributes to 13% of a total fish production of the country from different types of inland open water systems. There are 14 major rivers and 44 medium rivers and innumerable small rivers of the country which can provide one of the richest fish faunistic resources of the world. No data are available for some riverine systems, but the major rivers showed the yield ranging from 0.64 to 1.64 tones per km with average of 1 tone per km. The reservoirs form the largest inland fisheries resources in terms of resource size with 56 large reservoirs, 180 medium reservoirs and 19,134 small reservoirs covering water area of 1.14 million ha, 0.527 million ha and 1.485 million ha respectively. Floodplain wetlands are other potential fisheries resources in the states of Assam, West Bengal and Bihar, which offer tremendous scope for both culture and capture fisheries. These water bodies play vital role for recruitment of fish stocks of the riverine system and provide nursery grounds for commercially important finfishes and shellfishes. It has been estimated that these floodplain possess potential to yield higher production levels of as much as 1000-1500kg//ha/yr., while the present level remains at only 100-150kg/ha. It is evidenced that the availability of live food organisms make this water bodies ideal for culture-based fisheries leading to higher growth of stocked fish species compared to the reservoirs. The scientific management and practice can enhance the production to great extent in Uttar Pradesh and in Kerala. The increased pollution and over exploitation of the resources leading to the stagnation of fisheries production, so that the culture based practice in reservoirs and floodplain wetlands holds the key for future of the inland fisheries development in India.

FRESHWATER AQUACULTURE

The aquaculture is an emerging fish producing system in India. After the government initiatives for active culture practices, Fish Farmers Development Agencies (FFDA) were set up in each district for transferring the technologies for practices, training and extension besides financial assistance to the beneficiaries. Indian aquaculture has shown appreciable growth rates than those of capture fisheries with production increasing from 1.01 million tones in 1990 to 2.10 million tones in 2000. Further the freshwater aquaculture has form a major share of the aquaculture production. The freshwater aquaculture has been contributed dominantly by Indian major carps which share as much as 1.6 million tones. Around the India there are about 0.65 million ha of water area under fish farming and about 1 million beneficiaries were trained. The present annual production is around 2.2 t/ ha. There are about nineteen thousand million fry per year were produced in India for culture practices. The necessary feed also available in India. The carps accounts for over 80% of farmed fish comprised with rohu (*Labeo rohita*), catla (*Catla catla*), mrigal (*Cirrhinus mrigala*), grass carp (*Ctenopharyngodon idella*), common carp (*Cyprinus carpio*), silver carp (*Hypothalmitcthysmolitrix*), catfish (*Clarius batrachus*), singi (*Heteropneustes fossilis*), rainbow trout (*Oncorhynchus mykiss*). Giant freshwater prawn (*Macrobrachium rosenbergii*) has emerged as a new species for farming with promising results. In recent years, the freshwater aquaculture has made remarkable strides in India with appreciable growth trend with annual growth rate of about 6% during the last decade and the sector possess higher growth rates than other food producing sectors. The freshwater aquaculture started for own need in West Bengal and Orissa, in recent years it become commercialized, in states like Andhra Pradesh, Punjab, Haryana, Maharastra, etc. taking up freshwater aquaculture as a trade. After the technological inputs, entrepreneurial initiatives and financial investments, the productivity has gone up from 500-600kg/ha/yr to over 2000kg/ha/yr. There are some farmers who are achieving higher production levels of 6-8 t/ha/yr. The Indian major carps such as Catla, rohu and mrigal contribute share with production of 0.546, 0.567 and 0.517 million tones respectively recorded during the year 2000. Further, other groups such as catfishes, freshwater prawns and pearl culture have also been brought into the farming. In addition, the non-conventional culture systems like sewage-fed fish culture, integrated farming system, cage and pen culture, raceway fish culture and so the freshwater aquaculture made on increasing activity across the country. Recently the organic farming has become more popular in India which is enabling eco-restoration.

CULTURE OF CARPS

Research and development efforts during last five decades have greatly enhanced the average fish yields in India making carp culture an important economic enterprise. The three Indian major carps were the principal species cultured by the farmers in ponds. Introduction of exotic species viz., silver carp, grass carp and common carp into carp polyculture system during early sixties also added new dimension to the aquaculture development of country. Central Inland Fisheries Research Institute (Cuttack) has initiated the development of scientific carp culture in India. With the adoption of technology of carp polyculture or composite carp culture production levels of 3-5 tones/ha/yr could be demonstrated in different regions of the country. The technology of intensive carp culture has demonstrated higher production levels of 10-15 tones/ha/yr.

CULTURE OF CATFISHES

The catfishes have potential commercial importance but their culture in India is yet to make any mark. *Clarias batrachus* (magur) and *Heteropneustes fossilis* (singhi) are the two air-breathing catfishes which are most suitable for induced breeding and seed production. The catfishes stocked at the density of 20,000-50,000 fingerlings/ha, production levels of 3-5 tones/ha of magur, which attain 100-200 g in 6-8 months. These are successfully cultured in both monoculture and polyculture systems.

CULTURE OF FRESHWATER PRAWN

The giant freshwater prawn, *Macrobrachium rosenbergii* is the largest and fast growing species. The technology was standardized for commercial seed production with an average annual level of 60%. There are about 40 freshwater prawn hatcheries, established mainly in the states of Andhra Pradesh, Tamil Nadu and Kerala. During the last ten years, freshwater prawn farming sector has evidenced overwhelming growth with as much as 24,000 ha additional area achieving production over 30,000 tones. Monoculture of freshwater prawn at stocking densities of 30,000-50,000/ha has shown production levels of 1-1.5 tones/ha in a culture period of 7-8 months. The polyculture of freshwater prawn along with carps was found to be sound culture practice and economically viable option for enhancing the income of the farmers.

BRACKISHWATER AQUACULTURE

Shrimp Farming

The traditional brackishwater aquaculture has been practised as bheries in West Bengal and Pokkali in Kerala. But the systematic brackishwater aquaculture was started only a little over two decades ago but this is the booming export-oriented industry for the last five years. The brackishwater shrimp farming is the significant fishery sector in food security, employment generation and livelihood. A number of farms were set up in many places by multinational companies along the two coasts, especially on the east Andhra Pradesh and Tamil Nadu. The semi-intensive monoculture of tiger prawn was undertaken and a production as 9 t/ha/crop was obtained from an MPEDA supervised farm. The hatchery technology is available for about one dozen species of shrimps but only four species viz: *Penaeus monodon*, *P.japonicus*, *P.merguiensis* and *F.indicus* are used for aquaculture. Among these, *P.monodon* is the most promising candidate species with high market value. The country has the estimated area of brackishwater (available for aquaculture) is about 1.2-1.4 million ha, of which; about 14% area has been brought under farming. The shrimp production from coastal aquaculture during 2004 was approximately 1, 20.000 tones. Farmed shrimp accounts for about 60% of shrimp exported from the country. The shrimp is the single commodity that contributes almost the total production of the sector. The production levels of shrimp recorded marked increase from 28,000 tones in 1988-89 to 1, 27,170 tones in 2001-2002. The black tiger prawn contributes the major share. There are about 260 shrimp hatcheries in the country with production capacity of 11 billion seeds, of which 200 are in operation with an output of 7 billion shrimp larvae. There are 33 feed mills with a capacity of 1, 50,000 tpa. The sector is providing direct employment to 0.6-0.7 million people. After the outbreak of white spot syndrome virus (WSSV), the tiger prawn culture experienced a set back. The latest development is the establishment of PCR labs, at hatcheries and other places to detect whether the seed, spawners/broodstock or the animals in the grow-out facilities are carrying the infection.

Fish Farming

In recent times, the brackishwater fish farming has become an important activity in the country. Candidate species for brackishwater fish farming are the mullets (*Mugil cephalus*,

Liza tade, L.parsia and *L.macrolepis*), milk fish (*Chanos chanos*), pearl spot (*Etroplus suratensis*), seabass (*Lates calcarifer*), groupers (*Epinephelus* spp.) and snappers (*Lutjanus* spp.). So far the induced breeding and seed production technology was developed for seabass *L.calcarifer* by the Central Institute of Brackishwater Aquaculture (CIBA), Chennai in 1997. No other finfish species are reared for induced breeding and seeds are not available for farming. The CIBA transferred the hatchery technology to Rajiv Gandhi Centre for Aquaculture, where the fish are being spawned since 2001. A pilot hatchery facility was developed under an Indo-French collaborative programme with M/S COFREPECHE at CIBA field station, Muttukadu (Chennai) and the seabass seeds are successfully produced throughout the year by recirculation system. Experimental culture of seabass with *Tilapia* as forage has given some successful results in Maharastra with large size variations. The hatchery produced seeds of *L.calcarifer* attained an average weight of 1.2kg in a tide-fed pond in 7 months with no provision of artificial feed. The successful result was obtained by the RGCA in seabass under cage culture in earthen pond at demonstration farm located at Karaikal.

Mariculture

Intensive researches have been carried out during the last two decades by the Central Marine Fisheries Research Institute to develop the seed production technology for the culture of important marine crustaceans, mollusks and seaweeds. The culture technology has been standardized for green mussel, *Perna viridis* and brown mussel *P.indica* by rafts and long line methods. Mussels attained harvestable size of 70-80 mm in 6-7 months of culture period and obtained the production of 12-14 kg mussel/meter of rope. The economic analysis of the mussel farming made by CMFRI showed over 40% profit margins on investment of about Rs. 24,000 per raft of 8mx8m during the culture period of 6-7 months.

Edible Oyster Culture

The oyster, *Crassostrea madrasensis* is the only species that is found to be fit for commercial farming. The species reach harvestable size of 80 mm in a culture period of 7-8 months and production levels of 8-10 tones of shell on oyster/ha are obtained.

Seaweed Culture

Seaweeds form an important marine creature in the sea. Seaweeds and their byproducts are used as food, fodder, fertilizers and several industrial and pharmaceutical products. In India the seaweeds are found to be more common on the coasts of Tamil Nadu and Gujarat. Since 1972, CMFRI is involved in experimental culture of different seaweeds species and developed culture technologies for species like *Gracilaria edulis*, *G.corticata* and *Gelidiella acerosa*. The seaweeds are cultured by net and rope culture, the 1 kg of seed material can give an average yield of 3 kg in culture period of 60 days in *G.edulis*. 1000 nets of one hectare area can give production of 30 tones. With minimum four crops a total of 120 tones of fresh *G.edulis* can be produced in one year. Central Salt Marine Chemicals Research Institute (CSMCRI) also engaged in seaweed culture especially the species called *Kappaphycus alvarezii*. The species was successfully cultured in Palk Bay and Gulf of Mannar area of Tamil Nadu and the technology was transferred to the beneficiaries. Central and State governments can support the publics and women's shelf help groups for seaweed farming in and around Rameshwaram and Mandapam coastal area by providing financial support and subsidies along with technical guidance by CMFRI and CSMCRI.

CONCLUSION

The estimated area of our country is 3.29 million km that includes 8,118 km coastal length with an exclusive economic zone of 2.02 million km^2. But the area under the utilization is so meager. The population of our country is about 1.1 billion in 2004. The fishes used for direct human consumption per capita is 4.8 kg/year and it is expected to be increased to a level of 11 kg per annum for fish eating population for which production and distribution has to be scaled up appropriately. The over exploitation can leads the scarcity of fish potential in the wild and so no further fishery resources for our future need. Aquaculture is recognized as an important source for meeting future demands for protein food in the country. So there is an urgent need to develop appropriate technology for induced breeding and seed production for commercially important fin fishes and shrimp seed production and grow-out culture to meet the future demand.

REFERENCES

Ayyappan, S. and J.K. Jena., 2003. Indian fisheries and aquaculture: present status and future prospects, FISH FOR ALL: National Launch., 13-24.

FAO, 2008, World Fisheries.

FAO. 2004. Fishery and aquaculture country profiles (India).

Pillai, N.G.K., 2008. CMFRI Research Highlights 2007-2008 (Edt. K.S. Mohamed and K. Vinod), 36pp.

Tripathi, S.D., 2003. Inland fisheries in India, FISH FOR ALL: National Launch., 33-57.

Bioresources for Food Security and Rural Livelihood, Pages 35–46
Edited by : G.K. Kulkarni and B.N. Pandey

BIO-PROSPECTING OF BIO-RESOURCES FOR LIVELIHOODS OF THE RURAL PEOPLE

D.K. Belsare, S.D. Belsare and M. Murlidhar

ABSTRACT

The bio-resources and their sustainable use for livelihood of the people are considered as the major priority of the government's policy. Training in multiple uses of these resources can provide income and food security. Freshwater fish culture, apiculture, lac culture, sericulture and value added products like jams, juices, sauce, squash and pickles from wild edible plants are some of the bio-resources that can alleviate poverty of rural people through proper processing and marketing for income generation. These resources are critical especially for the rural poor in securing subsistence needs and are more important at time of hardship and emergencies when quick cash is required or when casual food stores run out.

Keywords : Bioresources, fish culture, apiculture, sericulture, rural proverty.

INTRODUCTION

India's population is directly dependent on its vast bio-diverse resources for their livelihoods. The combined population of 28 states and 7 union territories of India on March 1, 2001 were 1037 million. The decadal growth of population varies from state to state e.g. it is lowest in Andhra Pradesh and Kerala and highest in Bihar. Similarly the overall population density also varies e.g. 10/km^2 in Arunachal Pradesh to 904/km^2 in West Bengal . Thus there is uneven distribution of local bio-resources in relation to population. The country's tribal population overlaps with the forest areas and there is maximum concentration of poverty in these areas. 72 % of the India's population live in rural areas and is directly dependent on terrestrial and aquatic resources for their food, health and diverse livelihoods. These include forest dwellers, tribal communities, small and marginal farmers. Out of 5, 80,000 villages of India, 2, 00,000 villages depend on forest resources for their livelihoods. Women are the major gatherers of bio-diverse resources from common land and key actors in subsistence farming and livestock rearing. They have specific knowledge of bio-diverse plant and crop systems, including seed selection, domesticated livestock breeds and wild fruits, food, and medicinal plants, acquired by them based on their interaction with local ecosystems for ensuring household food security. The' green revolution' and modern technology reduced the value of women's specific knowledge about local biodiversity. The solution to achieve their sustainable livelihoods and employment is to find out specific bio-resources and bio-prospecting them in the form of finished products for more income generation and to sustain their livelihoods. This was the main aim of this presentation.

1. FISH CULTURE AS BIO-RESOURCE

The freshwater fish diversity can serve as a pathora of livelihood and biodiversity of conservation values. They serve as source of protein and food for increasing human population in developing countries (Belsare 1984). The extensive farming of fish is more sustainable for livelihood of the rural poor (Belsare 1986), because it serves as security of food, health and earned income. When natural disasters like flood and earthquake occur, our first response is obviously to address the immediate needs of the survivors in terms of employment or earned income and food. In the longer term, however, the loss and recovery of livelihoods require in depth analysis by a wide range of expertise, but driven by the real needs of the local people. The utilization and proper management of freshwater fish diversity resource serves as good source for livelihood of the people. The multiple uses of fish for food, raw material for industrial finished products, sports, ecotourism, and bio-manipulation of water quality can serve as means of sustainable development.

The polyculture or composite fish culture are the general practices in India.. The culture of freshwater prawn, *Macrognathus rosenbergii* is newly introduced aquaculture practice in India, in which common carp (*Cyprinus carpio*) is replaced by prawns. The demand of catfish seed for aquaculture in India will be met by recent technology developed at Powarkheda Center of CIFE. This center raised 20,000 magur fry this year, 2009. These practices mainly aim at increased fish production for meeting increasing demand for food. The fish farmers are earning more income through these practices and training imparted to them for their sustainable livelihoods. Educating them regarding use of new synthetic hormone synthetic agonists of GnRH (GnRHa) create a surge in the use of hormones to control reproductive processes in aquaculture. The most recent development is the incorporation of GnRHa into polymeric sustained release delivery systems, which release the hormone over a period of days to weeks. These delivery systems alleviate the need for multiple treatments and induce (a) long term elevation in sperm production and (b) multiple spawning in fish with asynchronous or multiple batch group-synchronous ovarian physiology. The compounds used as matrix can be: cholesterol and cellulose, lactic acid and glycolic acid, or co-polymers of dimer fatty acid and sebasic acid, or acetate of ethylene and vinyl (EVAc). The farmers then would develop fish seed at any time of the year and stock them to replenish old fish stock. What is required is intensive training to them.

India has major (Indian) carps (rohu, catla and mrigal) and silver carp as the dominant species in composition, culture intensity, productivity and marketing. High stocking density and low use of supplementary feed and fertilizer characterize cultured fisheries in India. Cultured ponds make use of relatively lower inputs and thus are semi-intensive or improved extensive. It is taking shape on commercial scale now. Yields, costs and returns vary considerably, reflecting variations in intensity, production environments and systems, and culture practices. Rates of return and break-even prices are lowest in India, which suggests that farmers from this country are more cost-effective. Feed costs account for most of the variable costs for pond operation. The cost of fingerlings is not taken into account of the total cost. . Women's participation is comparatively less in India.

Indian water bodies are much altered by anthropogenic influence on them, such as clustering of human population around them, intensive aquaculture of a few fish species, eutrophication, soil erosion on the adjoining land where natural vegetation is destroyed or altered and pollution which created drastic ecological imbalance, especially of aquatic food web. The fisheries departments and corporations claim that the freshwater fish production has gone up (based on major carps & common carp and in few cases of catfish, freshwater prawn), the actual fish landings are depleted. The indigenous fish species are replaced by a few species and bottom fish fauna is drastically altered. Exactly similar situation occurred in African natural and man-made lakes (Worthington and Lowe-Connell 1995). Pandey (2003) pointed out that the North Bihar has high potential of fish production that may feed the poor, but his presumption lacks proper analysis of ecological changes in water bodies due to human influence. The poor people of North Bihar still depend on *Amphipnus cuchia* , catfishes and river major carps for their food and some earnings by transporting human population across the Ganges as well as makhana & trapa farming. *In situ* conservation of freshwater fish diversity and livelihoods of the people in India and elsewhere requires in depth analysis of aquatic ecology, if we want to avoid this disaster and think of the future generation's welfare.

It has been pointed out that within 30 years, an average rate of extinction of freshwater fish species is 50% (Finlayson, 2005). This loss is due to various factors such as habitat loss, introduction of alien species for aquaculture, climate change and several anthropogenic activities. The habitat loss is the main cause of loss of fish species diversity. The majority of wetlands are under threat due to anthropogenic activities like conversion of wetlands to agriculture use, riverine projects, mining activities, dams, over-fishing, target fishing, fishing in banned water bodies, eutrophication of water resource and reclamation of wetlands etc. (Scott & Poole 1989) The food chain of fish is disturbed and breeding ground is lost, as a result of which fish population is declined. The introductions of alien species to freshwater wetlands also result in decline of indigenous fish species because the former become competitive to native species in obtaining aquatic food source and breeding. In Kashmir Valley and the Loktak Lake the loss of *Schizothoracine* species is due to introduction of common carp. Similarly introduction of *Tilapia* sp in wetlands of Orissa affected native carp and catfish population.

Under Bhoj Wetland Project of Bhopal, the Grass Carp (was introduced to eradicate weeds, but the net result was depletion of the native fish fauna like *Mystus vittatus* (the major catch from the Upper Lake in the past). The range and abundance of fish species are changed dramatically due to climate change and they may migrate to new locations. For economic security and achieving livelihood, a large scale intensive aquaculture practices in village ponds and reservoirs also affected native fish fauna. The dam constructions prevented fish migration to upstream and hence affect their breeding potential. The aquaculture of hybrid varieties has severe impact on fish diversity. The run-off water from agriculture fields carries pesticides, which result in fish kills. Similarly heavy metal pollution from industrial effluents causes fish toxicity and loss of their population.

Certain freshwater fish species such as Mahaseer (*Tor putitora*), snake-heads (*Channa marilius, C. striatus, C. punctatus*), cat-fishes (*Wallago attu, Clarias batrachus, Heteropeustes fossilis, Mystus seenghala* etc) are the sport fish of the anglers. The villagers can provide those services and can earn income. In Pawai Lake of Maharastra state, *Barilius bengalensis* from Madhya Pradesh was introduced by Tata Enterprise for the purpose of sport. Now it has become famous for tourists activity. The fish can be used for developing ecotourism-selecting sites for tourists and can be source of income for rural inhabitants for their services. The other species is *Osteobrama belangeri* (Pengva), which is recently bred at Pawarkheda Center of CIEF , Mumbai, can serve as sport fish rather than fish for aquaculture for food. 50,000 spawn obtained in this Center in 2008, are being raised up to fingerling stage and would be distributed to fish farmers, but they may be educated for its use as sport fish rather than food fish to obtain more income. Due to its shining silver colour and swift running nature, it can serve as 'anglers paradise' like *B. bengalensis.*

2. APICULTURE AS BIO-RESOURCE

The apiculture (bee-keeping) is important bio-resource technology for income generation and livelihoods of the people. It is widely practiced in Bihar, Uttar Pradesh, Tamil Nadu and other states of India, but production of honey from this resource is poor in these states. The entrepreneurs do not get proper return from the sale of bee products. This is mainly due to unorganized marketing and traditional technology. It involves rearing of bees in scientifically manufactured wooden or polystyrene boxes (hives). The place where the bee colonies in hives are kept is known as 'apiary '. It is a subsidiary occupation and does not require much time and capital. It is a suitable activity to augment the income of the farmers and for livelihood of the rural people.

In India several million ha of land is under cultivation of oilseeds, pulses, fruits and other crops. Bees are the natural cross pollinators of these crops. Similarly India has 20% of its land as forests, which have good bee-keeping potential and can sustain millions of bee colonies. To fulfill the needs of pollination of agricultural and horticultural crops, several times more number of bee colonies can be established in non-forest areas. As per available data, the state-wise production of honey from bee colonies established by farmers for the year 1992 is indicated in table .1.

Table 1 : Statement Showing Statewise Production of Honey During 1992

States	Production of honey (tons)	Major Plant Sources
A.P	104.50	Coriander, Tamarind, mixed forest.flora, Mango
Arunachal Pradesh	0.30	Mixed forest flora
Assam	401.15	Mustard, Citrus, Mixed forest flora.
Bihar	360.97	Litchi, Mustard, Mixed forest flora
Goa	0.16	Cashew and other plantation, Mixed forest flora.
Gujarat	2.02	Fruits, Forest flora
Haryana	6.96	Mustard, Eucalyptus
Himachal Pradesh	60.00	Orchards and Forest flora
Jammu & Kashmir	96.70	Orchard, False acacia.
Karnatak	662.59	Rubber, Coffee, Orchards, Forest flora.
Kerala	3658.52	Rubber, Cardamom, Coconut, Forest flora.
Madhya Pradesh	86.32	Cultivated and forest flora
Maharashtra	78.24	Jamun, Mango, forest flora
Manipur	207.50	Citrus, forest flora
Meghalaya	57.92	Orange, Forest flora
Nagaland	19.22	Forest flora.
Orissa	590.00	Mustard, Forest flora
Punjab	540.00	Mustard, Eucalyptus
Rajasthan	13.96	Cultivated and forest flora
Sikkim	1.46	Orchards & forest flora
Tamil Nadu	1946.42	Rubber, Cardamom, forest flora.
Tripura	2.72	Rubber, Forest flora
Uttar Pradesh	97.33	Mustard, Litchi, Drumstick, Forest flora.
West Bengal	496.96	Mustard, Litchi and Forest flora.
Total apiary honey	**9494.12**	
Total forest honey	**18000.00**	
Grand Total	**Rs. 27494.12**	

Being a natural food collected by bees from nectar of flowers honey has medicinal values. Viscous supersaturated sugar in honey is easily assimilated in the blood stream and gets deposited in liver as glycogen and provides energy to our body. The calorific value of one kg of honey and other products is; honey (3150 to 3350), milk (620), fish (620), mushroom (270), ray-bread (2040), apple (400), orange (290), and cucumber (140) . Honey is used in manufacture of agro-processed foods either as a substitute of sugar or for giving flavours to the agro-processed recipes, cosmetics, medicines, wax preparations, confectionery etc. Honey bee pollination increases the seed or fruit yield of various crops viz., Mustard, Carrot, Coriander, Grapes, Cotton, Horse-gram; Soy bean etc., processed by them, honey contains several vital constituents having nutritional importance.

3. SERICULTURE AS BIO-RESOURCE

Sericulture is the technique of silk production. It is an agro-industry playing an eminent role in the rural economy of India. Silk fiber is a protein produced from the silk glands of silkworms. There are four types of sericulture practices in India i.e. Mulberry, Tassar, Eri & Muga silk cultures. There are five varieties silk worms reared in this country. *Bombyx mori*, the silkworm, feeds on the leaves of *Morus*, to produce the best quality of fibre among the different varieties of silk produced in the country. *Antherea assama* produces the famous muga silk. This variety is confined only to the Brahmaputra valley of India in the world. Tasar silk is a product of *Antherea mylitta*, which feeds on *Terminalia tomentosa* grown in the thick jungles of Bihar, Madhya Pradesh and Orissa. *Antherea royeli* and *Antherea perniyi* has been recently introduced to produce the oak tasar silk, *Phylosamia ricini*, the eri silkworm, which feeds on *Ricinus communis,* is raised in Assam and Orissa.

In India sericulture has become the most promising rural activity due to certain reasons like minimum gestation period, less investment, maximum employment potential and quick turnover for the investment. Sericulture as a whole, by its very nature of activity, creates large-scale employment and income generation opportunities in the rural and semi-urban areas accelerating the economic growth of these areas.

In recent years sericulture has achieved enormous progress in evolving suitable mulberry varieties and techniques to bring about new silkworm races suitable for tropical climatic conditions. With the evolution and introduction of more productive silkworm races, the productivity has increased and sericulture has become a highly remunerative activity. Attracted by these advantages many more farmers have taken up sericulture and the industry has spread to almost all the states in India

Karnataka has been the leading producer of mulberry silk accounting for more than 50 percent of its production in the country and is now regarded as the "Silk Bowl of India". Andhra Pradesh comes next to Karnataka in producing mulberry raw silk. In Andhra Pradesh, sericulture has proved to be a money-spinner for many middle class families. Except for Hyderabad district, almost all the districts in the State have taken to sericulture activity. In 1956, mulberry cultivation in the State was only in 1212 hectares and it went up to 90,800 hectares by the end of 1993-94.

ECONOMICS

S. No.	Details of crop	No. of crops in a year	Annual income per acre (in Rs.)	Expenditure (in Rs.)	Net income per acre (in Rs.)
01.	Groundnut	One to two	7,500	5,243	2,257
02.	Ragulu	Two to three	12,000	3,600	8,400
03.	Paddy	Two to three	20,500	6,075	14,175
04.	Mulberry	Five to six	49,500	21,500	26,000

Thus from mulbery crop the net income is 26,000 Rs / acre.

4. LAC CULTIVATION AS BIORESOURCE

Lac is produce by insect species, *Laccifer lacca* of the Laccaferinae sub-family of the Lacciferidae. In India, two strains are cultivated, "kusiumi" and "rangeeni", which differ in their seasonal cycle and preferred host trees. Strains in other lac producing countries are less well defined. Lac cultivation is a seasonal, part-time agro-forestry activity which may be based on cultivated or wild host trees. In order to obtain maximum yields of stick-lac, the insects are cultured, the host trees are managed and attention is given to their control. Lac cultivation has a number of advantages like low cost, small investment, quick and regular income, easy transportation from remote areas and sustainability, because of perennial host trees can be used for many years. The female cell is shorter and more rounded in shape, and remains fixed to the twig. She retains her mouthparts but fails to develop any wings or eyes. Although she develops rudimentary antennae and legs, she becomes an immobile shell-like organism with little resemblance to an insect; increasing in size to accommodate her growing number of eggs. Lac resin begins now to be secreted at a faster rate and a continuous layer coalesces or grows into one body, covering the entire swarm. After fourteen weeks the female contracts, allowing light into the cell, and lays her eggs. When the eggs hatch they emerge as larvae and the whole process begins again. Her ovaries contain a crimson fluid called lac dye, which resembles cochineal (a colouring used mainly in the food industry and derived from dried bodies of coccus insects). After the cycle has been completed and around the time when the next generation begin to emerge, the resin encrusted branches are harvested. They are scraped off, dried and processed to form shellac. A portion of broodlac is retained to tie to more trees and so produce the next crop.

The first operation is pruning of the host tree in order to stimulate the growth of young shoots which provide sap as food for the insects. This is done four to six months prior to inoculation of the tree with "broodlac", a cocoon containing mature females at a stage just prior to emergence. Eggs laid on the host develop into larvae and form a resinous cocoon ("sticklac"). Harvesting is undertaken approximately six months later and the tree is subjected to a repeated treatment of pruning and inoculation.

Economic Gain

Procurement and supply of more brood lac.........24.3 tons, Service providers...15; Investment for working capital... Rs. 0.75 million

Number of trees pruned, marked and inoculated....11,000; Estimated value of output ...Rs. 4 million; Estimated net income to families (3000)...Rs 3 million

Average per capita income (over 8 months)....Rs. 10,000

5. WILD EDIBLE PLANTS AS BIO-RESOURCE

The interest in wild edibles has grown significantly with the increasing awareness in linking forest conservation with rural development and poverty alleviation. There are about

18 wild edible plants in Hoshangabad district of Madhya Pradesh which may play a dominant role in poverty alleviation of people with transfer of appropriate biotechnology. Their raw material is sold in the market at throw-away price. If they are trained in technology of value added products from these edible plant resources, like jams, juices, sauce, squash and pickles., it can serve as a source of income for them, especially women folk. These edible plants are: *Ziziphus mauritiana* (Indian plum), *Syzygium malacceuse* (Malay apple), *S. jambos* (Rose apple), *S. cumini* (Jamun), *Limonia acidissima* (Elephant apple), *Madhuca latifolia* (Mahua), *Aegla marmalos* (Wood apple), *Annona sqamosa* (Custard apple),*Emblica officianalis* (Emblic myrobalan), *Diaospyrus ebenum* (Indian abony), *D. peregrina* (Wild magasteen, Tendu), *Mangifera indica* (Mango), *Citrus reticulate* (Orange) and *Tamarandus indica* (Tamarind) etc.

For livelihoods security the following policy outlines are to be For livelihoods security the following policy outlines are followed; (i) access to food security through food production, food supply, food price, food subsidies, food aid and food research, (ii) access to resources through land reforms and development, forest regeneration and management and water planning and management; (iii) access to non-farm rural activities (iv) access to special nutrition programme; (v) access to basic services and healthy environment; (vi) access to credit services; (vii) access to population education and (viii) access to legal and constitutional rights. These outlines must be followed in utilizing bio-resources of the region. The present paper discusses some of the bio-resources like fish culture, apiculture, lac culture, sericulture and bio-prospecting of wild edible plants for sustainable livelihoods and rural employment.

Freshwater aquaculture can make a significant contribution to bridging the widening gap between demand and supply of fishery products in India in the face of declining capture fisheries production and growing populations. Sustainable development policies are needed that address use of natural resources, research, inputs and outputs and pricing, as well as problems in marketing, credit, trade, investment and exchange rates. Plurality of ownership for closed water bodies that is hindering fish culture should be settled and policy measures are needed to ensure equitable access to land and water resources by various groups. Protecting cultured water bodies from floods is very important for India. Fertilizers and agro-chemicals should be used judiciously on crops to protect the natural habitats of fish. The extensive and traditional fish culture practices should encourage intensive use of inputs to move to semi-intensive fish culture. Considerable public investment in infrastructure, capacity building and institutional strengthening is needed for sustainable development of aquaculture, with special attention to easing credit constraints. Marketing and related infrastructure improvements required include more physical space in fish markets, standardization in weighing; open-price bidding (auctioning); quality control measures; grading and pricing; and display of prices. Refrigerated transport is required for transport of fish, as well as adequate icing facilities at markets, to provide good quality fish for consumers and to help producers obtain better market prices. Research on fish nutrition should be undertaken in order to develop cheap but good quality feed. Identification of potential future constraints to spectral growth is an equally important area of research. Changing attitudes to decentralization implies more need for networking in the form of collaboration between research institutions and policies to this end need to be formulated and nurtured. Policy guidelines involving the participation of the private sector in research areas like fish marketing and processing should also be drawn up and followed through. Research findings on fish reproductive physiology in the laboratory should be applied to boost up aquaculture under field conditions and thus gap between laboratory and field should be bridged. More emphasis should be given on captive breeding of cultivable species, using synthetic gonadotropin-releasing analogues with suitable vehicle, so that fish seed is available throughout the year.

22 million fisher-folk of India depend on aquatic habitat for their livelihood (Kocherry, 2001). The Wetlands International and the IUCN Species Survival Commission (IUCN-SSC) have made collaborative venture towards the conservation and sustainable use of freshwater fish diversity globally (Gordon, 2005). Under the leadership of Marcel Silvius, they brought out publication on Wetlands, Water and Livelihood which was presented at the upcoming Ramsar Conference of Parties (Awoyinka , 2005). The strategy is on micro-credit, gender empowerment, chains of communication, needs-versus-rights ('Bio-rights') based approach and cooperation with the national government. This approach should be applied to food security. The recent field surveys and taxonomic work by Wetlands International in the Pacific Islands has revealed that how little we actually know about the freshwater fishes of this region (Jenkins, 2005). The hill stream fish fauna of the eastern Himalayan region has diverse species that are used by the local people for food and sports. Exploration of fish diversity in hill streams of the peninsular region of India is not taken seriously and awaits discovery of new species. The recent report of ZSI (Jabalpur Station) has brought out list of freshwater fishes of Madhya Pradesh, but could not identify new species. There are several villages on the bank of Narmada, which are mainly villages of fishermen. Our major task is now to survey their livelihood depending on local fish catch. The Central Institute of Fisheries Educations (CIFE) Center at Powarkheda is giving training to rural folk for aquaculture of fin-fish and prawn (*Macrobrachium rossenbergi*) and educating them regarding latest technologies for more income generation from fish culture.

The common constraints faced by freshwater fish farmers, especially in India, include: pluralities of ownership, lack of credit facilities, and lack of technical know-how, illegal poaching, deliberate poisoning and inadequate marketing opportunities. In addition, problems like multiplicity of pond ownership, non-recognition of aquaculture as a land-based activity, the absence of long-term leasing policies and non-assurance of seed supplies at appropriate times constrains the receipt of credit in India. In addition, the country is to some extent characterized by a social taboo against stocking community ponds. In general, common water bodies have been shrinking through degradation, encroachment and siltation. Industrial pollution is another problem for fish culture.

Marketing is organized almost entirely by the private sector, with some degree of government intervention.. Market structure, conduct and performance are poor in India. Similarly the markets are poorly competitive. Marketing channels are generally short due to the perishable nature of the product. Marketing infrastructure is poor, producers' bargaining power is low and flow of institutional credit for production and marketing of fish is low in this country, as it is dominated by the private sector.

In India, the Indian bee, *Apis cerana indica* and Italian bee *A. mellifera ligustica* are domesticated and recommended for commercial rearing. The honey yield per annum per colony of an Indian bee is about 6.5 to 9.5 kg, whereas that of Italian bees varies from 40 to 100 kg per colony per annum. Italian bees have been imported from Europe and successfully reared in the Kangra valley of Himachal Pradesh and now practiced in the states of Punjab, Himachal Pradesh, Jammu & Kashmir, Bihar and Uttar Pradesh. Italian bees are mild in temperament, swarm less and are very good gatherers of honey, but these are also highly susceptible to European foulbrood disease. There is a growing demand among the beekeepers in Northern India to adopt Italian bee. Artificial beehives are kept in the vicinity of flowering crops or trees to facilitate the cross pollination of crops or collection of nectar from blooming flowers. The site may change as per the blooming season and from field to field. The Khadi and Village Industries Board are promoting bee-keeping activity in the country. Recently bee-keeping has been identified as bankable project for loan to the farmers. The high-tech bee-keeping , using Apis mellifera ligustica, can yield not only more production of honey, but other valuable products of industrial importance (Belsare 2009).

Sericulture is a growing enterprise in India, although the problems are bound to be many and sometimes unavoidable. The problems faced by sericulturists mainly relate to the insufficient financial support from the government agencies, climatic hazards, wide fluctuations in cocoon prices, and also, to some extent, inadequacy of extension services (Ghosh & Ghosh, 1995). The other problems are rearing sheds, lack of awareness, low female literacy, absence of non-governmental organizations in the area, etc. There are no formal banking services for capital generation and economic organization. The ratio between land, labour and organization is based on caste and other social factors than economic factors.

Rural people have not yet come up with the scientific development for lac, They also lag behind because they have no marketing strategies (Rao & Singh 1990) Unlike the agricultural sector, this one is unorganized and the middleman makes the most profit. The transfer of technology looks after this need of the lac grower and interested entrepreneurs. Lac cultivation is a traditional occupation of tribal in Jharkhand and other parts of the eastern Indian plateau. It has a large untapped potential to generate livelihoods. In Jharkhand alone, more than 40,000 families (all from tribal communities) have unutilized lac host plants. According to survey conducted in Torpa block of Ranchi district, more than 3,000 families can earn Rs 5-7,000 from existing host plants. The key step is to speed up land allocations to farmers for lac culture. Assistance should be given for them to establish their own farms and raise compactly planted lac hosts which can be conveniently managed and protected. Loans should be given to the farmers for purchasing brood lac. There is an urgent need to train the farmers in lac cultivation techniques, such as growing host plants, rearing of lac insects, and harvesting of sticklac and broodlac. Farmers, who are former workers of state, lac plantations should be encouraged to establish their own lac farms. Adequate training in farming broodlac and financial assistance are essential for them. Extension of lac cultivation to the states like Jharkhand and Madhya Pradesh are necessary .Adequate funds should be provided to establish several broodlac farms in this area. Research should be conducted on developing bush-like host trees and on the introduction of new hosts, particularly dwarf forms, for intensive lac growing (Vishwanath *et al.*, 1994). Assistance should be given to establish an association of lac cultivators, traders, processors and consumers. This association would contribute to promoting lac production and stabilizing the market. Technology transfer of improved lac cultivation methods should be encouraged in states where lac is produced. Lac production should be encouraged in national economic and agro forestry programs for the mountainous areas and for other forestry programs supported by international organizations.

Total monitory output from products of edible plants is very high. The fruits of these plants can easily be collected where input is less. To create awareness among the people about their highly nutritious and medicinal value as a healthy food, not only help in solving malnutrition problem among the rural as well as semi-urban areas, but also for economic development of the region. These resources are critical especially for the rural poor in securing subsistence needs. They are more important at time of hardship and emergencies when quick cash is required or when casual food stores run out. The use of this wild edible plants for livelihoods of the rural people has not been taken up in India , except in Uttarakhand state (Maikhuri *et al.*, 2007).

REFERENCES

Awoyinka. K. 2005. Wetlands and livelihood working groups. Wetlands. **13:** 14.

Belsare, D.K. 1984. Fish production in tropical waters as protein and food source in developing countries. In UNEP Bulletin on Ecological Approaches to Resources (ed. E. Seidel), Dresden. **12:** 81-162.

Belsare, D.K. 2009. Introduction to Apiculture. Shivneri Pub. Amaravati pp 246.

Belsare, D.K. 1986. Tropical Fish Farming. Environmental Pub. Karad (Maharastra State).

Finlayson, M. President's Report. Wetlands. **12:** 3.

Ghosh, G..K. and S. Ghosh. 1995. Indian Textiles (Past and Present)., 306 p.

Gordon, R. 2005: Freshwater fish specialists group. Wetlands. **12:** 14.

Hawang, G.H. (1990). Uses of the lac insect in industries. *Chinese J. Entomology* (Taiwan), Special Publication No. 5, 147-152.

Kocherry, T. 2001. Biodiversity and communities that live on subsistence. A note prepared for the livelihood. TWG under NBSAP., National Fish Workers Forum, Thiruvanthapuram.

Maikhuri, R.K., V. Negi, L.S. Rawat & V.K. Purohit. 2007. Ecofriendly appropriate technologies for sustainable development of rural ecosystems in central Himalaya. G.B. Pant Institute of Himalayan Environment & Development. 38 pages.

Pandey, B.N. 2003. Prospects and potential of inland fisheries with reference to North Bihar – adoption of modern technologies of aqua-farming for enhancing aquaculture production. In 'Biodiversity Conservation, Environmental Pollution and Ecology.' (eds B.N. Pandey, R.K. Choudhary & B.K. Singh) APH Publishing Corporation, New Delhi. Vol. II: 3-13.

Rao, A.R. and Singh, P. (1990). Lac cultivation and marketing. *Indian Forester*, **116(5):** 459-463.

Scott, D.A. and Poole, C.M. 1989. A Status Review of Wetlands. AWB, Kaula Lumpur, Malaysia. 140 pp.

Vishwanath, T. H, S. *et al.* (1994). The butea tree for lac and rice production in India. Agroforestry *Today*, **6(2):** 10.

Worthington, E.B. and Lowe-MacConnell, R. 1995. African lakes reviewed: creation and destruction of biodiversity, Environ. Conser. **21:** 199-211.

Bioresources for Food Security and Rural Livelihood, Pages 47–58
Edited by : G.K. Kulkarni and B.N. Pandey

BREEDING TECHNIQUES OF INDIAN MAJOR CARPS: ADVANCES AND FUTURE STRATEGIES

P.D. Prasada Rao, P. L. Paul and T.K. Ghosh

ABSTRACT

The major carps of India, that contribute a considerable proportion of farmed fishes, fail to breed in ordinary perennial tanks, but breed profusely in flooded waters having altered environmental conditions during monsoon season and establish challenges with respect to breeding in captivity. This review outlines the initial application of hypophysation technique for fish breeding, analyses the involved shortcomings and traces the improvements made thereafter. To avoid spoilage and achieve longer shelf-life, acetone-dried pituitary powder and other powder-forms like, 'Sumach' have been evolved. Emergence of knowledge about the hypophysial gonadotropins (GTHs) has led to devising new spawning agents using this hormone. Partially purified GTH and human chorionic GTH were also used for breeding. Demonstration of gonadotropin-releasing hormone (GnRH), a decapeptide in fish brain that is involved in GTH regulation has been a land-mark discovery. Additional knowledge about dopaminergic inhibition of GTH, blockage of dopamine action by domperidone, supplemented by *in vitro* and *in vivo* studies has resulted in formulation of a well-known spawning agent, ovaprim. Its excellent performance has led to its introduction into India and extensive usage with commendable success. Later, other allied forms (ovatide, wova-FH) were commercialized based on the same ovaprim composition. In order to provide fish seed during periods beyond monsoon months, multiple breeding techniques were developed after subjecting the fish to additional procedures like priming them with carp pituitary injections, altering photoperiods and providing specific diet to the fish etc. Newer approaches for improved breeding deserve attention. Several synthetic GnRH anlogues (GnRH-a) were developed and *in vitro* studies using of pituitary fragments perfusion system revealed that some superactive GnRH-a such as [D-Arg^6]-GnRH was more potent than salmon GnRH. Future investigations aimed at utilization of superactive GnRH analogues, and attempts to exploit GTH inhibiting hormone (GnIH) for improved breeding are warranted. All the above breeding techniques are of invasive nature involving some degree of stress to the brood fish, and it would be valuable to develop non-invasive procedures. Electrical stimulation of olfactory system leads to release of neurosecretory material (NSM) from the hypothalamo-hypophysial system. This classical NSM includes release of several neuropeptides that partake in GTH release. It is hypothesized that providing appropriate chemical stimuli through sprays in water could trigger release of (neuro)hormones essential to induce breeding and any attempt along these lines would be a valuable step that might lead to a break-through.

Keywords : Major Carps – Gnrh - Ovaprim – Synthetic Gnrh Analogues - Sensory System - Environmental Cues

INTRODUCTION

With immense possibilities for both open water fisheries and fresh water aquaculture, India ranks second only to Japan in inland sector fish production. The inland-sector, which has a growth rate of 6 per cent, contributes around 55 per cent of the yearly total fish production of about 6.57 million metric tones. Carps contribute as much as 87 per cent of the total aquaculture production. The resources in the inland sector comprise of river and canals 1,91,024 km, reservoirs 3.15 million ha, tanks and ponds 2.85 million ha, other types of lakes and derelict water 1.3 million ha. Thus there is a vast scope for culture of fishes including major carps of India *viz., Labeo rohita, Cirrhinus mrigala* and *Catla catla.* Since these carps do not breed in ordinary tanks, other approaches are essential to obtain required fish seed. Control of reproductive function including breeding in captivity is one approach to achieve the fish seed target for the sustainability of commercial aquaculture production, Administration of different spawning agents to induce fish breeding has been adopted successfully and methods like prior priming the fish with pituitary injections prior to actual breeding, subjecting them to altered photoperiod, temperature, and providing specific feed have proved to be advantageous in repeat spawning. Despite excellent performance in major carp breeding, there are certain aspects that need to be explored for improved breeding and there exist possibilities for evolving non-invasive methods of breeding as the practiced methods are not stress-free.

Reproductive Patterns in Fishes

In fishes, the reproductive activity is mostly cyclic but the frequency of spawning is variable. The rainbow trout breeds only once in lifetime, while a few others bred once in a few years, once in a year, once in a month or even every few weeks. Interestingly, the black bream *(Acanthopagrus butcheri)* and greenback flounder *(Rhombosolea tapirina)* are multiple-repeat spawners probably on a daily basis (see Hobby *et al.*, 2000; Suresh *et al.*, 2008). However, most cyprinids inhabiting tropical countries and all Indian major carps do not breed in ordinary perennial tanks, but breed only in rivers, inundated fields and certain types of tanks called the "bundh" type of tanks during monsoon period. Amazingly, aquatic species particularly those in the tropical zone, have evolved strategies to breed at a time when there is abundance of water, *e.g.* during the monsoon season in order to ensure maximal gamete survival, fertilization and survival of their young ones.

Environmental Conditions and Gonadal Maturation

The major carps of India show gonadal recrudescence from March to June when both photoperiod and temperature exhibit an increase. It is generally believed that these two environmental factors play an important role in initiating gonadal recrudescence. Fishes integrate their physiological process with environmental cycles, and the endogenous periodicities of physiological mechanisms are responsible, in part, for seasonal reproduction. In addition, certain environmental factors which act as cues for the approaching favourable season for reproduction impinge on the sensory organs and through them affect the central nervous system, the hypothalamus and the pituitary gland and finally the gonads. It is through such environmental factors that the endogenous rhythm is brought into a state for the precise breeding time. Photoperiod, temperature and seasonal rainfall are important environmental factors that regulate reproductive cycles in fishes. Spawning is initiated by environmental factors that prevail during the monsoon season. Monsoon rains bring about several changes in water conditions including an increase in water levels, velocity of flow, flooding of shallow areas, dilution of certain ions and concentration of others, changes in smell and taste of water as well as changes in biotic parameters such as growth of algae and others green vegetation. Environmental conditions required for initiation of oocyte maturation, ovulation and spawning are much more critical than those for spermatogenesis. In the majority of fishes, maturational processes are initiated only when thermal conditions are optimum. In others, a precise combination of day length and temperature is required. In still others, radically different environmental stimuli such as rainfall and inundation precipitate maturational events.

Important Factors for Spawning

The major carps of India do not breed in ordinary perennial tanks, though the males and females are sexually mature, exist in proper male-female ratio and healthy. On the other hand, they profusely breed in inundated rivers and certain tanks probably induced by a combination of several environmental factors in the monsoon period (Hora, 1945). Several chemical, physical and environmental factors have been suggested to trigger breeding. These comprise increased pH, turbidity, flow of water, concentration of dissolved minerals, depth of water, minerals, nature of vegetation and several environmental factors such as cloudy weather and rainy conditions, fall in temperature, flooded condition, even lunar periodicity, change in photoperiod etc. However, the nature of control they exercise needs to be evaluated. Furthermore, it is not known which one or combination of which of the suggested factors, in fact, partakes in or stimulates the spawning process. The role of pheromones in stimulation of spawning process is also not understood.

Pathways Involved in Activation of Spawning

Specific combination of some or all of these factors may be perceived through sensory system and conveyed to the hypothalamus, which then activates the pituitary to release gonadotropins (GTHs) and other hormones to induce maturation, ovulation and spawning. Generally, the reproductive activity is influenced by the brain-pituitary-gonadal axis, involving mediation of forebrain gonadotropin-releasing hormone (GnRH), pituitary GTH and sex steroids (Maruska *et al.*, 2007).

Gonadotropins in Teleosts

Gonadotropins are of pivotal importance in fish reproductive physiology as these hormones regulate various reproductive processes in carp breeding. Since major carps do not breed in confined waters, one has to depend upon induced breeding of fish, which mostly involves utilization of GTH for this purpose. Before going into the details of neural regulation of GTH release, the following is a brief description about the current knowledge about gonadotropins and localization in the teleostean pituitary gland. Initial studies on the pituitary gland of fishes did not reveal two different hormones comparable to mammalian follicle stimulating hormone and luteinizing hormone. However recent studies have revealed that both types of gonadotropins, *viz*., GTH1 and GTH2 occur in fishes (such as tuna, goldfish and eel) and that these are structurally similar to those in tetrapod pituitaries (see Trudeau, 1997)

INITIAL FISH BREEDING PRACTICES

The pioneering discovery of Houssay (see Chondar, 1970) that fishes can be induced to spawn by injecting pituitary homogenates has somewhat mitigated the problem. The principal advantage of this technique is referred to as hypophysation in aquacultural practice. A particular species of fish may be injected with the hypophysial extract of the same species or with the pituitary extract of a different species and these extracts are referred to as homoplastic and heteroplastic pituitary extracts respectively.

Effect of Hypophysation

The precise combination of environmental factors required for maturation, ovulation and spawning, when available, brings about an accelerated release of GTH from the pituitary into the blood stream. As a result, GTH which was hitherto being secreted only at a tonic level for the maintenance of yolky oocytes, suddenly appears as a surge. The elevated titres of gonadotropin bring about maturational changes culminating in the act of spawning. However, quite often, under farm conditions, the requisite environmental factors are either not available or do not persist for sufficient length of time for spontaneous maturation to occur. By simulating the natural GTH surge, it bypasses, to some extent, the environmental variables of photoperiod, temperature, rainfall, etc.

Maturation and ovulation have been induced in a variety of fishes including many cultivated fishes such as rohu *Labeo rohita,* mrigal *Cirrhina mrigala,* and catla *Catla catla,* grass carp *Ctenopharyngodon idella*, silver carp *Hypophthalmichthys molitrix,* common carp *Cyprinus carpio* and also in the murrel *Channa punctatus*. The technique of hypophysation is an established method for inducing spawning in cultivated fishes that mature under confined conditions (Chaudhuri and Alikunhi, 1957; see Chondar, 1970). Nevertheless, the collection of pituitary glands is not only time-consuming but also involves the sacrifice of a potential broodfish for the collection of glands from fish to be marketed, which often results in considerable devaluation of the product.

Drawbacks of Hypophysation

A serious drawback of the conventional hypophysation technique is the lack of standardization of the pituitary material, the demand for which generally exceeds supply. The pituitary extract is a mixture of several hypophysial hormones, some of which may not contribute to the physiology of gonads and reproduction. The gonadotropic potency of the injected material depends on the stage of sexual maturity and physiological state of the donor, as well as on the method of pituitary collection and preservation, while its effect in the recipient depends on its sexual maturity and on donor-recipient phylogenetic relationships. These extracts cannot be stored for a long time.

Inexpensive Agents with Long Shelf-Life

In view of the above-mentioned limitations, there have been efforts to identify potential fish spawning agents in order to breed fish in a more effective manner. Investigations have been conducted to uncover agents that are inexpensive, potent, with predictable activity, long shelf-life and effectual in simplifying the process of fish breeding. Consequently, several products were augmented to achieve the goal. Some important ones include acetone-dried salmon pituitary powder termed "Sumach", human chorionic gonadotropin (hCG), Ovaprim and Ovatide, Wova-FH. The last three products show some definite advantages over the pituitary gland extracts.

Acetone-dried Salmon Pituitary-Powder

As indicated above this pituitary - powder has the same drawbacks as the pituitary gland extracts, except for the fact that this has the benefit of long shelf-life.

Sumach

This product is also a dried pituitary gland extract but had the same limitations cited above. Some reports mention that this power is prepared using human chorionic gonadotropin. With the availability of newer spawning agents, the use of "Sumach" has become diminished.

Role of Partially Purified and Purified GTH

The mammalian gonadotropic hormones *viz*., luteinizing hormone (LH) and human chorionic gonadotropin (hCG), were found to be effective in inducing maturation and ovulation in fishes. Partially purified or purified piscine pituitary GTH isolated from carp and salmon has been extensively used for inducing ovulation and spawning in a number of fishes.

Maturation, ovulation and spawning have been induced in major carps using hCG, LH, partially purified salmon gonadotropin (SG-G100) (Chaudhuri *et al*., 1977). Similarly in the catfish, *Heteropneustes fossilis* SG-G100 along with hCG or even deoxycorticosterone were found to be effective in inducing spawning. Human chorionic gonadotropin was known to form antibodies in the fish and was not found suitable. Consequently continued research has helped in development of new agents that have been proved to be superior to the agents discussed so far.

ADVANCES IN FISH BREEDING

Single and Multiple Spawning

Generally, the major carps are bred only once through hypophysation and earlier investigators have not attempted to apply any experimental approach to enable to attain another chance of maturation or to use them for a subsequent induced breeding. However, recent years have witnessed attempts to attain multiple spawning of carps for enhanced seed production. Successes have been achieved to prepone and postpone induced breeding periods, without confining to monsoon months. Nowadays carps have been domesticated to breed well ahead of the monsoon and beyond, ensuring seed availability during the pre- and post-monsoon months. Priming the fish, *L. rohita* with purified carp GTH has led to enhanced oocyte maturation in premonsoon season to such an extent, that no marked seasonal differences in the gonadal condition was evident. Dasgupta *et al.* (2009) have further shown that the unresponsiveness of oocytes to hormonal induction was the primary cause for reduced spawning performance in pre-monsoon rohu. This problem was deciphered through injections of purified carp gonadotropin to achieve better spawning response in rohu during the pre-monsoon period. These authors have shown that the rohu could be induced to breed successfully as early as in May. Multiple spawning has been shown to be possible two or more times in a season through administration of combination of gonadotropin releasing hormone (GnRH) and domperidone (see below for details of this combination) opting the routine hypophysation technique. Major carps have been bred as many as four times between March and September. The brood fish carps used for multiple spawning have been referred to as professional broods (Gupta *et al.*, 1988; Dasgupta *et al.*, 2009). Nandi *et al.* (2007) have shown that enrichment of fish diet with polyunsaturated fatty acids not only improved the quality of eggs and larvae, but also enhanced the physiological response of broodfish to induced spawning. Oil enrichment in the broodfish diet should therefore be mandatory to ensure the optimum polyunsaturated fatty acid content for warm water carp for quality seed production. Exposure of fish to altered photoperiods has also permitted attainment of sexual maturation during periods that are different from normal maturation times through involvement of pineal hormones (Dey *et al.*, 2007).

Gonadotropin-Releasing Hormone Peptide

It has been known for a long time that hypothalamus regulates the pituitary hormones through various hormone-releasing and/or hormone-inhibiting factors or hormones. However, recent advances involving application of high performance liquid chromatography (HPLC), immunocytochemical and radioimmunoassay procedures have brought to light a wealth of new information that necessitated alterations in previous concepts about regulation of different hypophysial hormones. The teleostean GTH is known to be under a dual (stimulatory and inhibitory) control. The stimulatory control is exercised by several neuropeptides, the most important being GnRH. Gonadotropin-releasing hormone was originally referred to as luteinizing hormone - releasing hormone (LHRH) and it was isolated and synthesized in 1971 using procine and above hypothalamic extracts (Matsuo *et al.,* 1971). Although it was thought to be a unique form, biochemical studies have revealed that GnRH molecule occurs as a family of structurally similar peptides. To date, ten distinct GnRH molecular forms have been characterized from vertebrates and two from a protochordate. In all the ten forms, aminoacids 1,2,4,9 and 10 resides that are important for receptor binding and/or biological activity, are conserved (cf. King and Millar, 1982; Sherwood, 1994). Fish GnRH was first isolated from chum salmon (Sherwood *et al.*, 1994). The salmon GnRH (sGnRh) is not different from the mammalian GnRH (mGnRH); both have 10 aminoacids, 80% sequence similarly and the same N-terminal (pGlu) and C-terminal (NH_2) modifications. Following the first discovery of sGnRH structure, several distinct GnRH peptides have been isolated and sequenced from different species of fish. The ten distinct forms of GnRH peptides show sequence similarly of 50% or greater. Recent studies revealed that more than one type of GnRH molecule exists in the same species. So far, as many as five molecular forms of GnRH have been identified *viz*., seabream (sbGnRH), chicken (cGnRH-II), salmon (sGnRH), catfish (cfGnRH) and mammalian (mGnRH) in the brain of a cichlid and pumpkinseed fish. It was also shown that three forms of GnRH are widespread in the order Perciformes. In the goldfish we have shown that there are three types of receptors in the brain (Baby *et al*., 2002). Five different types of putative GnRH receptor gene are expressed in the brain of Masu Salmon (Jodo *et al*., 2003).

GTH Release by GnRH

Release of GTH from the pituitary is regulated by GnRH from the hypothalamus and other extra hypothalamic sites of the brain, although several neuropeptide participate in this function. The information about the effect of GnRH on GTH has primarily stemmed from *in vitro* studies that involved maintenance of pituitary fragments in a perifusion system, challenging these fragments with GnRH and measurement of GTH levels in the medium using radioimmunoassay method. Several studies, conducted at the laboratory of Prof. Peter, have amply elucidated the role of GnRH and its analogues on the release of GTH from pituitary fragments maintained in medium. However, it must be noted that the results of *in vivo* studies on GnRH effects on serum GTH in the goldfish are not very similar. Injections of superactive mammalian GnRH agonist [D-Ala^6-Pro^9-NEt]-LHRH (LHRH-A) and salamon GnRH agonist [D-Arg^6-Trp^7-Leu^8-Pro^9-NEt]-LHRH (sGnRH- A) stimulate only a modest increase in serum GTH and are ineffective in inducing ovulation and spawning in these species. This has led to investigations to unravel this perplexing difficulty for induced breeding. *In vivo* studies revealed that certain hypothalamic neurotransmitters are involved in inhibition of GTH release. In the goldfish, this inhibition was found to be exercised by dopamine and this important clue has opened avenues to tackle the problem of induced breeding in a successful manner.

Dopamine Inhibition of GTH and its Blockage

The knowledge about the inhibitory role played by dopamine in hypophysial GTH releasehas brought to light the necessity for the identification of dopamine blockers. Reserpine depletes all catecholamines and is highly effective in potentiating the action of GnRH; however, this drug has side effects and hence not useful. Domperidone and pimozide are most effective dopamine blockers. The former drug does not cross blood brain barrier and hence the side effects are minimal. Therefore, this is the most effective dopamine blocker and hence used in the preparation of ovaprim.

"Linpe" Method and Ovaprim

Administration of dopamine receptor antagonist pimozide (PIM) or domperidone (DOM) greately potentiates the action of LHRH-and sGnRH-A on GTH release. Administration of GnRH and of domperidone proved to be highly effective in inducing ovulation and spawning in several species of fishes. This finding by Prof. Lin and Prof. Peter that is useful for induced ovulation and spawning of cultured fish is referred to as is called the "Linpe" method after the names of these two investigators. The "Linpe" method has been commercialized by Syndel Laboratories Inc., Canada, under the commercial name Ovaprim (cf. Lin and Peter, 1991). Ovaprim was extensively tested on Indian major carps and other fishes and was found to be an excellent agent for induction of ovulation and spawning in fishes. Currently, it is marketed in our country by Glaxo, India and this has been in extensive use in India for induction of spawning in Indian major carps. Subsequently, other spawning agents have been developed in India, some of which are mentioned below.

Ovatide

Subsequent to ovaprim's availability in India, a cheaper spawning agent, termed as Ovatide was prepared and marketed in India by Hemma Pharma. Composition-wise it appears to be similar to that of ovaprim. This is extensively used and claimed to yield good results.

Wova – FH

Another spawning agent, termed as Wova – FH, introduced and marketed by Biostadt in India, is also used for breeding Indian major carps. According to the manufacturer's information the composition seems to be a replica of the original ovaprim.

Efficacy of Different GnRH-based Spawning Agents

It was reported that injections of ovaprim, ovatide, and wova – FH yielded about 90-100% breeding response (mean spawning fecundity, mean spawn recovery rate), whereas carp pituitary extract injections yielded about 77 - 84% breeding performance (Rath *et al.*, 2007). However, it would be important to evaluate which of three GnRH-based spawning agents is more effective in inducing the fish to breed.

FUTURE STRATEGIES

Other GTH Regulators

Although ovaprim, ovatide and Wova-FH are currently considered as the best induced spawning agents, there are several hormones and peptides/neurotransmitters that cause GTH release. Serotonin, neuropeptide Y, norepinephrine, gamma amino butyric acid, endogenous opioid peptides, and gaseous nitric oxide (see Baby *et al.*, 2000) are known to play a role in GTH release. Attempts to examine the usefulness of these either independently or alone would be valuable.

Superactive Synthetic GnRH Analogues

Discovery of the primary structure of mammalian LHRH (pGlu^{1}-His^{2}-Try^{5}-Gly^{6}-Leu^{7}-Arg^{8}-Pro^{5}-Gly^{10}-NH2 was a milestone in neroendocrine research. Several synthetic GnRH analogues are available and the efficacy of these analogues with respect to their ability to stimulate the release of GTH in teleostean species has been investigated employing in vitro perifusion experiments by Prof. Peter and his collaborators. *In vitro* studies using pituitary fragment perifusion system have shown that [D-Arg^{6}]-GnRH was more potent than sGnRH which was itself significantly more potent than [Arg^{6}]-GnRH. [DAla^{6}]-and [D-Leu^{6}]-GnRH were as potent as sGnRH, whereas the corresponding L-aminoacid-containing analogues, [Ala^{6}]-and [Leu^{6}]-GnRH were significantly less potent. [D-Arg^{6}-Trp^{8}]-GnRH and [D-Lys^{6}-Trp^{7}-Tyr^{8}]-GnRH had a high potency relative to GnRH. Although [His^{5}]-GnRH had a lower potency, [His^{5}, D-Trp^{6}]-GnRH had a higher potency than sGnRH. These findings indicate that substitution of selected D-aminoacids in position 6 is essential for development of superactive analogues of GnRH. Generally, [D-Arg^{6}] - or [D-Lys^{6}]-GnRH in combination with constituent hydrophobic aminoacids of salmon GnRH and chicken GnRH-II provides analogues of highest potency in the goldfish (cf. Peter *et al.*, 1995). Despite the superior ability of superactive analogues to trigger the release of GTH, no attempt has been made to test these analogues to induce breeding. Studies in this direction may prove rewarding.

Gonadotropin Inhibitory Hormone

A novel peptide, that inhibits gonadotropin release, was identified in quail in the year 2000 and termed as gonadotropin-inhibitory hormone (GnIH). GnIH acts on the pituitary and gonadotropin-releasing hormone (GnRH) neurons in the hypothalamus via a novel G protein-coupled receptor for GnIH to inhibit gonadal development and maintenance by decreasing gonadotropin release and synthesis. The occurrence and functional nature of comparable peptide is yet to be identified in the carps. If present, then the extent to which it plays a role in GTH inhibition needs to be investigated. Furthermore, it would be useful to explore the possibilities to block its release during appropriate age of the fish with a view to achieving better spawning in the major carps.

Non - Invasive Breeding Technique Possibilities

We have shown that the teleostean pituitary gland receives neuronal processes from several nuclei in the hypothalamus (Prasada Rao *et al.*, 1993). We also showed that the nucleus preopticus (NPO) is the major source of neuronal processes in the neurohypophysis. Although the NPO is composed numerous classical neurosecretory neurons, recent immunocytochemical studies have shown that this nucleus contains several neuropeptides and some of these are co-expressed by the same neuron. Several neurpeptides/ neurotransmitters such as GnRH, neurophysin, neurotensin, galanin, growth hormone-releasing hormone, cholecystokinin, somatostatin, gastrin, endorphins, melanophore concentrating hormone, urotenisn I, gamma aminobutyric acid, neuropeptide Y, arginine vasopressin, arginine vasotocin, corticotropin-releasing factor, endorphins etc. have been identified in the NPO neurons. Electrical stimulation of olfactory tract has resulted in release of neurosecretory material from the preoptico-hypophysial complex (Prasada Rao and Dabhade, 1973). It would be worthwhile to consider application of this principle of olfactory stimulation to augment the release of hypophysial GTH to result in the release of eggs in the process of spawning. With the commencement of monsoon season, a combination of complex environmental factors is generated and some or all of these seem to trigger the process of spawning. The environmental cues seem to stimulate the olfactory system that signals the brain-pituitary complex, acting as a link between the environment and hypophysial GTH. It is not known which of the neuropeptides are released as a result of olfactory stimulation and the levels of GTH in the circulating blood. Investigations along these lines might yield rewarding information that could be used in evolving non-invasive fish breeding techniques. Non-invasive procedures would also be useful in avoidance of stress while handling of broodfish for injection of spawning agents, and other experiments connected with induced breeding.

ACKNOWLEDGEMENTS

One of the authors (PDPR) was supported by "INSA Senior Scientist" Scheme, New Delhi. The authors are thankful to the Director, National Environmental Engineering Research Institute, Nagpur, for encouragement and help. PLP is grateful to the Principal, Hislop College, Nagpur, for help and encouragement.

REFERENCES

Baby, S.M., Ueck, M. and Prasada Rao, P.D. (2000) Gonadotropin-releasing hormone-immunoreactive neurons and associated nicotinamide adenine dinucleotide phosphate-diaphorase-positive neurons in the brain of a teleost, *Rhodeus amarus*. Gen. Comp. Endocrinol., **120:** 44-54.

Chaudhuri, H. and Alikunhi, K.H. (1957) Observation on the spawning in Indian carps by hormone injection. Curr. Sci., **26:** 381-382.

Chaudhuri, H., Bhowmick, R.M., Kotwal, G.K., Jana, R.K. and Gupta, S.D. (1977) Efficacy of partially purified salmon pituitary gonadotropin (SG-G100) in induced spawning in Indian major carps. J. Inland Fish. Soc., India **9:** 162-164.

Chondar, S.L. (1970) Hand book on breeding of Indian major carps by pituitary hormone injection. Satish Book Enterprise, Agra, India.

Dasgupta, S., Sarkar, S.K., Sarangi, N. and Bhattacharya, S. (2009) Variation in spawning responses, egg and larvae productions from induced rohu (*Labeo rohita*) during pre-monsoon and monsoon seasons: Relationship with hormonal changes and oocyte responsiveness during final maturation. Aquaculture, **290:** 320-326.

Dey, R., Bhattacharya, S., and Maitra, S.K. (2004) Temporal pattern of ovarian activity in a major carp *catla catla* and its possible environmental correlate in an annual cycle Biol. Rhythm Res., **35:** 329-353.

Gupta, S.D., Reddy, P.V.G.K., Pani, K.C. (1988) Advancing maturity and spawning in Asiatic carps through brood stock management. Proc. Carp Seed Production Tech. Asian Fisheries Society, Indian branch, pp. 29-34.

Hobby, A.C., Geraghty, D.P. and Pankhurst, N.W. (2000) Differences in binding characteristics of sex steroid binding protein in reproductive and nonreproductive female rainbow trout (*Oncorhynchus mykiss*), black bream (*Acanthopagrus butcheri*), and greenback flounder (*Rhombosolea tapirina*). Gen. Comp. Endocrinol., **120:** 249-259.

Hora, S.L. (1945) Analysis of the factors influencing the spawning in carps. Proc. Natl. Inst. Sci., India, **11:** 303-312.

Jodo, A., Ando, H., and Urano, A. (2003) Five different types of putative GnRH receptor gene are expressed in the brain of masu salmon (*Oncorhynchus masou*). Zool. Sci., **20:** 1117-1125.

King, J.A. and Millar, R.P. (1982) Structure of chicken hypothalamic luteinizing hormone-releasing hormone. II. Isolation and characterization. J. Biol. Chem., **257:** 10729-10732.

Lin, H.R. and Peter, R.E. (1986) Hormones and spawning in fish. Asian Fisher. Sci., **9:** 21-33.

Matsuo, H., Arimura, A., Nair, R.M. and Schally, A.V. (1971) Synthesis of the porcine LH- and FSH-releasing hormone by the solid-phase method. Biochem. Biophys Res. Commun., **45:** 822-827.

Maruska, K.P, Mizobe, M.H, Tricas, T.C. (2007) Sex and seasonal co-variation of arginine vasotocin (AVT) and gonadotropin-releasing hormone (GnRH) neurons in the brain of the halfspotted goby. Comp. Biochem. Physiol. A. Mol. Integr. Physiol., **147:** 129-44.

Nandi, S., Routray, P., Gupta, S.D., Rath, S.C., Dasgupta, S., Meher, P.K., Mukhopadhyaya, P.K. (2007) Reproduction performance of carp *Catla catla* (Ham.) reared on a formulated diet with PUFA supplement. J. Appl. Icthyol. **23:** 684-691.

Peter, R.E., Nahorniak, C.S., Bastos, V.L., Murthy, C.K., Prasada Rao, P.D., Milton, R.E., de L., Millar, R.P. and Rivier, J.E. (1995) 5th Internat. Symp. Reprod. Physiol., The University of Texas, Austin, pp. 52.

Peter, R.E., Prasada Rao, P.D., Baby, S.M., Illing, N., and Millar, R (2003) Differential brain distribution of gonadotropin-releasing hormone receptors in the goldfish. Gen. Comp. Endocrinol., **132:** 399-408.

Prasada Rao, P. D. and Dabhade, K. B. (1973) *In situ* demonstration of changes in preoptico-neurohypophysial complex after electrical stimulation of the olfactory tract in the catfish, *Clarias batrachus* (Linn.) Z. Zellforsch., **139:** 95-110.

Prasada Rao, P. D., Job, T. C., and Schreibman, M.P. (1993) Hypophysiotropic neurons in the hypothalamus of the catfish, *Clarias batrachus:* A cobaltous lysine and HRP Study. Brain Behav. Evol., **42:** 24-38.

Rath, S.C., Sarkar, S.K., Gupta, S.D. and Sarangi, N. (2007) Comparative account of induced breeding of Indian major carps with ovaprim, ovatide, wova-FH and carp pituitary extract. Ind. J. Anim. Sci., **77:** 1057-1060.

Sherwood, N.M., Harvey, B., Brownstein, M.J., Eiden, L.E. (1984) Gonadotropin-releasing hormone (Gn-RH) in striped mullet (Mugil cephalus), milkfish (Chanos chanos), and rainbow trout (*Salmo gairdneri*): comparison with salmon Gn-RH. Gen. Comp. Endocrinol., **55:** 174-181.

Suresh, D.V.N/.S., D.V., Baile, V.V. and Prasada Rao, P.D. (2008) Annual reproductive phase-related profile of sex steroids and their carrier, SHBG, in the Indian major carp, *Labeo rohita*. Gen. Comp. Endocrinol., **159:** 143-149.

Tsutsui, K. and Osugi, T. (2008) Evolutionary origin and divergence of GnIH and its homologous peptides. Gen. Comp. Endocrinol., **161:** 30-33.

Trudeau, V.L. (1997) Neuroendocrine regulation of gonadotrophin II release and gonadal growthin the goldfish, *Carassius auratus.* Rev. Reprod. **2:** 55-68.

Bioresources for Food Security and Rural Livelihood, Pages 59–73
Edited by : G.K. Kulkarni and B.N. Pandey

IMPROVEMENT OF BIORESOURCES THROUGH BIOTECHNOLOGICAL APPROACHES

M. Bhaskar

ABSTRACT

The demands for urbanization and production have created a great challenge to the sustainability of bioresources in India. Conserving the precious bioresources has become the greatest challenge of balancing productivity and development. It is well known that the diversity of organisms is the product of millions of years of evolution that have shaped the natural communities, species and gene pools. The number of genes, species and ecosystems in India are not completely known so for and an effective conservation mechanism and balanced productivity can provide over all development in future. The researchers need to emphasize biotechnological approaches for rapid biodiversity assessment, natural resource and conservation management, environmental monitoring, impact assessment and exploration of new biological resources.

The tools of Biotechnology has become the major preoccupation of biologists and industries interested in commercial exploitation of bioresources for identifying new processes targeted to a variety of medical, agricultural and industrial applications. Bioresources form inexhaustible source for novel products of immense commercial value, can make life comfortable. Our life style, culture and economy are becoming increasingly dependent on our management, conservation, and exploitation of biological resources and products.

An attempt has been made to focus on current status of bioresource, under the impact of natural calamities and anthropogenic activities on homeostatic diversity of bioresource and to explore the biotechnological approaches role in improving bioresource production to meet the food challenges in 21st century.

Keywords : Biotechnology, Food challenges, Bioresources, Biomass, Bioenergy, Forest resources.

INTRODUCTION

Richness and diversity of bioresources are the biological wealth of a given nation. India is one of the few fortunate countries in the world endowed with plenty of bioresources of varied diversity from the lofty Himalayan mountains to the confluence of great oceans at Kanyakumari. India is home to 8 % of the recorded species in the world and is the first country of 'mega' diversity to have enacted legislation in 2002 that will govern extraction and use of biological resources. Bioconservation, bioresources and biotechnological approaches are the important corner stones of sustainable development.

Scientists and technologists can play essential roles in the selection of technological alternatives that are appropriate to people's long-term needs. It is essential to design a simple and reliable mechanical systems for underdeveloped regions, can have relevance today in an increasingly interdependent, crowded, and polluted world. Biotechnologists can help in two ways to promote technologies that make sense, providing for future wellbeing, and minimizing risks. First, can we exercise personal judgment in our work, as we pursue technological progress. We need to consider whether the likely products of our work are compatible with the world that we want to leave for our grandchildren. Second, we can provide guidance to our fellow citizens, as society grapples with the political and economic choices associated with progress.

Biotechnology tools have most often been used in the context of aid to third-world countries. Equipment and processes to be used in different parts of the world need to be suited to the local conditions, including the availability of electricity, running water, *etc*. It doesn't make sense to send a lot of plug-in refrigerators to a region that lacks a power grid. Likewise, expensive farming equipment will become useless in a land where there are no spare parts or people able to do the repairs. The concepts of appropriate technology also are needed most urgently in the developed world. The idea is not new, but we have to list out the ways in which modern societies can benefit from technologies that are inherently simpler, cheaper, safer, or less centralized than the current state of the art. Present lifestyles are becoming vulnerable due to the pressures of population growth, climatic change, and depletion of resources. Hence, it is essential to pay attention on the following aspects to improve bioresources which helps to meet the challenges of food scarcity in 21st century.

Use of Communication Technology

Information and Communication Technology (ICT) and other relevant technologies enhance the creation of data sets, sample designs and experimental designs for adaptive management in production, conservation and disturbed or polluted, environments.

Use of Biodiversity as a Tool to Improve Bioresources Production

Diversity and richness of organisms generate the ecosystem processes that regulate air, soil and water quality, maintain soil fertility, and breakdown domestic, industrial and agricultural wastes. Biodiversity underpins our most important industries agriculture, forest and tourism and many aspects of daily life. The disciplines that understand biodiversity - evolutionary biology, ecology, genetics, systematic and natural history - are yielding arrays of new concepts and products from a wide variety of unexpected sources. Applications of various molecular biology techniques and bioinformatics tools are used to enhance the bioresources production.

Use of Enterprise Technology

Bioenterprise primary objective is to identify, exploit and commercialize bioresources and biological processes of animals and plants as an enterprise of a biologist and or industrialist. Traditional entrepreneurial activities like aquaculture, apiculture, sericulture and vermiculture need to be integrated with ICT and biotechnology to meet the ever increasing demand for food and fodder. Human resources trained in life sciences need to be sensitized to these issues and should be equipped with skills to commercialize bioprocesses and optimize productivity through training in entrepreneurial skills to exploit biological resources. Biofertilizers, biopesticides and integrated pest management should form the main training areas for the farmers to educate them for the improvement of crop production.

The loss of forest's and bioresources is occurring at an alarming rate, a consequence of increasing population pressure, agricultural land degradation, urbanization and neglect. There is a growing recognition worldwide that conservation and management of bioresources are pressing priorities in the world today. The choice of conservation methods and techniques depend on the objectives of the particular conservation effort, the breeding system and behaviour of the species in question as well as the available resources including funds, trained personnel, infrastructure and technologies. The use of biotechnological tools and "bioprospecting" will open new vistas in medicine, agriculture, silviculture, horticulture, environment and other important issues.

WHAT IS BIORESOURCES?

Bioresources refers to the total biological variation manifested as individual plants, animals or their genes, which could be taken by man for use as drugs, food, livestock feed, construction materials for shelter, environmental protection, etc or in the development of improved crops and animals for higher yield and tolerance to biotic and abiotic stresses. Man depends on these bioresources for his continued existence and, therefore, he must use and preserve them for future generations yet unborn. The concept of sustainable development indicates that economic and environmental protection are inextricably linked and that the qualify of present and future life rests in meeting basic human needs without destroying the environment on which life depends.

Use of Forest Resources

Deforestation and forest degradation are large-scale problems in developing countries. Since we cannot do without exploiting the available bioresources to our advantage, there has to be a balance between uses of resources and their conservation. In this way, we would preserve an ecosystem, which although altered would still be rich in bioresources and at the same time would provide food and other needs as well as perform vital environmental functions on a long term basis. It is necessary to conserve time naturally endowed resources which apart from their direct usage by man, also serve several ecological functions.

Use of Modern Biotechnology

Biotechnology can be defined as any technique that uses living organisms or parts thereof to make or modify a product, improve plants or animals or develop microorganisms for specific uses. It is invaluable in research on conservation of bioresources.

Use of Genetic Resources

Some of the conservation strategies applicable for genetic resources includes, 1. Micropropagation; 2. *In vitro* conservation; 3. Cryopreservation; 4. *In vitro* production embryos; 5. Embryo Culture and Transfer; 6. Artificial insemination; and 7. Intracyioplasmic sperm injection.

In view of solving the problem of challenges by using biotechnological approaches the following measures has been employed to improve the bioresources production.

1. Sustainability
2. Sustainable, ecosystem and agricultural.
3. Biomass as a bioresources
4. Bioenergy potential
5. By-products from agriculture

1. What is Sustainability?

The Sustainability associates with different aspects such as,

(i) Resource sufficiency – anthropocentric view:
(ii) Functional integrity (ecocentric view):
(iii) Sustainability and components:
(iv) Sustainability and Ecosystem:

(i) **Resource sufficiency (anthropocentric view) :** Emphasis should be given on supply of food, materials, energy etc. for present and future generations using biophysical and technological resources. If current or foreseeable supplies meet or exceed the amount required in a certain timeframe the practice is sustainable. In addition we have to focus on the relationship between input and output/availability of resources. Concept usefulness also help for understanding the role of non-renewable resources in economic development and acknowledges the rights of future generations.

(ii) **Functional integrity (ecocentric view) :** Assumes systems to have crucial elements that are reproduced over time in a manner or at a rate that is depend on previous system states (e.g. soil fertility, natural enemies, the family). Reproduction of critical systems elements may be impaired with a substantial time lag as consequences of change due to an unsustainable practice and the system can not recover. It is essential to have fully awareness about the long-term consequences of many practices. Acknowledging the complex interactions between ecological, economic, and social values, systems view is required.

(iii) **Sustainability and components :** It is a complex interaction and balance between environmental, economic and social part. It is important to look at the entire system. Most of industiialized agroecosystems are overusing and degrading the soil, water, etc. The conversion to sustainable agrosystems should become the goal. "The definition of sustainable agriculture must move beyond the narrow view of agriculture that focuses primarily on the development of practices or technologies designed to increase yields and improve profit margins. These practices and technologies must be evaluated on their contributions to the overall sustainability of the farm system."

(iv) **Sustainability and Ecosystem :** Ecosystem is the functional system of complementary relations between living organisms and their environment, delimited by arbitrarily chosen boundaries, which in space and time appears to maintain a steady yet dynamic equilibrium. Such an equilibrium can be considered to be sustainable in a definitive sense. A well-developed, mature natural ecosystem is relatively stable, self- sustaining, recovers from disturbance, adapts to change, and is able to maintain productivity through using energy inputs of solar radiation alone.

2. Sustainability, Ecosystem and Agriculture

We can expand the ecosystem concept to agriculture – then we will have basis for looking beyond a primary focus on traditional system outputs (yield or economic return). We can instead look at the complex set of biological, physical, chemical, ecological and cultural interactions determining the processes that permit us to achieve and sustain yields. Agroecosystems are often more difficult to study than natural ecosystems because they are complicated by human management. An agroecosystem is created when human manipulation of an ecosystem takes place for the purpose of establishing agricultural production.

(a) **Ecosystem converted into sustainable agroecosystem :** The energy flow, and nutrient cycles are two important pathways play an important role in converting the eco-nutrients into useful suitable products. The model assumes that nutrients and leftover energy are returned to the agroecosystem as reusable materials, and that the use of non-renewable human inputs is minimized. The following are the four key emerging qualities of ecosystems (which can be used as indicators for agroecosystem sustainability).

(i) Energy flow – very often it becomes a through-flow system, with a high fossil fuel input and the energy direct out at the time of each harvest. For sustainability to obtain, renewable sources of energy must be maximized.

(ii) Nutrient cycling – small amounts of nutrients continually enter an ecosystem through several hydrogeochemical processes. In mature ecosystem small losses of nutrients are replaced by local inputs. In agro-systems the considerable quantities are lost with the harvest or a result of leaching or erosion due to a great reduction in permanent biomass levels. Modern agriculture rely heavily on nutrients inputs derived from petroleum based fertilizers.

(iii) Sustainability requires that the leaks of nutrients will be reduced to a minimum and recycling mechanisms will be reintroduced and strengthened.

(iv) Population regulating mechanisms – due to human directed genetic selection and overall simplification of agro-systems, population of crop plants is rarely self reproducing or self regulating. Sustainability requires the reintroduction of the diverse structures and species relationships that permit the functioning of natural control and regulation mechanisms.

(b) **Conversion to sustainable agrosystems design and management :** Sustainable agrosystems design and management play a crucial role in improving the bioresources and it is essential to follow the below aspects systematically to manage the system effectively.

- Shift from through-flow nutrient management to recycling of nutrients, with increased dependence on natural processes, such as biological N fixation.
- Use renewable sources of energy instead of non-renewable.
- Eliminate the use of non-renewable off-farm human inputs that have the potential to harm environment or the health of farmers, farm workers or consumers.
- When materials must be added to the system, use naturally occurring materials instead synthetic, manufactured inputs.
- Manage pests, diseases and weeds instead of "controlling" them.
- Reestablish the biological relationships that can occur naturally on the farm instead of reducing and simplifying them.
- Make more appropriate matches between cropping patterns and the productive potential and physical limitations of the farm landscape.
- Use a strategy of adapting the biological and genetic potential of agricultural plant and animal species to the ecological conditions of the farm rather than modifying the farm to meet the needs of the crops and animals
- Value most highly the overall health of the agroecosystem rather than the outcome of a particular crop system or season
- Emphasize conservation of soil, water, energy and biological resources
- Incorporate the idea of long-term sustainability into overall agroecosystem design and management

3. Biomass as a Bioresources

What is Biomass?

Biomass is nothing but a wide range of products and by-products from agriculture and forestry as well as municipal and industrial waste streams. It includes: arable crops, perennial crops and short rotation forestry, trees, agriculture and forest residues, effluents, sewage sludge, manure, industrial by-products and the organic fraction of municipal solid waste. After the conversion process, biomass can be used as a fuel to provide heat, electricity or as transport fuel, depending on the conversion technology and the type of primary biomass.

Bioenergy

The growing demand for bioenergy may create further competition for land and water between existing agricultural activities, energy production and the use of agricultural land for nature conservation and urbanization needs. The following are the two important aspects need to be emphasized.

(i) Most agricultural land use in EU is already intensive, therefore increased agricultural biomass production for energy purposes could cause additional pressures on agricultural biodiversity as well as on soil, water and air resources.

(ii) Future biomass production on agricultural land should not impose any additional pressure on farmland biodiversity and environmental resources. If possible, environmental pressures from farming need to be reduced to achieve current EU environmental objectives in the water, air quality or biodiversity areas. (EEA 2007).

Current Bioenergy Patterns

In 2005, an estimated 3.6 mill ha of agricultural land in the EU-25 was directly dedicated for biomass production for energy use. The appropriate percentage of bioenergy sources has been given below.

83% - oil crops, 11% - ethanol crops, 4% - biogas production, 2% - short rotation forestry.

"Whether bioenergy feed stocks are imported or not, there are clear environmental risk associated with industrial scale installations. It is important to develop criteria that support an environmentally sustainable approach to energy production from agricultural biomass".

Environmentally Compatible Bioenergy Production

- How can we produce bioenergy from agricultural resources without harming the environment?
- Is it a difficult task?

The following are the key factors for an environmentally compatible bioenergy production.

(i) Bioenergy crop with low environmental impacts should be used – both perennial and annual.

(ii) **Soi l:** to maintain year round soil coverage, creating wind brakes, etc. E.g. sugar beets provide little protection against soil erosion, whereas perennial biomass crops are particularly efficient in soil coverage.

(iii) **Water :** the choice of crops, especially in arid areas, should aim for crops with low water demand; therefore, some perennial crops can perform better then the conventional arable crops used for biomass production.

(iv) **Biological and landscape diversity :** selection of crop mixes can help increase the biological and landscape diversity.

(v) **Risk of fire :** it is important to choose crop with low fire spreading characteristic, especially in dry parts of Europe.

(vi) "**Maintain or reach a 30%** share of environmentally oriented (both organic and high nature) – farming".

(vii) Set aside 3% if intensively used farmland as ecological compensation areas by 2030" e.g. it would be possible to combine conversation objectives with a certain removal of biomass from such areas (grassland cut – 2nd gen. biofuels).

(viii) "Maintain extensive land use categories in their current land cover" – e.g. permanent grasslands could be used for lignocellulose crop, but in general it should be avoided, so as to maintain the environmental quality of these land use types

(ix) "Choose environmentally optimal crop mixes and cropping practices" – perennial biomass crops generally exert less environmental pressure than arable crops (benefits to farmland biodiversity, high energy yield, lignocellulose feedstock has higher net CO_2 emission reduction.

(x) "Develop advanced biomass conversion technologies and bioenergy pathways" – from 2010 it is expected that biogas will become widely implemented and used in conventional cogeneration systems. Biogas crops – including those grown in double cropping farming practices delivering the whole crop as biomass, obtain a more prominent position in the optimal crop mix.

(xi) Willow plantations for waste water treatment.

(xii) The nutrient content in municipal wastewater (an annual wastewater load of 600mm equivalent: 100 kg N, 20 kg P, 65 kg K) usually corresponds quite well to the nutrients requirements in willow cultivation – it will also supply water and other macro-nutrients. Application of digestate from biogas plant is also possible.

(xiii) Utilizing cuttings from permanent grasslands for bioenergy production (maintaining the diversity, but it should fit the requirements of maintaining high nature-value grassland) – there are already project for biogas production. but also fibers, proteins, sugars (bioethanol), mineral concentrates (fertilizers).

The following are the some Examples of environmentally compatible bioenergy cropping systems:

- Double cropping – growing two or more crops in the same space during a single growing season
- reduced ploughing
- near all-year rounds crop coverage
- green crop harvesting (silage) reduces the growing period of one crop leaving time for production of an extra (biomass) crop
- closed cycle of nutrients by using fermentation residues (digestate)
- reduced herbicide application (weeds can also grow on the field)
- diversity on the field
- negative impact due to double harvesting
- the system is restricted to regions with sufficient water ability and long enough growing season
- Multiple cropping – growing of two or three crops simultaneously on the same land
- crop diversity and structural diversity
- crops has higher stress tolerance – more yield security
- lower need for pesticides and herbicides (lower pest pressure – no monoculture)
- better soil cover and lower crop quality levels need to be reached
- Row, strip or alley cropping
- increase landscape diversity (biodiversity in farmlands)
- help to prevent erosion and lower nutrient losses

Intercropping is the good example for environmentally compatible bioenergy cropping systems

- **Intercroping: It consists the**
    - The simultaneous cultivation of several species on a field
    - Well-known in grass mixtures
- **Benefits**
    - improved resource use efficiency
    - greater biomass productivity
    - regulation of weed, pest and disease infestation
    - yield stability and resilience to stress
    - crop quality
    - lower nutrient losses
    - impact on landscape quality

4. Bioenergy Potential

Global food requirement for three diets: vegetarian: 2388 kcal cap-1 day-1 of which 166 kcal cap-1 day-1; from animal products; moderate: 2388 kcal cap-1 day-1 of which 554 kcal cap-1 day-1 from animal products; and an affluent: 2746 kcal cap-1 day-1 of which 1160 kcal cap-1 day-1 from animal products. The actual population size in 1998 (5.9·109 people) and the estimated population size in year 2050 (9.37·109 people), as expressed in grain equivalents 109 tons dry weight per year.

New EU energy plan include a cut in CO_2 emissions by at least 20% by 2020. The commission will propose increasing the use of renewable energy sources to 20% of the total demand, to limit global temperature changes to no more than 2°C above pre-industrial levels. It also wants to improve the EU's energy efficiency by 20%. This would make Europe the most energy-efficient region in the world.

Methane Production from Energy Crops

- Economic biogas production depends on high biogas yields
- Biogas can be produced a wide variety of energy crops, animal manure and organic wastes. Due to high flexibility of this technology, it can be adapted to the local needs.
- The digestate can be used as a fertilizer to grown crops.
- In order to obtain high biogas yield from energy crops following factors has to be fulfilled:
- optimum yield
- optimum harvesting time
- optimum nutrient composition
- optimum conservation
- and optimum pre-treatment technology
- The key factor is to optimize the methane yield per hectare.
- But we need to remember to manage arable land and grasslands according to sustainable principles
- According to some researchers to achieve integrated crop rotations that offer the supply of food and feed, the production of raw materials and energy and to maintain sustainable development of landscape, following strategies are proposed:
- food non-food switch – alternation of crops for the production of food, feed and raw materials
- cascade utilization of different parts of the same crop for different option: e.g. starch from maize corns and biogas from the res of the plant
- mixed cultivation of several energy crops: e.g. sunflower and maize
- choice of the optimum variety and genotype: energy crops for biogas production must produce high biomass yields and contain optimum nutrient patterns
- choice of the optimum harvesting time

Sustainable Crop Rotation

Example of biomass and methane yields from a sustainable crop rotation in Lower Austria that integrates food, feed and energy crop production.

Year	Crop	Biomass yield ($t\ VS\ ha^{-1}$)	specific CH_4 yield ($l_N\ Kg^{-1}\ VS$)	CH_4 yield per hectare ($m_N^3\ ha^{-1}\ a^{-1}$) Crop only	Crop rotation
1.	Maize (whole crop silage)	1512	390	5897	1179
2.	Winter wheat (straw)	5.44	189	1028	206
	Intercrop (clover grass)	2.71	335	906	181
3.	Summer barley (straw)	3.81	189	720	144
4.	Sugar beet (leaves)	7.20	210	1512	302
	Pressed beet pulp silage	14.36	4.30	6173	1235
5.	Sunflower (whole crop silage)	11.02	300	3300	660
	Intercrop (luocrne)	3.61	335	1208	242
Methane yield of the whole crop rotation ($m_N^3\ ha^{-1}\ a^{-1}$)					4149

It was assumed that crop rotation covers food and feed demands. It is important that the intensity of the production is adapted to the energy crops, food and feed are grown.

5. By-products from Agriculture are Important Bioresource

They possess

(i) Several suitable agricultural by-products can be used for bioenergy and/or biochemicals.

(ii) Many of waste from agriculture can be used as feedstock in another process – by-products.

(iii) The advantage of such a feedstock is the low cost (in many cases no much more then the cost of transport).

(iv) Restriction on landfilling and increasing costs of waste disposal cause the negative cost of some waste.

Different Types of Agriculture By-products

Brown and green juice – water which is removed from the grass before drying it to produce fodder pellets from such a crops as perennial rye grass, Italian rye grass, clover grass and alfaalfa. The by-product is called green juice.

When the crops are steamed in a cooker to a temp up to 80ºC before processing, the process result in coagulation of the part of protein and damage the plant cell – the by product is called brown juice. The dry matter (DM) of those by-products is around 6%. The green juice contains 13% DM carbohydrates, 35% DM proteins. The brown juice contains 33% DM carbohydrates, 26% DM proteins.

Corn Steep Liquor

By-product from the wet corn milling industry. Steeping of the corn before the wet milling process is essential for high yields and high yield quality. Corn steep liquor contains approximately 47% protein and is valuable nutrient source in the fermentation industry

Potato Waste Water

Water form operations like washing, peeling, etc. The effluent contains starch (19.5 g/L) and some protein (2.9 g/L)

Molasses

By-product from sugar manufacturing. In production of sugar from beets, sugar is extracted with hot water (70ºC), the pulp and liquid are separated. The molasses (thick brown liquid) is used as animal feed but also in fermentation industry, it consist of 50% of sucrose and some proteins (6-9%). The following waste materials are highly useful as bioresources which helps the people meet the needs of them.

- Wheat straw – only in Denmark app. 6 million tons is produced each year. It has a content of app. 40% cellulose and 30% hemicellulose
- Corn stover – by-product from cultivation of corn. It contains app. 50% cellulose and 28% hemicellulose.
- Rice straw and hull – rice straw is probably most abundant agricultural by-product. The composition: 40% cellulose and 18% hemicellulose. Rice hull contains 26% cellulose and 13.5% hemicellulose.
- Wine shoots – the lignocellulose residues are usually burned in the field, causing environmental problems. Wine-trimming waste contain 34% DM cellulose and 19% DM hemicellulose.
- Sugar cane bagasse – by-product from sugar production from sugar cane (40% DM cellulose and 21% DM hemicellulose).
- Sorghum bagasse – sweet sorghum is used for production of syrups, brown sugar, and forage. The leftover contains 37% DM cellulose and 26% DM hemicellulose.
- Olive cake – it can be used as a fuel, but in many cases is discarded as waste in the environment. (10 %DM cellulose, 16% DM hemicellulose, 5% DM protein)

- Citrus waste – large amount of residue consisting of washing waters, peel, membranes, seeds and pulp.
- Banana waste – banana is the most consumed fruit in the world. Each hectare of banana crop generated nearly 220 tons of plant residue (mainly lignocellulose material).
- Hydrocarbons which are one of the major causes of environmental pollutants and also degradation of oil-degrading microbes.
- Mosquito control using modern synthetic insecticides which are highly toxic and persistent in the environment; in order to minimize their use, botanical/herbal insecticides are to be used from alternative sources.
- Methane production from termite guts. The influence of temperature will also play important role on production of methane from the bacteria, methanobacteria and methanosarcina.

Our goal is to facilitate the movement of products from the manufacturer to the consumer. Bioresources was founded in 1994. We've produced great results for our clients - so we've grown and expanded steadily since then.

Biotechnological Approaches to Improve Bioresources

Researchers are working together with the aim of using biotechnology to explore novel high value products such as drugs and enzymes. The following areas of expertise highly useful for the development of bioresources.

- Taxonomy of microorganisms
- Developing biological assays
- Natural product chemistry
- Purification and elucidation of chemical structures using nuclear magnetic resonance (NMR)
- Pre-pilot scale production of commercial valuable bio-products.
- Enzyme discovery
- Recombinant expression systems
- Biodiversity and ecology of plant and microorganism
- Database and GIS establishment
- In addition to research and development, the Biotechnology programmes will provide a selected range of special services to the private and public sector in order to facilitate R&D in the following biotechnological approaches such as 1. Bioactive compound screening, 2. Animal Cell Service, 3. Enzyme Screening Service, 4. Technical Service and 5. BIOTEC Culture Collection.

1. **Bioactive compound screening induces the assay :**
    - Anti-proliferative assay for cancer cell lines (KB, MCF-7, NCI-H187)

- Anti-*Candida albicans* assay
- Anti-herpes simplex type 1 assay
- Anti-inflammatory assay
- Anti-*Mycobacterium tuberculosis* assay
- Anti-*Plasmodium falciparum* assay
- Cytotoxicity assay against Vero cell line

2. **Animal Cell Service consists of :**
   - Mycoplasma testing of cultured cells
   - Cytotoxicity testing
3. **Enzyme Screening Service: it possess the following assay :**
   - Xylanase assay
   - Cellulase assay
   - b-glucosidase assay
   - Phytase assay
   - Protease assay
   - Amylase assay
4. **Technical Service: it includes the following techniques :**
   - NMR techniques (1D and 2D NMR)
   - LC-MS and MS Analyses
   - Technical service for IR (Infrared) Spectrophotometer
5. **BIOTEC culture collection : helps to improve the production of bioresources :**
   (i) Culture preservation.
   (ii) Microbial cultures.
   (iii) Lyophilization and preparation of cultures for storage by freezing in which cultures are expertly preserved and prepared before returning to customers for storage.
   (iv) Purification and identification of fungi, yeasts, and bacteria.

Sustainable use of bioresources demands that while utilizing the resources so generously placed by nature at our disposal, we should try not to be cruel to the environment and our children yet unborn. Biotechnology provides methods through which a balance between the economic exploitation of bioresources and their conservation for the future can be achieved. Finally, it is suggested that extensive biotechnological oriented training is required for the improvement of bioresources which helps to meet the food challenges of 21st century.

REFERENCES

An introduction to Genetic Analysis by Anthony, J.F. J.A. Miller, D.T. Suzuki, R.C. Richard Lewontin, W.M-Gilbert, W.H. Freeman publication, 1998.

Benjamin Lewin, Gene IX, 9th Edition, Jones and Barlett Publishers, 2007.

Biology of the Gene, 6th Edition, Benjamin Cummings Publishing Company Inc, 2007.

Biotechnological Innovations in Animal Productivity, BIOTOL Series, Butterworth – Heineman Ltd. Oxford, 1992.

Brown TA, Genomes, 3rd Edition. Garland Science 2006

Freifelder D., Physical Biochemistry, Application to Biochemistry and Molecular Biology, 2nd Edition, W.H. Freeman & Company, San Fransisco, 1982.

Griffiths, A. J. F., Miller, J. H., Suzuki, D. T., Lewontin, R. C., Gelbart, W. M. (1997). An Introduction to Genetic Analysis, W. H. Freeman & Company, New York.

Jackson AT., Bioprocess Engineering in Biotechnology, Prentice Hall, Engelwood Cliffs, 1991.

Shuler ML and Kargi F. Bioprocess Engineering: Basic Concepts, 2nd Edition, Prentice Hall, Engelwood Cliffs, 2002.

Bioresources for Food Security and Rural Livelihood, Pages 75–81
Edited by : G.K. Kulkarni and B.N. Pandey

BYPRODUCTS OF SERICULTURE INDUSTRY

A.K. Tripathi

ABSTRACT

Importance of *Antheraea mylitta* as a tasar producing insect is known to us from time ancient. In Bihar it is reared by tribal people of old south Bihar (now Jharkhand) to obtain cocoons for tasar silk yarn. After taking out the silk thread from the cocoons, the dead pupae are thrown out. The adults also die (Males after mating and females after oviposition) and are discarded as waste matter which create pollution in the area.

An attempt has been made to investigate whether or not these wastes can be utilized as fish food and manure. Tests on fresh water fish *Channa punctatus* have shown the utility of the material as assessed by significant increase in weight and energy content (Calory) against control fish of the same weight group maintained on a standard laboratory diet. Secondly these wastes when used as manure the bioassay test clearly indicate, their utility assessed by increase in the root and shrot of *oryza sativa.* This finding is supported by estimation of mitotic index in the root of *Alium cepa.*

This finding will help the poor tribal people for their betterment as by rearing of *A. mylitta* they will not get only tasaryarn but also a good amount of money from the wastes. Secondly by this way the environment also will be pollution free.

Keywords : *Antheraea mylitta*, Pollution, Fish Food, Weight, Calorific value, Manure, Mitotic Index.

INTRODUCTION

Antheraea mylitta D belonging to the family Saturniidae of order Lepidoptera is reared by poor tribal people of old south Bihar (now Jharkhand) to obtain cocoon for tasar silk yarn. After taking out the Silkyarn from the cocoons, the dead pupae are thrown out. The adults also die (male after mating and female after oviposition) and are discarded as waste matter. So when sillk is produced commercially, there is an accumulation of a huge mass of dead pupae and adults which have no further use in silk industry rather these create pollution in the area where good quantity of fat, protein and carbohydrate in these are thrown. Good quantity of fat, protein and carbohydrate in these wastes have been observed by Tripathi *et al* (1990, Table 1) and therefore, it was considered worthwhile to find out whether or not these discarded wastes could be advantageously used as fish food and manure. The present communication is a report on the feasibility of using these wastes as fish food in piscicultural practice and manure in the agriculture.

Table 1 : Gross chemical composition of *Antheraea mylitta* D. Each figure is mean of 25 replica

SL. No.	Stage	Water % wet weight	Lipid % Drywt	Protein % wet Dry	Carbohydrate % Dry wt.	Ash % Drywt
1	Vth instar	79.69± 1.08	23.79± 0.23	59.60±1.02	10.16±0.25	07.00±0.23
2	Pupa	77.77±1.10	19.54±0.28	59.21±1.09	10.28±0.24	10.12±0.26
3	Adult male	73.33±1.05	18.34±0.37	60.55±1.10	10.16±0.24	10.44±0.26
4	Adult female	75.93±1.10	19.20±0.38	60.86±1.07	09.45±0.25	10.76±0.23
5	Male After copulation	72.10±1.11	16.60±0.28	68.20±1.10	08.50±0.26	12.80±0.26
6.	Female after copulation	78.30±1.12	19.92±0.33	60.10±1.09	08.76±0.25	11.43±0.26
7.	Female After oviposition	64.63±1.13	16.34±0.28	60.98±1.10	08.13±0.25	14.60±0.26

1. **Fish food :** Specimens of *Channa punctatus* (6-7g) were weighed and placed in two large size aquaria (25 per aquarium), each containing equal volume of water, which was suitably aerated by aerator pump. The fishes of both the aquaria were fed on pellets of a standard diet prepared by mixing up equal amount of cotton see powder, wheat flour, milk powder and fish meal. The fishes in one of the two aquaria (experimental group) were also allowed to eat pillets of a mixture of dried and powdered pupal and adults of *Antheraea mylitta*. After 30 days, the fishes were weighed and % increase in weight was calculated. The energy content (Catories / g dry weight) of the fishes at the beginning and end of 30 day experimental period with and without the experimental food was also determined by using a Bomb calorimeter. The experiment was done four times (January, February, March and April) using fresh fish every month and the data so collected was subjected to students 't' test to assess the effect of tasar insect meal in relation to growth and energy content.
2. **Manure :** To see the effect of wastes, the Bio-assey test was performed for which 6%, 3%, 1.5%, 0.75% and 0.37% (weight / volume) aqueous solution of pupae, moths, faeces and 1:1:1 mixture of pupae, moths and faces of *A. myllitta* were made separatly and used as manure along with 2% aqueous agar solution and the length root and shoot of paddy (*oryza sativa*) was measured after 3 days and the result

was compared with control (2% Agar solution). To explain the grwoth of root the mitotic index of onion (*Alium cepa*) was calcualted after using the same manure. To calculate the MI following formula was employed.

$$M1 = \frac{\text{Number of dividing cells under one focus}}{\text{Number of total cells under one focus}} \times 100$$

OBSERVATION

1. **Fish food :** The data on the change in weight and energy content of fish maintained on standard and experimental diet are presented in Table 2. The control fishes shown 61.47% increase in weight, whereas it was found to be 76.24% in the fishes on experimental diet. The difference was observed to be highly significant ($t = 4.726$; $p > 0.01$). There was also a significant ($t = 6.914$; $p > 0.001$) increase (28.05%) in the energy content (Cal/g dry weight) of the experimental fishes.

Table 2 : Effect of feeding dead pupae and adults of *Antheraea mylitta* D on weight gain and energy content of the fish *Channa punctatus*. Each figure is mean of 25 replica.

Sl. No.	Months	% Fresh wt. gain		Cal/g dry weight		%calory gain
		Control	Experimental	Control	Experimental	
1	January	57.36± 2.85	70.09±4.80	3200.23±27.58	4321.25±29.22	35.03±2.12
2	February	58.20±1.06	78.79±3.08	3156.16±25.23	4102.18±17.38	29.97±2.23
3	March	66.39±3.43	76.00±2.42	3122.35±33.88	3973.22±35.57	27.25.±1.96
4	April	63.96±2.65	80.09±6.61	3099.82±26.49	3718.45±30.67	19.97±1.86
Mean ISE		61.47±2.19	76.24±2.22	3144.25±21.96	4028.53±12.59	28.05±3.14

Result: The effect of food is positive in both, the weight gain as well as the calorific gain.

2. **Manure :** The data on the change in length of root and shoot of *Oryza sativa* produced by the control (2% Agar solution) and different concentrations of wastage are presented in Table 3 which indicate that 1.5% mixture gives the best result (11.97 + 0.16 cm root and 20.87 +0.16 cm shoot) followed by adult (18.45 + 0.14cm shoot) and pupa (15.95 + 0.12 cm shoot) respectively in comparison to control (14.05 + 0.17cm shoot) where as all the concentrations of faeces give worst result. The data related to mitotic index of onion (*Allium cepa*) roots after using 1.5% aqueous solution of different stages of wastage as manure against control (distilled water) are presented in Table 4. It also reflects the same pattern as shown by the length of root and shoot of *Oryza Sativa*. In addition it is also observed that more cels under go mitosis within 24 hour and after that reduces with lapses of time (24 hour - 12.52 + 0.10; 48 hour - 9.87 + 0.08; 72 hour - 7.70 + 0.16 and 96 hour – 4.40 + 0.18).

Table 3 : Length (cm) of root and shoot of paddy (*oryza sativa*) after using different stages of *Antheraea mylitta* as manure against control (2% Aquous Agar solution). Each figure is mean of 25 replica.

Manure	Control		Experimental										Critical
	Root	Shoot	6%		3%		1.5%		0.75%		0.37%		Difference (CD)
			Root	Shoot	Root	Shoot	Root	Shoot	Root	Shoot	Root	Shoot	
Pupa	11.94	14.05	0.00	1.61	6.41	7.28	10.43	15.95	5.66	9.48	10.80	8.86	Root-0.228
	±0.90	±0.17	±0.00	±0.10	±0.14	±0.12	±0.08	±0.12	±0.06	±0.08	±0.11	±0.06	Shoot-0.225
Adult	Do	Do	1.43	4.40	5.43	7.65	6.48	18.45	4.500	16.46	6.97	10.91	Root-0.246
			±0.15	±0.14	±0.18	±0.05	±14	±0.05	±0.07	±0.05	±0.07	±0.06	Shoot-0.528
Faeces	Do	Do	0.00	0.66	0.58	1.76	1.16	2.47	1.15	0.05	4.58	6.58	Root-0.237
			±0.00	±0.04	±0.08	±0.08	±0.06	±0.08	±0.06	±0.06	±0.06	±0.11	Short-0.239
1:1:1	Do	Do	6.41	4.01	8.98	14.42	11.97	20.87	10.51	13.54	9.96	8.53	Root-0.234
mixture			±0.10	±0.7	±0.08	±0.08	±0.16	±0.16	0.07	±0.04	±0.09	±0.11	Short-0.210

Resut : 1. 1.5 % Mixture gives better result

2. Faeces gives the worst result.

Table 4 : Mititoitc index of onion (*Allum cepa*) roots after using 1.5% aquous solution of different stages of *Antheraea mylitta* as manure against control (Distilled water). Each figure is mean of 25 replica.

Time (Hours)	Control	Pupa	Adult	Larval faeces	1:1:1 Mixture	Critical Difference (CD)
24	9.87±0.15	9.87±0.15	9.56± 0.10	4.93± 0.10	12.52±0.10	0.392
48	7.10±0.12	6.29±0.11	6.81±0.10	2.44±0.12	9.87±0.08	0.346
72	5.38±0.17	5.31±0.19	4.97±0.05	0.99±0.10	7.70±0.16	0.436
96	2.33±0.13	2.37±0.13	1.96±0.05	0.00±0.00	4.40±0.18	0.367
Mean± SE	6.17±0.10	5.96±0.09	5.82±0.02	2.09±0.06	8.62±0.09	0.326

Result : 1. 1.5% mixture gives the best result
2. Faeces gives the worst result

An exhaustive review of information on all aspects of applied fish nutrition (energy requirements, protein and lipid requriements, feed formulation and evaluation has been presented in a treatise edited by Halver (1972). Useful information on fish is also available in relation to food (Das and moitra, 1956a, 1956b), food and growth (Paloheimo and Dickie (1966) and digestion, absorption and conversion of food (Pandian 1967). Laboratory experiments provide precisely calculated data based on prepared diets of known composition and energy content put in nature, fishes ingest a complex of materials providing a balanced mixture of proteins, carbohydrates, and lipids. Artificial diets based on laboratory experiments on fish are expensive. It is therefore, always advantageous and practicable to provide extra feeds from natural sources, if readily available at a low cost. The results of the present work are strongly indicative of the usefulness of the feed prepared from dead pupae and moth of *A. mylitta*, which are otherwise rejected as waste. Tripathi *et. al* (1990) have observed the gross chemical composition of pupae and moths of *A. mylitta*. On a dry weight basis, the pupae have around 59% protein, 21.5% lipids and 9.5% carbohydrates and adults have 61% protein, 19% lipids and 9.5% carbohydrate. Thus there is little difference in the chemical composition of pupae and adults and averaging of figures would give 60% protein, 20% lipid and 9.5% carbohydrate. The corresponding figures on a fresh weight basis can be worked out to be 14.6% protein, 4.9% lipid, 2.4% carbohydrate, 2.5% ash and 75.6% water. It appears that the feed prepared from pupae and adults contain balanced amounts of different food ingradiatents and therefore, it promotes better growth and more calories in the test fishes. Therefore, dead pupae and moths of tasar silk insect should not be rejected as wastes as those can be safely used as fish food.

Tremendous works on biofertilizers, are carried out by Dahama (1998); Bagyaraj (2003); Gupta (2004); Meshram *et al* (2004) Kokade *et al* (2005); Gahutkar (2006);

Ravindran *et al* (2007); Sandhu (2008); Gupta (2009) and Panigrahi *et al* (2009) out of which Gahutkar (2006); Ravindran *et al* (2007) and Sandhu (2008) have observed enhanced plant height, more number of compound leaves, more root nodules, larger grain size with better shining texture, better taste and better flavour when they used biofertilizers in paddy crop which resutled into better marketing price of the crop. These scientists have opined that when organic manure is added to soil, it provides and keeps all the materials and micro-organisms required for plant growth in the soil intact. There application is easy and cost effective and they do not cause pollution. In present project different concentrations of dead pupae and adults along with faecal pillets of larvae of *A. mylitta* are used as manure out of which 1.5% concentration has shown the better result (in the form of root and shoot length of *Oryza sativa* and more value of mitotic index in *Allium cepa* roots).

One very striking observations is related with faecal pillets as they have shown inhibitory effect on the length of root and shoot and the mitotic index. Hence it is a matter of further investigation to explore the chemical conditions of faecal pillets and the acutal physiology behind it. In my opinion if these faecal pillets are used for eradication of unwanted grasses such as parthenium in the field, they may be very effective.

On the whole it is very clear that rearing of *A. mylitta* is very helpful for betterment of livelihood of poor tribal people not only by producing tasarsik yarn but also by the use of wastes in the form of (i) Fish food (ii) manure (iii) use for eradication of unwanted grasses from the field and (iv) to make pollution free environment.

REFERENCES

Bagyaraj, D. J. (2003) Biofertilizer news letter, **11(1) :** 3-6.

Dahama, A. K. (1998) Agro Annual review of crop ecology **1(1) :** 37-43.

Das, S. M. and Moitra, S. K. (1956a) studies on the food of some common fishes of Uttar Pradesh, India Part II. Proc. Natn. Acad. Sci. India, **26 :** 213-223.

Das, S. M.and Moitra, S. K. (1956b) Studies on the food of some common fishes of Uttar Pradesh, India. Part III. Proc. Natn. Acad. Sci. India, **26 :** 224-233.

Gahutkar, R. T. (2006) Potential and use of biofertilizer in India : Every man's Science, XL **(5) :** 354-361.

Gupta, P. K. (2004) A hand book of soil, fertilizer and manure (2nd ed), Kisan Printing Press, Kankhal, Haridwar.

Gupta, R. K. (2009) पर्यावरण संरक्षण एवं उत्तम जैविक खाद के लिए केंचुआ पालन, पर्यावरण **62:** 40-41

Halver, J. E. (1972) The vitamins. In "Fish nutrition" Academic press, New York and London, 26-103.

Kakade, T. D., Wadaskar, R. M.; Siddhabhatii, P. M.; Nemade, P. W. and Tajne, V. S. (2005) Pestology, **29 (4) :** 32-38.

Meshram, S. V.; Pande, S. S; Shanware, A. S.; Kamdi, R. R. and Tajne, V. S. (2004) Biofertilizer news letter, **12(1) :** 7-10.

Paloheimo, J. E. and Dickie, L. M. (1966) Food and growth of fishes III. Relations among food, body size and growth efficiency, J. Fish Res. Bd. Can. **23 :** 1209-1248.

Pandian, T. J. (1967) Intake, digestion, absorption and conversion of food in the fishes *Megalops Cyprinoides* and *ophiocephalus striatus.* Mar. Biol. **1 :** 16-32.

Panigrahi, A. K.; Sahoo, T. R.; Behera, H. S. and Swain, N. K. (2009) Organic soil fertility management for enhanced paddy production : Model paddy fields in orissa, Every man's Science XLIII **(6) :** 368-374.

Ravindran, K. C.; Venkatesan, K.; Balasubramanian, T. and Balkrishnan, V. (2007) Studies on the effect of Halophytic compost in combination with farmyard manure and phosphobacteria on grwoth parameters in *Arachis hypogaea* linn. Proc. 94th Indian Science Congress part II (Abstracts) Section XIV : Plant Science pp. 25.

Sandhu D. K. (2008) Impact of Chemical and organic manures on crop production J. Env. Bio. Sci. **22(1) :** 87-92.

Tripathic, A. K. ; Sinha D. P. and Singh, S. B. (1990) Gross Chemical Composition of different stages of *Antheraea mylitta* (Saturniidae : Lepidoptera). Proc. Natn Acad. Sci. India. **60 (B)** : 25 - 30.

Bioresources for Food Security and Rural Livelihood, Pages 83–95
Edited by : G.K. Kulkarni and B.N. Pandey

FRESHWATER BIORESOURCE AND THEIR BIOGEOGRAPHY IN THE HIGHLAND RIVERS OF INDIAN SUBCONTINENT

Prakash Nautiyal, Rachna Nautiyal, V. P. Semwal, Asheesh Shivam Mishra and Jyoti Verma

ABSTRACT

Organisms in sea live in continuous media and species adjust their ranges as climate or ecological conditions change, but freshwater habitats are relatively discontinuous. Freshwater biodiversity is most threatened by hydropower schemes and faces threat from climate change. Major aquatic bioresource; diatom, benthic macroinvertebrate and fish were compared in the rivers of Himalaya, Central & Eastern Highlands and Western Ghat ecoregions. Himalaya was taxonomically richer than Peninsular ecoregions but the richness was not proportionate to its area, owing to harsh climate. Similarity was low for species but high for family and genera in diatoms and fish, and low for macroinvertebrates attributed to historical factors.

Keywords : species richness/freshwater/mountain/harshness.

INTRODUCTION

Biodiversity is the extraordinary variety of living creatures and ecological communities growing and interacting with each other all over the world. The loss of biodiversity in ecosystem has important implications, including diminished resistance and resilience to disturbance, system simplification, and loss of ecological integrity. Critical to preventing losses of biodiversity is to know the patterns in taxa richness (a common measure of biodiversity) at a variety of spatial scales. The freshwater biodiversity is distributed in a fundamentally different pattern from that in marine or terrestrial systems. Organisms on land or in the sea live in media that are more or less continuous over extensive regions and species adjust their ranges to some degree as climate or ecological conditions change. But freshwater habitats are relatively discontinuous, and many freshwater species do not disperse easily across the land barriers that separate river drainages into discrete units [1].

The Indian subcontinent has three distinct units, Plateau, Mountains (Himalaya) and Plains. These two highlands constitute a large area of India and the major rivers of India emerge from them, especially the Himalaya, Western Ghats, and Central and Eastern Highlands. These mountain chains are separated by the Gangetic Plains. The Himalaya, Central and Eastern Highland and Western Ghat differ in extent of mountain area, topography, elevation and climate by virtue of their location. The streams/rivers also vary due to glacier (resulting in ice cold water), in the Himalaya compared to Central and Eastern Highland and Western Ghats. Among freshwater aquatic biota, fishes and macroinvertebrates have attracted much attention, while diatoms, a major producer group in highland streams/rivers, are less known.

However, the large-scale patterns of richness of these groups of organisms remain unknown in the Indian subcontinent. The present study compares the level of richness in the above said highlands, examines taxonomic similarities among them and discusses possible reasons for the present state of richness and similarities in these ecoregions.

STUDY AREA

Himalaya : The Himalaya (H) is located between latitudes 26°20' and 35°40' North, and between longitudes 74°50' and 95°40' East, for a distance of about 2,500 km with an average width of 240 km[1]. The Himalaya comprises two nearly parallel ranges separated by a wide valley in which the Indus and Sutlej rivers flow westward and the Brahmaputra flows eastward for long distance and flow southwards on either extremes of the Himalaya. The northern range is called the Trans-Himalaya. The southern range has three parallel zones: the perpetually snow-covered Great Himalaya (average elevation 6,100 m); the Lesser Himalaya with 2,130–4,570 m elevations; and the Outer Himalaya 610–1,520 m. The Ganga bisects these southern ranges.

Central & Eastern Highlands (C&EHL) : This ecoregion comprising Vindhya, Satpura and Aravali occupy a large part of central India from east to west. These highlands extend in east-western direction. The Eastern Highland extends to east of Vindhya towards the State of Jharkhand and Madhya Pradesh. The river flow east, e.g. Damodar. This and the Vindhya region were considered for the present study. Vindhya, is a low mountain range, from Vârânasi to Gujarât State (103,600 km^2, from Sasaram and Rohtas in Western Bihar to Chitorgarh on the Aravali's, with the exception of a central tract in Bundelkhand) for a distance of about 1090 km. Elevations range from 455 to 910 m and reach a maximum of 1113 m (3651 ft). This part of eco-region is drained by the north flowing rivers Chambal, Parbati, Kali, Sindh, Betwa, Dhasan, Ken and Tons.

Western Ghats (WG) : The Western Ghat or Sahyadris extend unbroken to the extreme south of Malabar. The western ghat as the name Ghat denotes are down to Malabar, steep-sided, terraced, flat-topped hill or cleft facing the Arabian Sea cost and running with the general parallelism. The Western Ghat has a latitudinal range of more than 10 degrees and form a nearly unbroken relief parallel to the Western Coast of Indian

Peninsula for almost 1600 km. They extend from the mouth of river Tapti (21°N) to the South of India (about 8°N), the only gap in the chain being Palghat Gap. The Vindhya-Satpura ranges lie to north of these Ghats. Biogeographically, the hill chain of the Western Ghats constitutes the Malabar province of the Oriental realm. Rising up from a relatively narrow strip of coast at its western border, the hills reach up to a height of 2800 m before they merge to the east with the Deccan plateau at an altitude of 500-600 m. The average width is about 100 km.

In case of fish fauna information exists to species level for the entire Himalaya, Central & Eastern Highlands and Western Ghat compared to the family level for macroinvertebrates from the West Himalaya, Western Ghat (Kerala) and Vindhya (Central Highlands). In case of diatoms species level information exists for parts of Himalaya (northwest, west, northeast) and the Central & Eastern highlands (Vindhya region, Damodar basin) while genera level from lower part of Western Ghats.

The data presented here pertains to major research projects of the author[2, 3, 4, 5], studies made under D. Phil. research programme of the H.N.B. Garhwal University Srinagar[6, 7, 8], the University of Allahabad, Allahabad[9, 10] and related publications[11, 12, 13].

Taxonomic richness : Algae function as primary producers in aquatic ecosystems. Among algae diatoms play important role on global cycling of silica and carbon (probably 20 - 25 % of all organic carbon fixation of the planet is carried out by diatoms Round 1999, Mann 1999). They constitute a very large part of the algae in mountain river ecosystems. Their richness amounted to 702 taxa (species) in the Himalaya and 358 taxa in the Central and Eastern Highlands. Fifty eight genera were recorded from Central Highland, West Himalaya and the Western Ghats ecoregions of the Indian subcontinent, 51 each from the Himalaya (Garhwal region), as well as the Central & Eastern Highlands and 43 from Western Ghats. Thirty six genera were common to these 3 regions (Table 1). Within a region the estimates differed in the Himalaya and Central & Eastern Highlands (Table 2).

Though the number of diatom taxa was almost same in the Vindhya Rivers and total numbers amounted to 293 taxa only. The rivers differed considerably in species composition, as only 86 taxa were common among Ken, Paisuni and Tons[10]. Further, the number exceeded 200 taxa in each Vindhya river (202 to 211 taxa) but not in case of the Himalaya (99 to 160 taxa), as also in the Alaknanda-Ganga[11], Kumaun[14], Nepal[15, 16, 17, 18], eastern Himalaya[19] and even the Alps[20]. Further, the estimates from not only one river (Paisuni 3 stations and ca. 42 km, 202 taxa; Nagni 4 stations ca. 150 km., 130 taxa) but also one station (Paisuni P1 171 taxa; Ken K1 182 taxa; Nagni N1 104 taxa; Alaknanda 160 taxa) were high in the Vindhya. The Himalayan streams/rivers have poor electrolyte content and nutrients occur in traces[4, 20, 21]. Higher individual estimates in the Vindhya suggested more diversity in this region. However, more studies are needed to arrive at a conclusion.

Table 1 : Distribution of diatom genera in three highland eco regions of the Indian subcontinent.

	GENERA	H	CH	WG		GENERA	H	CH	WG
1.	*Cyclotella*	+	+	+	30.	*Surirella*	+	+	+
2.	*Diatoma*	+	+	+	31.	*Cymatopleura*	+	+	+
3.	*Fragilaria*	+	+	+	32.	*Epithemia*	+	+	+
4.	*Staurosira*	+	+	+	33.	*Rhopalodia*	+	+	+
5.	*Synedra*	+	+	+	34.	*Neidium*	+	+	+
6.	*Eunotia*	+	+	+	35.	*Amphipleura*	+	+	
7.	*Achnanthes*	+	+	+	36.	*Diploneis*	+	+	
8.	*Achnanthidium*	+	+	+	37.	*Fallacia*	+	+	
9.	*Planothidium*	+	+	+	38.	*Hippodonta*	+	+	
10.	*Cocconeis*	+	+	+	39.	*Placoneis*	+	+	
11.	*Amphora*	+	+	+	40.	*Denticula*	+	+	
12.	*Brachysira*	+	+	+	41.	*Tabellaria*	+	+	
13.	*Caloneis*	+	+	+	42.	*Aulacoseira*		+	+
14.	*Cymbella*	+	+	+	43.	*Diadesmis*		+	+
15.	*Cymbopleura*	+	+	+	44.	*Bacillaria*		+	+
16.	*Encyonema*	+	+	+	45.	*Ceratoneis*	+		+
17.	*Frustulia*	+	+	+	46.	*Encyonopsis*	+		+
18.	*Gyrosigma*	+	+	+	47.	*Reimeria*	+		+
19.	*Gomphonema*	+	+	+	48.	*Rhoicosphenia*	+		+
20.	*Navicula*	+	+	+	49.	*Melosira*	+		
21.	*Navicula sensu lato*	+	+	+	50.	*Meridion*	+		
22.	*Adlafia*	+	+	+	51.	*Fragilariforma*	+		
23.	*Luticola*	+	+	+	52.	*Peronia*	+		
24.	*Geissleria*	+	+	+	53.	*Nupela*	+		
25.	*Sellaphora*	+	+	+	54.	*Anomoeoneis*		+	
26.	*Pinnularia*	+	+	+	55.	*Gomphocymbelopsis*		+	
27.	*Stauroneis*	+	+	+	56.	*Mastogloia*		+	
28.	*Hantzschia*	+	+	+	57.	*Aneumastus*		+	
29.	*Nitzschia*	+	+	+	58.	*Scoliopleura*		+	

Similarly, 57 macroinvertebrate taxa (families) were observed in the Himalaya, 29 in the Vindhya and 39 in the Western Ghat. Estimates within a region were as follows: Gangetic drainage 31 taxa, Sutlej-Beas R. 18 taxa, Jhelum R. 51 taxa. In case of individual rivers the estimates for Uttarakhand part of Himalaya were; Yamuna, Bhagirathi, Alaknanda, Pindar basins 31 taxa, Ramganga and Mandakini basin 30 taxa, Bhagirathi 31 taxa (genera); Alaknanda at Srinagar 18 taxa, Khanda Gad 14 taxa, Vindhya; Paisuni 24 taxa, Ken and Tons 23 taxa.

Table 2 : Diatoms: Taxonomic richness and species diversity in the Himalaya and Central & Eastern Highla nds eco regions.

Biogeographical regions	Taxonomic richness	Diversity index (H)	Evenness (E)
Himalaya	419	4.32-4.69	0.19-0.39
Alaknanda (Ganga R)	251	4.69	0.38
Mandakini basin streams	200	2-3.5	0.96 – 1.22
Nagni Gad	130	4.47	0.39
Central & Eastern Highlands	**360**		
Vindhya	**293**	5.0-5.07	0.23-0.33
Ken	205	5.01	0.31
Paisuni	202	5.07	0.23
Tons	211	5.02	0.33
Damodar	122		

Total 346 fish species were recorded in the Himalaya, out of which 270 taxa from the east Himalaya, 174 from west Himalaya and 180 from the Central (Nepal) Himalaya. The number of fish species was less than the C & EHL (174) and WG (270). The estimates for individual rivers are: Alaknanda 61 taxa, Bhagirathi 54 taxa.

The taxonomic richness of diatoms was high in the Himalaya compared to the Vindhya, as also observed for the benthic macroinvertebrates and fish from the Western Ghat. The present pattern of richness and diversity in these mountain chains can be explained on the basis of species richness-area relationship and historical factors. This relationship suggests that the number of species increases with corresponding increase in the area[22], as also observed in case of highland fish fauna[23]. Considering this the Himalaya with over 2400 km length and largest area should be physically diverse than the WG and the C&EHL and hence support more richness. Though the richness of diatoms, benthic macroinvertebrates and fish increased with number of samples, locations and rivers and hence the area, the richness was not proportionate to the area of the Himalaya if compared with the WG and the C&EHL. The Himalaya (663,181 km^2; 516,000 km^2; 517540 km^2) is over four times of the WG (160,000 km^2) and six times the area of C & EHL (103,600 km^2), the proportions being 1:1.5: 6.4 in C & EHL, WG and H and the number of fish species should be 174, 278 and 1114, while the existing proportion of richness was 1:1.6:2[13].

Low richness in the Himalaya with respect to area is attributable to the extreme climate (harshness) in the Greater Himalaya (a large area) which is under perennial snow and which is inhibitive for the high biodiversity. These observations supplement the view that physical factors become important in the temperate and polar zones[24]. The richness considered here pertains to Lesser Himalaya which also exhibits extremes in topography and climate and has continental climate in the West and Central Himalaya adding to harshness of the habitats in this region also. The magnitude of harshness, however, diminishes as Lesser Himalaya experience rainfall compared to cold semiarid and arid conditions in the Greater Himalaya. Sub-Himalayan conditions prevail in the rest of the Himalaya where harsh continental conditions give way to monsoonal climate, especially the northeast India. Further, historically, the Himalaya has been tectonically active and hence unstable, a factor responsible for low richness. Thus despite large area the habitat diversity is low, consequently affecting the richness. Further, this area remains unexplored with respect to richness. However, if only the area of Lesser and Siwalik Himalaya is considered, the estimates can be real.

Another, intriguing fact is less number of macroinvertebrate and fish species in the C&EHL, located between the Himalaya and the Western Ghats (WG). The loss of diversity is considered to be secondary[14] and attributable mainly to historical events of tectonic nature[25] which not only led to changes in the drainage patterns but also climatic conditions. In the Western Ghats 'harshness' of mountain terrain is moderated by lack of extreme elevation, latitudinal position and oceanic climate, resulting in high richness. Geological history (Vindhya-Pre Cambrian, while Western Ghats and Himalaya-Coenozoic) is yet another important reason for the extent of richness, diversity and similarity and differences among these ecoregions.

With respect to the group of organisms considered herein, the number of diatom and benthic macroinvertebrate species are likely to increase with further investigations in the Central & Eastern Highlands, but not in case of fish where extensive information already exists. However, the number is not likely to be more than the Himalaya. Yet, the Central & Eastern Highlands though low in richness, the diversity was observed to be high in case of diatoms. Also the number of taxa per site was higher than the Himalaya as discussed above.

BIOGEOGRAPHY

The mountain chains considered herein differ in the elements of each group of organism. Only 20 to 25% diatom species were similar in the Himalaya and Central & Eastern Highlands ecoregions. Among diatoms the genera *Melosira*, *Fragilariaforma, Staurosirella*, *Peronia*, *Ceratonies*, *Encyonopsis*, *Reimeria*, *Rhoicosphenia* and *Cymatopleura* occurred only in the Himalaya while *Aulacoseira*, *Anomoeoneis, Anuemastus, Bacillaria, Diadesmis, Gomphocymbellopsis, Mastogloia* and *Scoliopleura* only in the Central & Eastern Highlands. Investigations will reveal such forms in the Western Ghats also. The Central & Eastern Highlands flora has many Himalayan genera, indicating exchange among these ecoregions. Central & Eastern Highlands had relatively more species of *Amphora*, *Cymbella*, *Cymbopleura*, *Navicula* and *Nitzschia* **(Table 3)**. Among the benthic macroinvertebrates few taxa were common in the Himalaya, Vindhya and Western ghats (**Table 4a**), as 17 families were restricted to the Himalaya, 11 in the Vindhya and 19 in the Western Ghat (**Table 4b**). Thus each region had specific elements.

Table 3 : Diatom flora: Comparison of the Himalayan and Central Highland eco regions. Vindhya is rich in *Cymbella, Navicula* both sensu stricto and lato, *Nitzschia*.

Genera	H	CH
Thalassiosiraceae		
1. *Aulacoseira*		1
2. *Cyclotella*	1	2
Melosiraceae		
3. *Melosira*	1	
Fragilariaceae		
4. *Ceratoneis*	3	
5. *Diatoma*	4	4
6. *Fragilaria*	4	2
7. *Staurosira*	3	2
8. *Synedra*	10	19
9. *Tabellaria*		1
Eunotiaceae		
10. *Eunotia*	1	6
Achnanthaceae		
11. *Achnanthes*	1	1
12. *Achnanthidium*	22	14
13. *Planothidium*	6	5
14. *Cocconeis*	4	6
Naviculaceae		
15. *Amphipleura*	1	1
Genera	H	CH
16. *Amphora*	5	12
17. *Anomoeoneis*		1
18. *Brachysira*	1	2
19. *Caloneis*	4	6
20. *Cymbella*	23	31
21. *Cymbopleura*	3	14
22. *Encyonema*	3	5
23. *Encyonopsis*	3	
24. *Reimeria*	2	
25. *Diploneis*	4	6
26. *Frustulia*	1	1
27. *Gomphocymbelopsis*		1
28. *Gyrosigma*	1	3
29. *Gomphonema*	10	15
30. *Mastogloia*		1
31. *Navicula*	24	32
32. *Navicula* sensu lato	3	5
33. *Adlafia*	2	2
34. *Aneumastus*		2
35. *Craticula*		5
36. *Diadesmis*		2
37. *Fallacia*	1	2
38. *Geissleria*	1	1
39. *Hippodonta*	2	2
40. *Luticola*	1	8
41. *Placoneis*	1	2
42. *Sellaphora*	1	7
43. *Neidium*		4
44. *Pinnularia*	2	7
45. *Rhoicosphenia*	1	
46. *Scoliopleura*		1
47. *Stauroneis*	1	2
Epithemiaceae		
48. *Epithemia*		1
49. *Rhopalodia*		2
Bacillariaceae		
50. *Bacillaria*		1
51. *Denticula*	1	1
52. *Hantzschia*	2	1
53. *Nitzschia*	19	32
Surirellaceae		
54. *Cymatopleura*	1	
55. *Surirella*	5	9
Total	**189**	**293**
Genera	**42**	**49**

Table 4(a) : Benthic macro invertebrate fauna: Distribution of families in the Himalayan, Central & Eastern Highlands and Western Ghats – Common elements.

Families	H	C&EHL	WG
Baetidae	+	+	
Caenidae	+	+	+
Ephemerillidae	+	+	
Heptageniidae	+	+	+
Leptophlebiidae	+	+	+
Hydropsychidae	+	+	+
Rhyacophilidae	+	+	
Glossosomatidae	+	+	
Brachycentridae	+	+	
Chironomidae	+	+	
Perlidae		+	+
Gomphidae	+	+	+
Agrionidae	+	+	

Table 4(b) : Benthic macro invertebrate fauna: Distribution of families in the Himalayan, Central & Eastern Highlands and Western Ghats – Native ? elements.

Himalaya	C&EHL	Western Ghats
Sipholonuridae	Neoephemeridae	Stenopsychidae
Philopotamidae	Hydroptilidae	Polycentropodidae
Psychomyiidae	Tabanidae	Nepidae
Dixdae	Heleidae	Sericostomatidae
Simulidae	Culicidae	Prosopistomatidae
Blepheroceridae	Dytiscidae	Notonectidae
Leptoceridae	Glossoscolecidae	Lepidostomatidae
Limnephilidae	Nephthydae	Notonectidae
Perlodidae	Salifidae	Coenagrionidae
Sialidae	Thiaridae	Euphaeidae
Corydalidae	Corbiculidae	Protoneuridae
Psephenidae		Corydalidae
Dryopidae		Lesticidae
Elmidae		Naucoridae
Planorbidae		Gerridae
Lymnaeidae		Belostomatidae
		Cordulidae
		Libellulidae
		Gyrinidae

Similarly, only 82 out of 520 fish species were common the Himalaya, C&EHL and WG, 193 species were found only in the Himalaya, 122 in the WG and only 19 species in the C&EHL. In case of families, 25 out of 45 families were common among these regions. Only 6 families were restricted to Himalaya, 1 to the C&EHL and 8 to WG. Families specific to each ecoregion were: Himalaya; Moringuidae, Ophichthinae, Psilorhynchidae, Chacidae, Olyridae, Chaudhuriidae; C&EHL; Cynoglossidae, W G Blenniidae, Oryziidae, Cyprinodontidae, Hemiramphidae, Lutjanidae, Pomadasyidae, Cichilidae, Soleidae, while WG and C&EHL Parapsilorhynchidae only.

The present level of similarity can be explained on the present day conditions as well as the geological factors. Apparently, the similarity though low, exists among these ecoregions as far more number of families and genera are similar than the species. As the Himalaya and Vindhya are connected by the Gangetic drainage, the floral and faunal exchanges seem obvious. The cause of this low similarity in the aquatic flora and fauna of these regions may be due to 'Gangetic Plains' which form a physical barrier for passively dispersing unicellular diatoms. The barrier can be more effective in case of diatoms rather than the insects which have wings and can be carried by wind also and fish which perform powerful locomotion. However, the distribution of all three groups is limited by the prevailing thermal regimes more so in case of poikilothermic fauna. The present day higher thermal regimes (ambient and aquatic) in the Gangetic plain are possible barriers to the exchange flora. Himalayan elements cannot easily proliferate for long distances in the hot Gangetic Plains and vis-a-vis. The effectiveness of the barrier gets enhanced due to great width of the 'Plains'. The 'Plains' are wide extending 6° N latitude. Besides, the river bed consists of soft sediments in the 'Plains' where only planktonic or epiphytic forms can flourish and the benthic forms will cease to exist due to lack of hard substrate, typical to highland streams/rivers. Thus, despite the Gangetic drainage connection similarity in the diatom flora is low among these highlands.

Possibly the Plains were narrow when their formation began following the tectonic rise of the Himalaya and as a physical and physiological 'barrier'. The climatic conditions were certainly different from today as monsoon is only few thousand years old. Further, creation of the Himalaya effected glaciations when the global climate was cooling. In view of these geological events the Indian subcontinent has remained unstable in the vicinity of the Himalaya. The factors related to geological events, disturbance, chance introductions, and dispersal ability govern distributions and are more important than environmental conditions[26].

The development of biota in the Himalaya and Central Highlands presently connected to each other by the Gangetic drainage and the Western Ghats with no apparent present connection has been envisaged in phases. In the **first phase** following the Himalayan upheaval and prior to the glaciations, the newly formed stream habitats in the sub-Himalaya must have been colonized from the north flowing rivers of the Indian Peninsula. The biota in the northern Peninsula must have developed through exchanges between west flowing rivers in historical past. The mechanism of colonization however, was active as species have to invade against the current in the rivers flowing down the mountain, thus leading to the development of mountain flora. From upheaval to glaciations there was enough time for adaptation and speciation. The species found only in the Himalaya may be the result of speciation within each genera, most of which were ancient by virtue of Peninsular ancestry.

In the **second phase** i.e. during the glaciations many of the Himalayan elements reached Peninsular India and some peninsular elements may have perished in the Central Highlands and the Western Ghats. Similarly, in the **third phase** i.e. 'post glaciations' some of the Himalayan forms retreated or perished from the Central Highlands and Western Ghats[25]. Thus, the Peninsular biota in Central India is a dilute mixture of the Himalayan and Peninsular floras which was further diluted in southern Peninsula. Thus, in past, the biota of the Peninsula and Himalaya, two distinct regions of the Indian subcontinent were possible more similar, which decreased after the end of recent glaciations as climate became warmer. The taxa unable to withstand desiccation perished and those which modified or adapted to sub tropical climate became new taxa, varieties and forms. The observation that most of the families and genera are common in these regions but their species and variety differ lends support to this opinion.

Bioresource as Livelihood and Protein Security: An Example of Indian Mahseers

It will be appropriate to discuss the role of mahseers in ensuring protein security of the forest inhabited primitive tribes of the Western Ghats in this context. *Kadar, Malayar* and *Muthuvar* of the Ghats are depending on the forest resources for their livelihood and the hill stream fishes serve as their major protein source[27] (Dinesh and Abraham, 2007; Dinesh *et al.,* 2007). In Nilambur forests of Kerala, *Cholanaikkan* (Manchery tribal colony) which is one of the most primitive tribal communities of Asia, depends largely on mahseer for their daily bread. *T. mussullah* which has a patchy distribution in the River Chaliyar is the major species caught. Tribal fishers go to the interior forest with their spouses accompanied by a dog and do fishing throughout the night. Though hook and line, gill netting and poisoning using plant based toxicants are the major fishing methods adopted by the tribes throughout the Western Ghats, mahseer is mainly caught by angling. As the fish has thick scales and tough meat, the quality degradation is comparatively slower in mahseer, which make it a preferable fish among the tribes both for consumption and marketing. Mahseers have been of considerable importance to the local fishermen[28] Kumar (2000) because of their large size, hardy texture, high commercial value and longer shelf life. Although organized marketing system, fish preservation techniques and value addition are yet to be introduced in these places, tribes usually earn approximately Rs.100/- for a day's catch. Interventions on the marketing of the produces, by the Department of Forests and Wildlife or other responsible agencies will help to reduce the exploitation by the middlemen during marketing. Another issue related with mahseer fishing and consumption is that local tribes believe that mahseer meat can not be consumed in certain seasons owing to the presence of some toxic material in it which is reported to cause severe vomiting problems in the consumers. Many reasons like ootoxicity, metal poisoning, physiological and biochemical reactions of the body can be attributed to the issue and hence further studies are warranted to resolve the problem. Tribal empowerment issues of the Western Ghats could be better addressed by incorporating fish-related activities as a component which may indirectly help the mahseer conservation programmes like angling, organized mahseer farming[29] (Kalita *et al.* (2007) and can become a vehicle for rural economic growth, apart from providing sustainable supplies of this fish caught during sport fishing and also through commercial farming by developing market linkages.

ACKNOWLEDGEMENTS

The academic support by the Head, Department of Zoology, and H.N.B. Garhwal University and University of Allahabad, is acknowledged.

REFERENCES

World Resources Institute (WRI), World Conservation Union (IUCN), and United Nations Environment Programme (UNEP). 1992. *Global Biodiversity Strategy: Guidelines for Action to Save, Study, and Use Earth's Biotic Wealth Sustainably and Equitably*. Washington, D.C.: World Resources Institute.

Nautiyal, P. (1996) The River Valley Projects in the Garhwal Region: Impact on the population dynamics of endangered Himalayan Mahseer with emphasis on bioconservation FTR (J110173/8/92-GPD/24.09.92) M O En & F New Delhi, p 96.

Nautiyal, P. (1997) Migratory phenomenon of the endangered Himalayan Mahseer *Tor putitora* in relation to the Ecology of the river Ganga.106 pp.Final Technical Report of Research Project No. 14/28/92- submitted to the Ministry of Environment and Forests New Delhi under Man and Biosphere Programme, p. 85.

Nautiyal, P. (2001) Diatom biodiversity in the Himalayan lotic systems. Final Technical Report of Research Project No.GBPI/IERP/98-99/04/565; March 23,1999, submitted to G.B. Pant Institute of Himalayan Environment and Development Kosi-Katarmal, Almora. [Ministry of Environment and Forests New Delhi], p. 75.

Nautiyal, P. (2005). Spatial and microdistribution of benthic macroinvertebrate community in tropical mountain streams of Garhwal region (Project No. SP/SO/C-42/98 dated 22 July, 1999, 18.01.2002 to 17.01.2005). Final Technical Report submitted to DST, New Delhi.

Nautiyal, R. (1992) Diatoms of the Cold Water River, Alaknanda. D. Phil. Thesis, H. N. B. Garhwal University, Srinagar, Garhwal.

Rawat, V. S. (2000) Studies on the abundance and species diversity of the Mayflies in Alaknanda and its tributary Khanda Gad (Garhwal). D. Phil. Thesis, H. N. B. Garhwal University, Srinagar, Garhwal, p. 197.

Semwal, V. P. (2002) Distributional patterns of benthic macroinvertebrate assemblages in Himalayan stream (Mandakini basin: In Garhwal region). D. Phil. Thesis, H. N. B. Garhwal University, Srinagar, Garhwal, p. 165.

Shivam, A. (2007) Longitudinal distribution of the benthic macro invertebrate community in the fluvial systems of Vindhyan Hills. D. Phil. Thesis, University of Allahabad, Allahabad.

Verma, J. (Unpublished) Biodiversity patterns of lotic (fresh water) diatoms in mountain chains (North India). D. Phil. Thesis, University of Allahabad, Allahabad.

Nautiyal, R. & Nautiyal, P. (1999) Altitudinal variations in the pennate diatom flora of the Alaknanda-Ganga river system in the Himalayan stretch of Garhwal region. In: *Proc.* 14th *Internat. Diatom Symp.* (S. Mayama, M. Idei and I. Koizumi, eds.), Koeltz Scientific Books, Koenigstein, p. 85-100.

Nautiyal, P., Nautiyal, R., Kala, K. & Verma, J. (2004) Taxonomic richness in the diatom flora of Himalayan streams (Garhwal, India). *Diatom,* **20**: 123-132.

Nautiyal, P. & Singh, H.R. (2008) Fish diversity in highlands of the Indian subcontinent. In: *Biodiversity and Ecology of Aquatic Environments*. Narendra Publishing House, Delhi, p. 105-160.

Jüttner, I. & Cox, E. J. (2001) Diatom communities in streams from the Kumaon Himalaya, north west India. In: *Proceeding of 16th International Diatom Symposium.* Athens and Aegean Islands. (A. Economou - Amilli, ed.), Amvrosiou Press, University of Athens, Greece, p. 237 248.

Ormerod, S. D., Rundle, S. M., Wilkinson, G. P., Daly, K. M. & Jüttner, I. (1994) Altitudinal trends in the diatoms bryophytes, invertebrates and fish of a Nepalese river system. *Freshwater Biol.,* **32**: 309-322.

Ormerod, S. J., Baral, H. S., Brewin, P. A., Buckton, S. T., Jüttner, I., Rothfritz, H. & Suren, A. M. (1997) River habitat surveys and biodiversity in the Nepal Himalaya. In: *Freshwater Quality: Defining the Indefinable* (P. J. Boon & D. L. Howell, eds.), Her Majesty Stationary Office, Edinburgh. p. 241-250.

Jüttner, I., Rothfritz, H. & Ormerod, S. J. (1996) Diatoms as indicators of river quality in the Nepalese Middle Hills with consideration of the effects of habitat-specific sampling. *Freshwater Biol.,* **36**: 475-486.

Rothfritz, H., Jüttner, I., Suren, A. M., & Ormerod, S. J. (1997) Epiphytic and epilithic diatom communities along environmental gradients in the Nepalese Himalaya implications for the assessment of biodiversity water quality *Arch. Hydrobiol.,* **138:** 465-482.

Rout, J. & Gaur, J. P. (1994) Composition of dynamics of epilithic algae in a forest stream at Shillong (India). *Hydrobiol.*, **291**: 61-74.

Cantonati, M., Corradini, G., Jüttner, I. & Cox, E. J. (2001) Diatom assemblages in high mountain streams of the Alps and the Himalaya. *Beiheft zur Nova. Hedw.,* **123**: 37-61.

Badoni, K., Nautiyal, R., Bhatt, J. P., Kishore, B. & Nautiyal, P. (1997) Variations in the epilithic diatoms community structure due to river valley projects on the Ganga between Rishikesh and Hardwar. *Proc. Ind. Nat. Sci. Acad.*, **63(6):** 523-536.

Begon, M., Harper, J. L. & Townsend, C. R. (1990) Ecology: Individual, Populations and Communities. 768 pp. Cambridge, Massachusetts, USA.

Nautiyal, P. (2005) Taxonomic Richness in the fish fauna of the Himalaya, Central Highlands and the Western Ghats. *Internat. J. Eco. Environ. Sci.*, **31(2)**: 73-92.

Krebs, C. J. (1994) Ecology 4th ed. Addision-Wesley Education Publisher Inc, USA. p. 801.

Mani, M. S. (1974) Ecology and Biogeography in India. Dr. W. Junk b. v. Publishers the Hague, p. 725.

Charles, D. F., Acker, F. W., Hart, D. D., Reimer, C. W. & Cotter, P. B. (2006). Large-scale regional variation in diatom-water chemistry relationships: rivers of the eastern United States. *Hydrobiol*, 561, 2757.

Dinesh, K. and Abraham, J. 2007. Protein security of the local tribes of the Western Ghats through aquaculture-interventions with special emphasis to conservation and utilization of fishery resources. In: Natural resource management and livelihood support systems. Compendium of selected Research papers and articles, Western Ghats Cell, Government of Kerala, Thiruvananthapuram. p. 323-327.

Kumar, K. 2000. Conservation and development of Golden mahseer (*Tor putitora* Ham.) in Himachal waters. *Fishing Chimes*. **20(9):** 26-27.

Kalita K., Bhagapati, S.K. and Sarma, D.K. 2007. Status of threatened fishes in Assam. *Fishing Chimes*, **26(10):**142-144.

Bioresources for Food Security and Rural Livelihood, Pages 97–110
Edited by : G.K. Kulkarni and B.N. Pandey

MARINE NATURAL RESOURCES AND FOOD PRODUCTION:
THE SAGA OF THE INDIAN "GREEN BULLET"

Jyoti D. Vora

ABSTRACT

In the sea, numerous species of marine organisms inhabit, adapting to the environment having a high salinity, high water pressure, low temperature and being a closed system of water. These environmental features lead to unique biological phenomena, metabolisms and metabolites of marine organisms. At the same time, although organisms originally occurred in the sea and current marine organisms are regarded as lower in the phylogenic classification, they have biological phenomena and constituents universally important for life. Thus, it is important to take particular note of both the specific and universal natures of marine organisms to utilize the diversity of marine bioresources.

Biodiversity encompasses the variety of all life on earth. India has rich Biological diversity and is one of the 12 diverse countries of the world. With only 2.5% of land area, India accounts for 7.8% of the recorded species at global level. India is also rich in traditional and indigenous knowledge both coded as well as informal.

INTRODUCTION

Biological diversity means the variability among living organisms from all sources, including inter alia, terrestrial, marine and other aquatic ecosystems and the ecological complexes of which they are part, this includes diversity within species, between species and of ecosystems.

Bioresources are important components for progress and economic activities of a nation. But bioresources management and utilization for human welfare is very important and critical components of the optimum utilization of the bioresources Awareness of the importance and implications of bioresources among common people as well as elite educated citizens for safeguarding and protecting the optimum and balance way of using the bioresources needs critical studies to focus the natural bioresources wealth for the benefit of not only the present generation of our people but also to our future generations for their better, healthy and peaceful living on the earth. The problems faced at present are the over exploitation of bioresources/natural resources which would not only have negative impact on the environment but also sometimes totally destroy and erode the important bioresources which are available at local level, regional level and national levels. Therefore, handling Bioresources in a proper manner in an appropriate way is important for the optimum use without over exploitation of our bioresources wealth.

Biological resources includes genetic resources, organisms or parts thereof, populations, or any other biotic component of ecosystems with actual or potential use or value for humanity.

Ecosystem management in a particular location is important and integral part for the conservation and protection of biological diversity of India. Ecosystem is a dynamic complex of plant, animal and microorganism communities and their non-living environment interacting as a functional unit and habitat is the place or type of site where an organism or population naturally occurs.

Biodiversity is being lost as on today more rapidly than at any time in the past several million years. Some Biologist believe that bout 60,000 of the world's 2,40,000 plant species perhaps even higher proportions of vertebrate and insect species could become extinct within the next thirty years if the same trends continue. But even a species at no risk of extinction can lose much of it potential through the loss of genetic material by reduction in range, numbers and varieties. The current losses to biodiversity can be attributed to direct causes including habitat loss and fragmentation, invasion of introduced species, over exploitation of living resources and modern agriculture and forestry practices. It is imperative, therefore to look towards other resources and hence, especially in the Indian context, the marine environment as a bioresource for food provision, is a very viable and significant option.

Given that in biology many phyla, families and genera have some species that live in the sea and others that live on land, marine biology classifies species based on the environment rather than on taxonomy. Marine biology differs from marine ecology as marine ecology is focused on how organisms interact with each other and environment and biology is the study of the animal itself.

Marine life is a vast resource, providing food, medicine, and raw materials, in addition to helping to support recreation and tourism all over the world. At a fundamental level, marine life helps determine the very nature of our planet. Marine organisms contribute significantly to the oxygen cycle, and are involved in the regulation of the Earth's climate. Shorelines are in part shaped and protected by marine life, and some marine organisms even help create new land. Marine biology covers a great deal, from the microscopic, including most zooplankton and phytoplankton to the huge cetaceans (whales) which reach up to a reported 48 meters (125 feet) in length.

The habitats studied by marine biology include everything from the tiny layers of surface water in which organisms and abiotic items may be trapped in surface tension between the ocean and atmosphere, to the depths of the abyssal trenches, sometimes 10,000 meters or more beneath the surface of the ocean. It studies habitats such as coral reefs, kelp forests, tidepools, muddy, sandy and rocky bottoms, and the open ocean (pelagic) zone, where solid objects are rare and the surface of the water is the only visible boundary.

A large amount of all life on Earth exists in the oceans though, exactly how large the proportion is still unknown. A lot of species living in oceans are still to be discovered. While the oceans comprise about 71% of the Earth's surface, due to their depth they encompass about 300 times the habitable volume of the terrestrial habitats on Earth.

Many species are economically important to humans, including food fish. It is also becoming understood that the well-being of marine organisms and other organisms are linked in very fundamental ways. The human body of knowledge regarding the relationship between life in the sea and important cycles is rapidly growing, with new discoveries being made nearly every day. These cycles include those of matter (such as the carbon cycle) and of air (such as Earth's respiration, and movement of energy through ecosystems including the ocean). Large areas beneath the ocean surface still remain effectively unexplored.

The marine ecosystem is large, and thus there are many subfields of marine biology. Most involve studying specializations of particular animal groups. (i.e. phycology, invertebrate zoology and ichthyology).

Other subfields study the physical effects of continual immersion in sea water and the ocean in general, adaptation to a salty environment, and the effects of changing various oceanic properties on marine life. A subfield of marine biology studies the relationships between oceans and ocean life, and global warming and environmental issues (such as carbon dioxide displacement).

Recent marine biotechnology has focused largely on marine biomolecules, especially proteins that may have uses in medicine or engineering. Marine environments are the home to many exotic biological materials that may inspire biomimetic materials.

Plant life is widespread and very diverse under the sea. Microscopic photosynthetic algae contribute a larger proportion of the worlds photosynthetic output than all the terrestrial forests combined. Most of the niche occupied by sub plants on land is actually occupied by macroscopic algae in the ocean, such as *Sargassum* and kelp, which are commonly known as seaweeds that create kelp forests. The non algae plants that survive in the sea are often found in shallow waters, such as the seagrasses (examples of which are eelgrass, *Zostera,* and turtle grass, *Thalassia).* These plants have adapted to the high salinity of the ocean environment. The intertidal zone is also a good place to find plant life in the sea, where mangroves or cordgrass or beach grass might grow. Macroscopic algae and plants provide important habitats for life, sometimes acting as hiding and foraging places for larval forms of larger fish and invertebrates.

India has a long coastline of about 8000 kms which is inhabited by over 6 million fisherfolk. Two island groups, Andaman and Nicobar Islands and Lakshadweep Islands, have been identified as part of Global 200 eco-regions (Marine Eco-regions) by WWF. It is estimated that the continental shelf alone has an area of 0.44 million square kms, with an estimated potential of 4.5 million tonnes of marine produce per year.

For centuries fishing community has exploited this vast ecological niche, home to a variety of marine organisms in the inshore waters, providing coastal and inland villages with an inexpensive source of protein, while maintaining an independent and self-sufficient life style.

India has declared a 200 mile Exclusive Economic Zone (EEZ) wholly in the Indian Ocean, with sovereign rights for exploring and exploiting the living and non-living resources of the sea. The largest mangroves in Sunderbans, Bay of Bengal, unique in its bio-diversity, act as a salt tolerant system and has been designated the status of a World Heritage Site by WWF in the Indo Malayan - Indo Pacific region. The country's marine biodiversity is outstanding with a wide variety of species found in the ocean. Coral reefs, which are regionally outstanding in terms of their spatial cover and diversity, are also found in the Andaman and Lakshadweep Islands and Gulf of Mannar.

Dugongs are found in large numbers in the Gulf of Mannar and Palk Bay, their residential population being the most vulnerable in the region. Some species of whales like Falsekiller whale, Balenoptera and Humpbacked whales have been documented by scientists.

Besides a large number of fish, India is home to 5 out of 7 species of marine turtles in the world and has a major mass-nesting site for Olive Ridley Turtle on the Orissa coast. All the five species of sea turtles along with marine mammals are severely threatened and endangered

Referred to as "green bullet", marine algae is considered as science and society¨ best hope for a clean bio-energy source that will help loosen broad dependence on fossil fuel, counteract climate warming, and power the vehicles of the future and it offers a sustainable solution to the food provision problem in the national context.

According to Greg Mitchell from the Scripps Institution of Oceanography at DC San Diego, marine algae are the most efficient organisms on Earth for absorbing light energy and converting it into a natural biomass oil product, the biofuel equivalent of crude oil. "Algae yields five to 1 0 times more bio-energy molecules per area, per time, than any terrestrial plant," he said. "Nothing else comes close," he added. From a sustainability perspective, algae hold the upper hand against other biofuel candidates, such as corn and soybeans. Algae can be grown on barren desert land using salt water, averting competition with agricultural cropland and the need for large amounts of precious fresh water for irrigation. Since they require carbon dioxide for growth, algae are inherently carbon neutral, and they can suck up CO2 directly from industrial pollution sources. Furthermore, algae can feed off the nutrients in discarded wastewater. Adding yet another layer to their allure, the rich protein left over from algae harvests can be converted to anima] feed. Since algae reproduce quickly (they can double their numbers in a single day), it is believed that they can more efficiently produce many more gallons of oil per acre than any other source.

"There is still a lot of work to do, but algal-derived biofuels have the potential to become a major source of transportation fuel," said Bernard Raemy, executive vice president of Carbon Capture Corporation.

According to Raemy, a string of challenges lies ahead, but with appropriate investment, he believes a new algal biofuelj industry, based on collaborations with public and private sectors, could be built within 10 years.

"Given their advantages, I believe marine algae are not only the most promising option for bio-energy fuel, but the only option that can scale up massively at the global level", said Mitchell.

"Most scientists who understand these processes are concluding that algae has the best chance. There is no silver bullet when it comes to energy, but there is a green bullet, or rather a green missile," he added. (ANI)

Although fish and other seafood products make delicious, healthy meals for people all over the world, many American children would not mind if they never had to eat tuna casserole again. But they would mind if suddenly there were no more cheese and chocolate milk, peanut butter and pudding, frozen desserts and fruit drinks. What could such different foods have in common? Along with hundreds of other common foods and household items, they contain the aquatic plants known as seaweed.

MARINE ALGAE ON THE INDIAN COAST

Seaweed or benthic marine algae are the group of plants that live either in marine or brackish water environment. Like the land plants seaweed contains photosynthetic pigments and with the help of sunlight and nutrient present in the seawater, they photosynthesize and produce food. Seaweeds an~ found in the coastal region between high tide to low tide and in the sub-tidal region up to a depth where 0.01 % photosynthetic light is available. Plant pigments, light, exposure, depth, temperature, tides and the shore characteristics combine to create different environment that determine the distribution and variety among seaweeds. They are basically classified according to colour into three main groups i.e. green (Chlorophyta), brown (Phaeophyta) and red (Rhodophyta). Seaweeds can also be classified by use (as food, medicine, fertilizer, industrial, etc.)

Types of Marine Algae

Cyanophycea (blue-green algae) comprise 67 genera with 287 species or intraspecific taxa. They are arranged according to the Geitlerian scheme as modified by Komarek and Anagnostidis, accompanied by a reconciliation with the Drouetian system. Their nomenclature is based on a 1753 Linnaean starting-point.

- Rhodophyceae (red algae) comprise 390 genera with 1810 species or intraspecific taxa.
- Phaeophyceae (brown algae) comprise 96 genera with 596 species or intraspecific taxa, of which 239 belong to *Sargassum.*
- Xanthophyceae (represented only by *Vaucheria)* comprise 11 species.
- Chlorophyceae (green algae) comprise 77 genera with 585 species or intraspecific taxa, of which 131 belong to *Caulerpa.*

The Indian Scenario

Seaweeds grow abundantly along the Tamilnadu and Gujarat coasts and around Lakshadweep and Andaman and Nicobar islands. Rich seaweed beds are found around Mumbai, Ratnagiri, Goa, Karwar, Varkala, Vizhinjam and Pulicat in Tamilnadu and Chilka in Orissa. Surveys carried out by the Central Salt and Manne and Chemical Research Institute (CSMCRI), Central Marine Fisheries Research Institute (CMFRI) and other research organisations have revealed vast seaweed resources along the coastal belts of South India. On the West Coast, especially in the state of Gujarat, abundant seaweed resources are present on the intertidal and subtidal regions. These resources translate into great potential for the development of seaweed-based industries in India. Despite the great number of sheltered bays and lagoons suitable for mariculture, large-scale attempts to grow seaweed have not been made so far.

Out of approximately 700 species of marine algae found in the Indian coast, 60 species are commercially important. Agar-yielding red seaweeds such as *Gelidiella acerosa* and *Gracilaria sp.* are collected throughout the year while algin-yielding brown algae such as *Sargassum* and *Turbinaria* are collected seasonally from August to January off the southern coast.

Importance of Shore Types in Marine Algae Ecology

The ecology of marine algae is dominated by two specific environmental requirements. These are the presence of seawater (or at least brackish water) and the presence of light sufficient to drive photosynthesis. A very common requirement is also to have a firm point of attachment. As a result, seaweeds are most commonly found in the littoral zone and within that zone more frequently on rocky shores than on sand or shingle.

1. **Sandy shore :** More than half of the Indian coast is sandy. This region is wholly sandy and bereft of any algal vegetation. The reason for this is the absence of any hard substratum suitable for the attachment of the marine algae. However this region shows the presence of many unicellular and filamentous algae.

Marine algae along the Indian coast:

Algae	Order	Family	Genus	Species variation	Total
Chlorophyta	7	19	43	179+ 37	216
Phaeophyta	6	13	37	159 + 32	197
Rhodophyta	16	36	136	406 + 28	434
Xanthophyta	1	1	1	3 + 1	4
Total	30	69	217	746 + 97	844
Coastline	**Chlorophyta**		**Phaeophyta**	**Rhodophyta**	**Total**
Gujarat	64		70	113	247
Maharashtra	52		47	99	197
Goa	22		29	32	83
Karnataka	13		20	23	56
Kerala	23		12	39	74
Lakshadweep	40		24	58	122
Total	214		202	363	779

2. **Shingle :** Shingles are shores with layers of stones and pebbles or fragments of coal. The shingle offers a suitable substrates for small algae which may be filamentous or leafy or crust forming. These may occur throughout the intertidal or may be confined to the low water level. There is no dominant species and consists of crust forms like *Hi/denbrandtia* and *Peyssonelia,* filamentous reds like *Polysiphonia,* filamentous browns like members of the *Ectocarpales,* green algae like *Cladophora, Enteromorhpa* and *Ulva* and various blue green algae.

3. **Rocky shores :** These may be varied in origin but 3 principal types can be distinguished: Silt covered rocky flats occur in various places along the Indian coast, particularly in the Gulf of Cambay and the Gulf of Kutch. The rocky flats are of recent calcareous sand stones eroded to various extent and showing a number of gullies and tide-pools. The algal growth is found in the gullies and tide-pools as the substrate does not offer a firm hold for the algae. The flora is limited to Viva and a few dwarfish filament types that attain their maximum development in tide-pools. Eg: Polysiphonia sp., Gracilaria verrucosa, Enteromorhpa sp. and Cladophora sp.

   - A more common type of rocky shore on the Gujarat coast consists of a bedrock of limestone which becomes thick with sand deposit in the upper intertidal region. The rock itself is often flat or gently sloping into the sea and occupies the lower intertidal region. In some places the sand cover does not exist and the algal growth is very poor in this region except for the presence of the blue green algae and a space growth of *Enteromorhpa sp.* In region covered with sand, there is abundant algal growth. In this region there is *Gracilaria* and *Hypnea* are present in the lower low-water level and *Enteromorhpa* and *Ulva lactuca* are present in the lower high- water level. This difference is caused by critical tide factors operating at lower low-water level and lower high- water level.

- Steep rocky shores characterise the major part of the west coast of India and are interrupted at frequent intervals by sandy areas associated with creeks, rivers and backwaters etc. Very little is known about the ecology of this region.

4. **Coral formations :** Coral formations are prevalent in the Gulf of Mannar in South India, the Gulf of Kutch in Gujarat and around the islands of the Arabian sea and the Bay of Bengal. The highly calcareous coral reefs form good substrates for the growth of marine algae. The coral reef although primarily built by the activity of anthozoans is developed as a result of the association of coralline algae which includes members of the Rhodophyta and few members of the Chlorophyta. However, very little is known of the coral reef in India.
5. **Estuarine lagoons :** The algal flora is composed of either floating or mud-inhabiting forms which have the capacity to withstand a wide range of salinity. *Enteromorhpa prolifera, Cladophora sp.* among green algae, *Rosenvingea* intricata among brown algae, *Graci/aria verrucosa* and *Hypnea nigrescens, Spyridia filamentosa* and *Polysiphonia sp.* among the red algae are the typical members of the Community.
6. **Mangrove swamps :** Members of Rhizophoracea and *Avicennia sp.* are predominant. Certain alga frequently associated with mangroves include *Cantella repens, Caloglossa sp. Bostrychia tenella, Herposiphonia insidiosa, Murrayella perlelados* and *Polysiphonia sp.* among the red algae some filamentous blue-greens belonging to the Oscillatoriaceae.

SEAWEEDS

Many kinds of seaweed are edible and rich in vitamins and iodine. They are as common in many Asian countries as green beans and carrots are in the United States. But until more people here develop a taste for sea vegetables, it is alginates, carrageenan, and beta carotene seaweed derivatives that act as stabilizers, thickeners, and colorants that end up on our dining room tables

Seaweeds are not really weeds but large forms of marine algae that grow in the coastal ocean waters of many countries. They include thousands of species ranging from microscopic plants called phytoplankton to giant floating or anchored plants.

The three main groups of seaweed are brown, red, and green algae, each providing important ingredients for the manufacture of food and other products. Carrageenan is a generic term for compounds extracted from species of red algae. Carrageenans are used in stabilizing and gelling food products. From brown algae come alginates. They make water-based products thicker, creamier, and more stable over extreme differences in temperature, ph, and time. For example, alginates prevent ice crystals from forming in ice cream. Beta Carotene, a natural pigment derived from green algae, is used as a yellow-orange food coloring and may help prevent certain types of cancers.

These seaweed derivatives represent only a small part of the many living and nonliving products we derive from ocean plants, animals, minerals, and sea water. Together, they provide an important reason to protect the oceans.

Naturally growing seaweeds are an important source of food, especially in Asia. They provide many vitamins including: A, B1, B2, B6, niacin and C, and are rich in iodine, potassium, iron, magnesium and calcium. In addition commercially cultivated microalgae, including both Algae and Cyanobacteria, are marketed as nutritional supplements, such as Spirulina, Chlorella and the Vitamin-C supplement, Dunaliella, high in beta-carotene.

Algae are national foods of many nations: China consumes more than 70 species, including *fat choy,* a cyanobacterium considered a vegetable; Japan, over 20 species; Ireland, dulse; Chile, cochayuyo. Laver is used to make "laver bread" in Wales where it is known as *bara lawr;* in Korea, gim; in Japan, nori and aonori. It is also used along the west coast of North America from California to British Columbia, in Hawaii and by the Maori of Newzealand. Sea lettuce and badderlocks are a salad ingredient in Scotland, Ireland, Greenland and Iceland.

The oils from some Algae have high levels of unsaturated fatty acids. For example, *Parletochloris incisa* is very high in arachidonic acid, where it reaches up to 47% of the triglyceride pool. Some varieties of Algae favored by vegetarianism and veganism contain the long-chain, essential omega-3 fatty acids, Docosahexaenoic acid (DHA) and Eicosapentaenoic acid (EPA), in addition to vitamin BI2. The vitamin BI2 in algae is not biologically active. Fish oil contains the omega-3 fatty acids, but the original source is algae, which are eaten by marine life such as copepods and are passed up the food chain.

Nori is commonly used as a wrap for sushi and onigiri. It is also a common garnish or flavoring in noodle preparations and soups. Nori is most typically toasted prior to consumption ("yaki-nori" in Japanese). A very common and popular secondary product is toasted and flavored nori ("ajitsuke-nori" in Japanese), in which a flavoring mixture (variable, but typically soy sauce, spices and sugar in the Japanese style or sesame oil and salt in the Korean style) is applied in combination with the toasting process. Nori is also eaten by making it into a soy sauce flavored paste *noritsukudani.* In Addition, nori is Sometimes used as a form of Food Decoration

Seaweeds are extensively used as food by coastal people, particularly in East Asia, e.g. Japan, China, Korea, Taiwan, Thailand, and Vietnam, but also in Indonesia, Belize, Peru, the Canadian Maritimes, Scandinavia, Ireland, Wales, Philippines, and Scotland, among other places. Tiwi, Albay residents discovered a new pancit or noodles made from seaweed, which has health benefits. It is rich in calcium and magnesium and the seaweed noodles can be cooked into pancit canton, pancit luglug, spaghetti or carbonara.

In Asia, Zicai in China), gim (in Korea) and nori (in Japan) are sheets of dried *Porphyra* used in soups or to wrap sushi. *Chondrus crispus* (commonly known as Irish moss or carrageenan moss) is another red alga used in producing various food additives, along with Kappaphycus and various gigartinoid seaweeds. *Porphyra* is a red alga used in Wales to make laver. Laverbread, made from oats and the laver, is a popular dish in Wales. Affectionately called "Dulce" in northern Belize, seaweeds are mixed uniquely with milk, nutmeg, cinnamon, and vanilla to be served as a common beverage throughout the country.

Seaweeds are also harvested or cultivated for the extraction of alginate, agar and carrageenan, gelatinous substances collectively known as hydrocolloids or phycocolloids. Hydrocolloids have attained commercial significance, especially in food production as food additives. [6] The food industry exploits the gelling, water-retention, emulsifying and other physical properties of these hydrocolloids. Agar is used in foods such as confectionery, meats and poultry products, desserts and beverages and moulded foods. Carrageenan is used in preparation of salad dressings and sauces, dietetic foods, and as a preservative in meat and fish products, dairy items and baked goods. Alginates enjoy many of the same uses as carrageenan, but are also used in production of industrial products such as paper coatings, adhesives, dyes, gels, explosives and in processes such as paper sizing, textile printing, hydro-mulching and drilling.

"Some algae are a delicacy fit for the most honorable guest, even for the king himself"

Edible seaweeds are algae that can be eaten or are used in the preparation of food. They may belong to one of several groups of multicellular algae: the red algae, green algae, and brown algae. Seaweeds are used extensively as food in coastal cuisines around the world. Seaweed has been a part of diets in China, Japan, and Korea since prehistoric times. Most edible seaweeds are marine algae as most freshwater algae are toxic. While marine algae are not toxic, some do contain acids that irritate the digestion canal, while some others can have a laxative effect.

CAULERPA

Caulerpa is a genus of seaweeds in the family Caulerpaceae (among the green algae). They are unusual because they consist of only one cell with many nuclei, making them among the biggest single cells in the world.

Food use: Some species (especially *Caulerpa lentillifera* and *Caulerpa racemosa)* are eaten under the names sea grape or green caviar or umi-budo in Okinawa. They have a peppery taste. Seagrapes are eaten in Indonesian cuisine, sometimes fresh, and other times coated in sugar. They are raised in Cebu, for domestic consumption in the Philippines as well as export to Japan.

GRACILARIA

Gracilaria is a genus of red algae (Rhodophyta) notable for its economic importance as an agarophyte, as well as its use as a food for humans and various species of shellfish. *Gracilaria* is used as a food in Japanese, Hawaiian, and Filipino cuisine. In Japanese cuisine it is called ogonori or ogo. In the Philippines it is called gulaman orguraman.

RED MARINE ALGAE

This therapeutic super food provides the body with a full array of nutrients including complete protein, complex carbohydrates, essential fatty acids, fiber, vitamins, minerals, trace elements, enzymes, and sulfated polysaccharides. Red Marine Algae's medicinal properties are thought to enhance the 'immune system's regulatory response, indicating that it is an immuno-modulatory/antiviral agent.

Red marine algae have been used by people as a food staple for thousands of years. In China, Japan, and the Indo-Pacific region; several dozen species of red algae are used. A nutritional analysis of this Red Marine Algae shows it to be composed of carbohydrates, proteins and fatty acids.

The unique combination of nutrients found in Red Marine Algae is important in the following ways:

- It is a complete protein with all the essential amino acids-unlike most plant foods. Complete proteins are crucial because they are involved in all major metabolic processes such as energy production, rebuilding, and enzyme production.
- It has a high carbohydrate content. It contains both simple and complex carbohydrates which affect a quick but sustained release, providing the body with an excellent source of additional fuel for its many constant energy needs. In particular, the sulphated complex carbohydrates are thought to enhance the immune system's regulatory response.
- It contains an extensive fatty acid profile, including Omega 3 and Omega 6. These essential fatty acids also playa key role in the production and availability of body energy.
- It has an abundance of vitamins, minerals and trace elements in a naturally-occurring synergistic design. For example, it contains an excellent magnesium to calcium ratio of 2: 1, and a potassium to sodium ratio of 4:3.

General health benefits show red marine algae useful in weight-loss programs and for lowering cholesterol and fat in the blood. It contains soothing, mucilaginous gels such as algin, carregeenan, and agar, which specifically rejuvenate the lungs and gastrointestinal tract. Once thought of as a liability that blocked assimilation, the tough cell wall in Dumontiaceae has been found to be invaluable. It binds with heavy metal pesticides, and carcinogens, and carries these toxins safely out of the body.

Red marine algae have long been known to assist the body's immune system to respond to virus attacks, and a regular intake can reduce the number of viral attacks and their severity. The polysaccharides are thought to stimulate the production of interferons, which work with the immune system as antiviral agents. They also improve the activity of T -cells and B-cells that destroy virus-infected cells in our body.

BROWN MARINE ALGAE : FUCOIDAN

Fucoidan is a sulfated polysaccharide (MW: average 20,000) found mainly in various species of brown seaweed such as kombu, limu moui, bladderwrack, wakame, mozulru, and hijiki (variant forms offucoidan have also been found in animal species, including the sea cucumber).

- One of the primary ways in which fucoidan can benefit your health is in its ability to interact with, and modulate the immune system. The ability of a substance to modulate the immune system is extremely useful. It has been shown that a dietary supply of certain saccharides from brown seaweed improves the integrity of glycoprotein production. Fucoidans provide an abundant supply of fructose, one of the necessary saccharides, and also provide smaller amounts of several other required sugars. Fucoidan supports the immune system in a number of ways. Some research suggests that it may cause an increase in the number of white blood cells, which are an integral part of the body's natural immune system.
- Fucoidan may also prove helpful in eliminating harmful cells through a process called apoptosis. Sulfated polysaccharides like fucoidan have been shown to assist in inducing appropriate programmed apoptosis, which is a natural process where living organisms eliminate harmful cells.

Brown algae are used to produce alginates, which are used as food additives and in industry. Common uses include food thickeners, stabilizers and fillers.

Besides their impressive nutritional profile, sea vegetables offer other health benefits. For centuries. Oriental medicine has recognized that sea vegetables contribute to general well-being and especially to the health of the endocrine and nervous systems. Diets high in sea vegetables have been associated with a lower risk for both colon and rectal cancers. Sea vegetables have also been shown to inhibit breast cancer. In summation it would not be inappropriate in enunciating that the Algae are truly Nature's gift to Mankind and in the Indian context represent the golden option for feed, food and fodder.

"What a country chooses to save, is what a country chooses to say about itself - Mollie Beattie"

REFERENCES

Abbott, Isabella A. and Hollenberg, George J. *Marine Algae of California.* Stanford CA : Stanford University Press, 1976. Nat Ctr Ref QK571.5.C2 M37 1976; Main QK571.5.C2 M37 1976.

Bear, F.E (1984) The Ftrman report (1948). Rutgers University

Beisel WR., Nutrition and immune function : Overview, J. Nutr., 126.2611S-2615S, 1996.

Bendich, A (1989) Carotenoids and the immune response. J. Nutrition **119:** 112-115

Benedetti S , *et al* (2004) Antioxidant properties of a novel phycocyanin extract from the blue-green alga Aphamzomenon flos-aquae. Life Sciences, Volume 75, Issue 19, pp. 2353-2362

Bland. J (1995) Food and nutrient effects on detoxitication Townsend letter (December) 40-44.

Bold, Harold Charles. *Introduction to the Algae: Structure and Reproduction.* Englewood Cliffs, N.J.: Prentice-Hall, 1985. Main QK566 .B64 1985.

Borowitzka, M.A. (1988) Vitamins and fine chemicals from micro-algae. In Micro-algal biotechnology, Borowitzke M.A. and Borowitzka J.L., 153-196, Cambridge Press.

Cannell. R.J.P (1990) Algal biotechnology. Appl Biochem & Biotech 85-105

Chao et.al., (1971). Purification and properties of glycogen isolated from a blue-green algae, Nostoc muscorum. J.Bacteriol., **105:** 331-338.

Commandeur, J.N.M and Vermeulen, N.P.E (1996) Cytotoxicity and cytoprotective activities of natural compounds. The case of curcumin Xenobiotica **16(7):** 667-680.

Connor, Judith. *Kelp Forests.* Monterey, CA : Monterey Bay Aquarium Foundation, 1989. Nat Ctr Ref QH541. 5 .K4 .C66.

Conquer, J A. and B.J Holub (1995) Effects of supplementation with an algae source of docosahexaenoic acid on risk factors for hean disease in vegetarians. Abstract of the 64th Congress of the European Atherosclerosis Society Utrecht Netherlands, June 10-13.

Conradson, Diane (ed.). *Natural History of the Fitzgerald Marine Reserve.* Moss Beach, CA : Friends of Fitzgerald Marine Life Refuge, 1999. Nat Ctr Ref QH 105.C2 N38 1999.

Darnall, D.W. (1991) Removal and recovery of heavy metal ions from wastewaters using a new biosorbent. AlgaSORB Innovative Hazard. Waste Treat Teclmol. Ser 365-72

Davis, Chuck. *California Reefs.* San Francisco: Chronicle Books, 1991. Nat Ctr QH 105 .C2 D378 1991.

Dawson, Elmer' Yale. *Seashore Plants of California.* Berkeley, CA : University of California Press, 1982. Nat Ctr Ref QK 149. D39 1982.

Dewille et.al., (1979) Effects of essential fatty acid deficiency, and various levels of dietary polyunsaturated fatty acids, on humoral immunity in mice. J.Nutr' 109(6) : 1018-1027.

Ebeling, W. (1981) The relation of soil quality to the nutritional value of plant crops. J. Appl Nutr., **33(1):** 24-34

Foster, Michael S. *Annotated Bibliography: Rocky Intertidal Communities of Central and Northern California.* Santa Cruz , CA : Kinnetic Laboratories, Inc., 1987. Main QH91.8 .L5 A55 1987.

Foster, Michael S. *The Ecology of Giant Kelp Forests in California: A Community Profile.* Washington, DC : U.S. Fish and Wildlife Service, 1985. Main QH 540 .U56 85:7:2, Nat Ctr Ref QH 541.5 .K4 E26.

Fuhr, U., Klittich, K., Staib, AH (1993) Inhibitoty effect of grapefruit juice and its bitter principal. naririgenin, on CYP 1A2 dependent metabolism of caffeine in man. Br. J.Clin Phannac **35 :** 431-436.

Gahan, E., Kline, P.R and Finkle, T.H. (1943) Chlorophyll in the treatment of ulcers Arch. of Dermatology and Svphilology. **49:** 848-851.

Garewal. H.S . et.al., (1995) Emerging role of beta-carotene and antioxidant nutrients in prevention of oral cancer. Arch. Otolaryng **111(1):** 141-144.

Bioresources for Food Security and Rural Livelihood, Pages 111–125
Edited by : G.K. Kulkarni and B.N. Pandey

PHYTOESTROGENS AND THEIR ROLE AS BIO-RESOURCES IN PROMOTING HEALTH AND CURBING DISEASES

C. Suresh

ABSTRACT

Phytoestrogens (PE) are naturally available biological compounds found in a wide variety of sources such as plant foods and are said to exhibit estrogen-like, antioxidant and anticancer properties. PE include mainly isoflavones (IF), coumestans, and lignans. These compounds are known to be present in fruits, vegetables, and whole grains which are commonly consumed by humans. IF are found in legumes - mainly soybeans, whereas flaxseed is a major source of lignans, and coumestans are significantly present in clover, alfalfa and soybean sprouts. Several studies for the past decade suggests that consumption of a phytoestrogen-rich diet is protective against breast, prostate, and bowel cancer and cardiovascular disease, amelioration of estrogen-deficiency symptoms in postmenopausal situations, osteoporosis, atherosclerosis and stroke, and even in neurodegenerative disorders. The ultimate actions of these compounds in specific cells are determined by many factors, including the relative levels of estrogen receptor (ER) alpha and ER beta and the diverse cocktail of co-activators and co-repressors present in any given cell type. The proposed mechanisms of these PE include estrogenic and antiestrogenic effects, induction of cancer cell differentiation, inhibition of tyrosine kinase and DNA topoisomerase activities, suppression of angiogenesis, and mainly antioxidant effects. Although, previously only the main source of isoflavones is from food, now being marketed in health-food supplements or drinks and tablets may soon be available over the counter as medicated 'natural' hormone-replacement therapy. The present paper highlights the possible beneficial and clinical effects of PE as a bio-resource material in detail and implications if any regarding the dietary and pharmacological intake of these compounds.

Keywords : Phytoestrogens, Plant sources, Consumption, Benefits Health, Disease

INTRODUCTION

Obesity and its related disorders, such as type 2 diabetes, high blood pressure, dyslipidemia, cardiovascular diseases (CVD), have recently been considered as major health problem causing health related disturbances in the population world wide (Engelgau *et al.*, 2004, Virgili, F., 2008). These diseases, commonly referred to as the Metabolic Syndrome (MS), are beginning to surpass malnutrition and other infectious diseases are said to be the most significant contributor to worldwide morbidity. In western societies, for the first time in modern history, life expectancy of newborns is declining as a result of these metabolic health complications (Olshansky *et al.*, 2005). Although the source of metabolic disorders is often the diet itself, nutrition can also form part of the solution, in fact providing health benefits. The rapid increase in obesity suggests that life-style factors such as high fat diets, physical inactivity and potentially environmental endocrine disruption, rather than genetics, are the most plausible causes. Normally, dietary intervention to control overweight, hyperglycemia and dyslipidemia has included low energy and low fat diets, but these are of limited use because of the strict and long-term commitment required. However, long term health benefits can be obtained from dietary and bioactive non-nutrients, called phytochemicals, which could be either supplemented into the diet or can be part of the food itself. These phytochemicals are biologically active plant-derived compounds, which structurally and functionally mimic estrogens (Dixon, 2004).

The importance of in plant derived estrogens—or phytoestrogens—has recently been increased by the realization that clinically hormone replacement therapy can produce side effects and are not effective as previously thought (Hays *et al.*, 2003). The presence of phytoestrogens in our diets and the biological effects that they exert need to be thoroughly examined. Estrogens, produced in the ovaries and testis, have many biological effects in the body beyond the reproductive system. The estrogen receptors which are localized in the nucleus can form dimers when bound to an estrogen. The dimers then interact with the estrogen response element (ERE), which regulates transcription of estrogen responsive genes. A small percentage for about 4-6% of estrogen receptors are located on the cell membrane and could contribute to non genomic effects of estrogen (Norfleet *et al.*, 1999, Razandi *et al.*, 1999, Xu *et al.*, 2003, Chen *et al.*, 2004). There are two known estrogen receptors, ERa and ERb. Although the two estrogen receptors can be localized within the same cell, the distribution in the tissue varies and can have different effects on mixed agonists and antagonists (Nilsson and Gustafsson, 2002). Both ERa and ERb function in vascular endothelia cells, normal ovarian follicular development, smooth muscle, myocardial cells and breast tissue (Nilsson and Gustafsson, 2002). ERa is more important in maintaining follicle stimulating and luteinizing hormone concentrations in blood, and ERb is involved in frontal lobe mediated learning and memory. ERa is involved in bone maturation in both males and females, however, only ERb plays a role in bone maintenance in females (Nilsson and Gustafsson, 2002). Although any compound that induces receptor dimerization and subsequent binding to the ERE, can be considered an estrogen, the dominant form of estrogen in the body is 17--estradiol (Fig. 1). Some compounds act as estrogen agonists and antagonists and are referred to as Selective Estrogen Receptor Modulators (SERMs). As an example, the antiestrogen tamoxifen acts as an estrogen antagonist in breast tissue but as an agonist in the uterus, bone and vascular system (Macgregor and Jordan, 1998). These agonist/antagonistic effects are said to be key factor for the differential effects of phytoestrogens compared to estradiol.

During the 1940s it was first realized that some plant derived compounds could cause an estrogenic effect (Bennetts *et al.*, 1946). It was observed that sheep that were grazing on pastures containing red clover had infertility problems. Immature animals were showing signs of estrus, ewes were unable to get pregnant and those pregnant were also often miscarried. The clover in these pastures had high amounts of the isoflavones, formononetin and biochanin A (Rossiter and Beck, 1966), which were among the first phytoestrogens discovered. Currently, four different families of phenolic compounds produced by plants are considered phytoestrogens: the isoflavonoids, stilbenes, lignans and coumestans (Fig. 1) are responsible for the estrogen-mediated response in different ways.

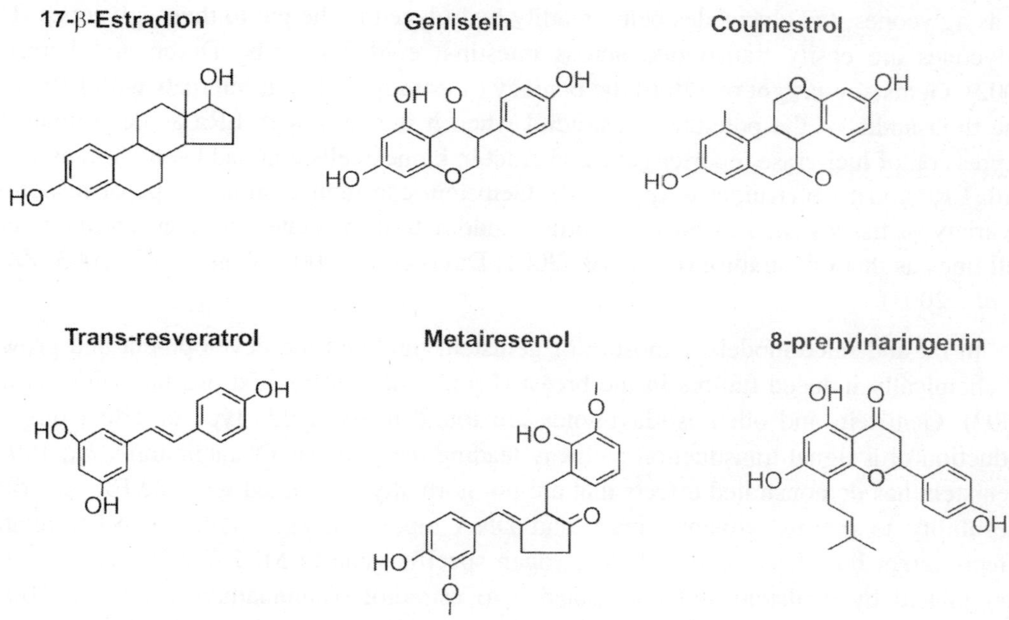

Fig 1 : The important chemical structures of phytoestrogen groups.

Classification of PE

The important dietary isoflavones are genistein, daidzein, glycitein, biochanin A and formononetin. These compounds are first found in legumes where they occur as glucosides. Soybeans and soy-based foodstuffs are a particularly rich source of isoflavones, especially genistein and daidzein and to a lesser extent glycitein, whereas biochanin A and formononetin are generally less prevalent in soy and are found mostly in clover and alfalfa sprouts (Bingham *et al*, 1998). The coumestans, of which coumestrol is the most common form, are structurally related to the isoflavones. They have been found in high concentrations in clover and fresh alfalfa sprouts (Humfrey *et al.*, 1998). Lignans are a class of phytoestrogens that exist as minor constituents in many cereals, vegetables and fruits, the highest concentration of these compounds (such as matairesinol and secoisolariciresinol), are found in foodstuffs such as grains, seeds and other fibre rich foods. Linseed (flaxseed) is the richest known source of lignans (Morton *et al*, 1997; Meagher & Beecher, 2000).

Isoflavanoids

The flavonoids are a important class of phytoestrogens that are formed through the phenylpropanoid-acetate biodemial pathway via chalcone synthase and condensation reactions with malonyl CoA. The isoflavonoids are a subclass of flavonoids, where one phenolic ring has migrated from C-3 to C-2. The isoflavonoids from legumes, including genistein and daidzein, are the most examined phytoestrogens. They can exist as glucosides or as aglycones, the glucosides being readily hydrolyzed in the gut to their aglycones. The aglycones are easily transported across intestinal epithelial cells (Dixon and Ferreira, 2002). Genistein has got one-third the potency of estradiol when it interacts with ERb, and one thousandth of the potency of estradiol when it interacts with ERa as determined by expression of luciferase reporter gene construct in kidney cells that had been co-transfected with ERa and ERb (Kuiper *et al.*, 1998). Genistein can induce similar type of responses I variety of tissues such as breast, ovarian, endometrial, prostate, vascular, and bone and cell lines as that of estradiol (Liu *et al.*, 2001; Davis *et al.*, 2002; Wang *et al.*, 2003; Zhou *et al.*, 2003).

In rat and mice models, exposure of genistein inhibited the development and growth of chemically induced tumors in the breast (Fritz *et al.*, 1998), and prostate (Fritz *et al.*, 2002). Genistein and other isoflavonoids are found to exert this type of effect through induction of a signal transduction pathway leading to apoptosis (Yanagihara *et al.*, 1993). Genistein has demonstrated effects that are not normally associated with the ER including the ability to inhibit tyrosine kinase and DNA topoisomerase . Other non-ER related effects might be related to the phytoestrogen specific gene in MCF-7 cells. This gene is upregulated by genistein and unresponsive to estradiol (Ramanathan and Gray, 2003). Genistein has differential effects on cell cycle progression in neoplastic (MCF-7) and non-neoplastic (MCF-10F) human breast cancer cells (Singletary *et al.*, 2002) by increasing cyclin B1 expression and phosphorylation of p34cdc2 in the non-neoplastic cell line and had no effect on the neoplastic cell line (Singletary *et al.*, 2002). Genistein inhibited cell proliferation and blocked the cell cycle at G2 in the non-neoplastic cell line with blockage of the cell cycle occurring only at specific concentrations (Singletary *et al.*, 2002). The ability of genistein to block cell proliferation of normal mammary cells may contribute to the preventive effect in breast cancer. A closely related and important compound to the isoflavonoids class is 8-prenyl naringenin , a flavanone, found in beer ingredient hops (Hummulus lupulus) is also considered to act as a protective agent against ovarian cancer.

Lignans

These are a class of phenylpropanoid dimers and oligomers. Secoisolariciresinol and matairesinol are two important lignan dimers that are not estrogenic by themselves, but are readily converted to the mammalian lignans, enterodiol and enterolactone, respectively, which are estrogenic in nature (Glitsø *et al.*, 2000). The conversion occurs by gut microflora and the mammalian lignans are readily absorbed and sources are flaxseed, whole grain breads, vegetables, and tea. Further they are very important constituents of higher plants (gymnosperms and angiosperms), such as whole grains, legumes, vegetables, and seeds, with exceptionally high concentrations of lignans found in flaxseed. Strawberries and cranberries has compounds possessing a 2,3-dibenzylbutane structure and exist as minor constituents of many plants, where they form the building blocks for the formation of lignin (as distinguished from lignan) found in the plant cell wall. Most of the structural changes occur in the colon, liver, and small intestine during enterohepatic circulation. Mammalian lignans differ in structure from plant lignans in that they have phenolic hydroxyl groups in the meta position only in their aromatic rings. Once in the colon, they are absorbed and then are conjugated with glucuronic acid or sulfate in the liver, re-excreted through the bile duct, deconjugated by the bacteria, and reabsorbed. Some reach the kidney and are excreted in the urine. Lignans are excreted in bile and urine as conjugated glucuronides and in feces in the unconjugated form. Originally, mammalian lignans were observed in urinary steroid profiles and thought to be new endogenous hormones. Enterolactone and enterodiol are metabolites of the plant lignans matieresinol and secoisolariciresinol, respectively. The higher the dietary intake of precursors, the higher the mammalian lignan production in the colon and the higher the excretion rate in the urine. Excretion of the lignans, enterodiol and enterolactone, was higher during a carotenoid (carrot and spinach) and cruciferous (broccoli and cauliflower) vegetable diet, than during a vegetable-free diet, suggesting that these vegetables may provide a source of mammalian lignan precursors.

Coumestans

Eventhough there are a large number of coumestans been reported only a small number have shown estrogenic activity, predominantly coumestrol and ethoxycoumestrol. The main dietary source of coumestrol, is legumes; however low levels have been reported in brussel sprouts and spinach (Knuckles *et al.*, 1976; Wang *et al.*, 1990; Franke *et al.*, 1994). Clover and soybean sprouts are reported to have the highest concentration, 28 and 7 mg/100 g dry wt., respectively; mature soybeans only have 0.12 mg/100 g dry wt (Knuckles *et al.*, 1976; Wang *et al.*, 1990; Franke *et al.*, 1994).

Phytoestrogens and Health

Diet

The research findings indicate that the diet can influence metabolism of phytoestrogens mediated by the gut microflora. As an example, consumption of less fat and more carbohydrate diet as a proportion of total energy intake was well correlated with greater equol production, particularly in women (Lampe *et al.*, 1998). The reason for this was unclear but there is always a possibility of complex carbohydrates stimulate fermentation in the large bowel, resulting in an increased breakdown of daidzein to equol (Rowland *et al,* 1999; 2000). Co-administration of dietary fructo-oligosaccharides with intragastric administration of isoflavone glucones was shown to modify the absorption pattern and entero-hepatic (Uehara *et al.*, 2001). In addition, low-fat diets are always associated with decreased ß-glucuronidase activity in the intestinal contents, which might impede absorption of more fatsoluble deconjugated compounds (Adlercreutz *et al.*, 1987) and can play a role in absorption and reabsorption of isoflavones (Tham *et al.*, 1998). High dietary fibre could result in the partial disruption of enterohepatic circulation of phytoestrogens and endogenous oestrogens. The effect of diet on the metabolism of phytoestrogens, both from the influence on gut microflora and differences in hepatic enzyme activities, can explain the biological response to phytoestrogens.

Role of Phytoestrogens in Disease

Phytoestrogens can show a favorable estrogenic effects on the risk of cardiovascular disease and are considered to be hypocholesterolemic, anticarcinogenic, antiproliferative, antiosteoporotic, and hormone altering agents. These health benefits could be an added advantage to a variety of potential mechanisms that are implicated in both the estrogen receptor dependent and independent nature.

Phytoestrogens and Cardiovascular Health

Recent epidemiological, clinical trial, and basic science reports suggests the possibility of a positive association between phytoestrogens and cardiovascular complications. High intake of dietary phytoestrogens in asian populations relative to those in other industrialized countries are consistent with a potential protective effect of phytoestrogens against CVD (Artaud-Wild SM *et al.*, 1993, Morisset, A.S. 2009). Estrogen-related studies forms a non-dietary line of epidemiological evidence that leads to a potential health benefit of phytoestrogen consumption. In a study soy product consumption has lowered non-high density lipoprotein (non-HDL) cholesterol and increased HDL cholesterol significantly, and therefore, both soy groups experienced an improved total cholesterol/low density lipoprotein (LDL) cholesterol ratio (Gooderham *et al.*, 1996) and reported no significant cholesterol changes in the subjects fed the supplements of 60 g/day of either soy protein or casein for 28 days despite increase in plasma isoflavone concentrations of genistein and daidzein during the study period. In another study, moderately hypercholesterolemic, 6.0 mmol/L postmenopausal women were given high (Murkies AL *et al* 1995) resulted decrease in cholesterol levels. There was no significant difference between the PE versus wheat flour groups in cholesterol concentrations despite significantly higher urinary concentrations of phytoestrogens for the women taking the diet rich of phytoestrogens. Two proposed mechanisms for the hypocholesterolemic effect of phytoestrogens are the up-regulation of LDL receptors and/or the inhibition of endogenous cholesterol synthesis. Phytoestrogens can stimulate the clearance of cholesterol, probably by up-regulating LDL receptors, and thereby increasing LDL receptor activity (Sanghvi A *et al* 1985). Lignans may also affect cholesterol homeostasis, as they have been shown to inhibit the activity of cholesterol-7a-hydroxylase, the rate-limiting enzyme in the formation of primary bile acids from cholesterol (Hirose N *et al* 1995). Lipoprotein(a) [Lp(a)] is a cholesterol-carrying particle in the blood that is structurally similar to LDL, with the addition of the apoprotein(a) moiety. The reports indicate that Lp (a) is an independent risk factor for coronary heart disease (Scanu AM, 1992). Estrogen treatment is one of the few factors that has successfully lowered Lp(a) levels although the effects of phytoestrogens on Lp(a) has not been studied directly, the structural and receptor binding similarities between phytoestrogens and endogenous estrogens suggest a potential benefit of the plant estrogens. Studies of cultured vascular cells have demonstrated that increased concentrations of isoflavonoids alter cellular processes associated with lesion development in the cardic cells and genistein was found to targets only proliferating cells, leaving quiescent, nondividing cells. This characteristic property is very important in designing the therapeutic applications by these compounds. Proteolytic degradation of the extracellular matrix by endothelial cells is controlled by angiogenic factors, such as basic fibroblast growth factor (bFGF), that induce the production of urokinase-type plasminogen activator and its physiological inhibitor, plasminogen activator inhibitor. Genistein was shown to reduced both bFGF-stimulated and basal levels of both plasminogen activator and plasminogen activator inhibitor-1 activity endothelial cells and genistein inhibited the bFGF-induced migration of endothelial cells in wounded confluent monolayers of endothelial cells. The mechanism is being the inhibition of production of proteolytic enzymes and migration of endothelial cells by genistein, ultimately acting on the events of angiogenesis. Genistein's role as a protein tyrosine kinase inhibitor is also responsible for an antithrombolytic effect and an increase in tyrosine phosphorylation at tyrosine residues of platelet proteins is associated with the platelet's activation, and protein tyrosine phosphorylation. This particular event is subsequently enhanced after thrombin stimulation, therefore, genistein can decrease tyrosine phosphorylation, which would lead to decreased platelet activation and have downstream events, such as a reduction in the deposition and aggregation of platelets and a decrease in the progression of atherosclerosis. Genistein and daidzein also plays and important role in blocking platelet response induced by collagen or thromboxane A2 is due to their ability to prevent thromboxane A2 from binding to its receptor *vs.* its ability to inhibit protein tyrosine phosphorylation, which is a key factor in the platelet aggregation.

Phytoestrogens and Cancer

Cancer prevalence differ in various places and populations of the world. Cancers related to hormones like breast, ovary, endometrium, and prostate has been indicated to vary atleast by several folds between populations. Incidences are mainly because of environmental factors rather than genetic predisposition. The increased rates of these cancers are typically observed in western lifestyles which include high fat, meat-based, low fiber diets, whereas the lowest rates typically observed in asian populations where plant-based diets with a high content of phytoestrogens are common (Parkin DM. 1989). A significant decreased incidence of common cancers like breast were observed in premenopausal and postmenopausal women who consume phytoestrogens. It was reported soy intake was inversely and animal product intake was positively associated with breast cancer. Dietary studies and urinary analysis of lignans in postmenopausal women have shown that lignan secretion is significantly lower in urine in women with breast cancer than in healthy omnivorous and vegetarian women (Adlercreutz H *et al.*, 1986). Diets that lower the androgen or estrogen levels are well associated with low risk of prostate, breast and ovarian cancer (Clarke R *et al.*, 1996, Zhao, R. 2009). There is a consistent relationship between reproduction, exogenous hormone use, and risk factors and plant estrogens are inversely associated with For breast, ovarian, endometrial, and colon cancers. In men, phytoestrogens elicits protection against prostate cancer (Adlercreutz H, 1997, Parker, L.P., 2009). Epidemiological data shows a relatively low incidence of prostate cancer in asian populations whose diet is rich of phytoestrogens (Donn AS, 1985). Substrates of lignans and isoflavonoid phytoestrogens were found to inhibit the conversion of testosterone to the more biologically active dihydrotestosterone. Genistein, biochanin A, and equol were able to inhibit 5a-reductase activity (Evans BA, 1995). Inflammation was drastically decreased in the subjects maintained with diet rich of phytoestrogens suggesting the fact that PE can play a protective role against the pathogenesis of prostatitis. Phytoestrogens inhibited the growth of cultured prostate cancer cells (Rokhlin OW, 1995) and can prevent prostate cancer by a variety of mechanisms, including reducing circulating androgen levels, increasing concentrations of sex hormone-binding globulin (SHBG), competitive binding to cellular hormone receptors, and apparent reduction in the production of dihydrotestosterone. The effects of a broad range of genistein concentrations on estrogen receptor (ER) binding, induction of the estradiol-regulated antigen pS2, and cell proliferation rate in ER-positive and ER-negative human breast cancer cells grown *in vitro* and genistein exhibited antiestrogenic effects and inhibited cell proliferation. A study reports that the inhibition of estrogen receptor-positive MCF-7 breast cancer cells was reversed by the addition of excess competing estrogen and the effect could be due to the antiproliferative effect of genistein, exerted via an estrogen receptor-dependent pathway (So FV 1997). In addition the antioxidant properties of genistein may also responsible for its anticarcinogenic effects. Genistein strongly inhibits tumor promoter-induced H_2O_2 formation both *in vivo* and *in vivo.* In addition, genistein moderately inhibited superoxide anion formation by HL-60 cells and scavenged exogenously added H_2O_2 under the same conditions as in cell culture. Phytoestrogens have been shown to inhibit enzymes that are associated with cell proliferation (*e.g.* ornithine decarboxylase, protein tyrosine kinase, and DNA topoisomerase) and enzymes involved in the production of estrone from the androgens (*e.g.* aromatase), thus denying the tumor a source of endogenous estrogen. Genistein is a specific inhibitor of tyrosine protein kinases, topoisomerase II, and protein histidine kinase. Protein tyrosine kinase activity is associated with cellular receptors for epidermal growth factor, insulin, insulin-like growth factor I, platelet-derived growth factor, and mononuclear phagocyte growth factor. The tyrosine kinases seem to play an important role in cell proliferation and transformation. Further, genistein inhibits DNA topoisomerase II and ribosomal S6 kinase, both of which may lead to protein-linked DNA strand breaks in cancerous cells, arrest of tumor cell growth, and induction of differentiation of several malignant cell lines into lines that may be benign (Wei H 1993).

Phytoestrogens and Osteoporosis

Osteoporosis is characterised by low bone mass and micro-architectural deterioration of bone tissue with a consequent increase in bone fragility and risk of fracture. Bone mass is lost when an imbalance between the cells responsible for the formation of bone (osteoblasts) and those responsible for the resorption of bone (osteoclasts) occurs in favour of bone resorption. This imbalance results in bone structural failure and an increased likelihood of fracture. Osteoporosis is a very common manifestation in postmenopausal women, due to the acceleration of bone loss associated with reduced levels of estrogen. Hormone replacement therapy (HRT) has been suggested to be effective and alternative therapy in preventing bone loss. Phytoestrogens can have a beneficial role in preventing bone loss following ovariectomy. Estrogen plays an important role in maintaining bone density by regulating the formation and resorption of bone (Nilsson and Gustafsson, 2002). Since lower circulating estradiol levels are found during menopause, calcium is lost from the bone into blood plasma, leading to osteoporosis (Yamaguchi, 2002). The main purpose of hormone replacement therapy (HRT) is to prevent or lower the incidence of osteoporosis in postmenopausal women and phytoestrogens are found to be effective in maintaining bone mineral density (BMD) in postmenopausal women (Kim *et al*., 2002). A double blind placebo controlled study of postmenopausal women showed significant increase in BMD at the femoral neck upon the supllementation genistein which was isolated from soy. In a 24-week study comparing isoflavone rich soy protein (80.4 mg aglycone isoflavones/ day) and isoflavone poor soy protein (4.4 mg aglycone isoflavones/ day) in perimenopausal women, both BMD and bone mineral content (BMC) were significantly higher with the diet high in isoflavones (Alekel *et al*., 2000). Studies indicate that phytoestrogen consumption decrased Bone alkaline phosphatase (BAP) leves. The women who had more bone loss had higher BAP levels. Therefore, BAP may be a better indicator of bone turnover than bone formation (Alekel *et al*., 2000). The lumbar spine seems to benefit the most from consumption of phytoestrogens. The studies on the effects of soy isoflavones in premenopausal women showed no effect on bone mineral density levels, while significant effects were observed in postmenopausal women.

Phytoestrogens and Central Nervous System and Immune System

Estrogen receptors are expressed in the central nervous system (CNS) (Kuiper *et al*, 1998). Oestrogens are known to be active in a number of areas of the brain and spinal cord, where they are thought to influence behaviour, movement, cognition, pain sensitivity and have a protective effect on the development of neurodegenerative diseases. Therefore phytoestrogens may also exert similar effects in the CNS. Phytoestrogens have been shown to modulate a number of biochemical processes in the CNS (Lephart *et al.*, 2002). Expression of brain-derived neurotrophic factor was increased in the brain of ovariectomised adult females consuming diet rich of phytoestroges or oestradiol supplemented diet Calcium binding proteins in the brain such as calbindin and calretinin bind calcium and are thought to play a role in preventing neuronal cell death. Lephart *et al* (2002) found the lower levels of brain calbindin and calretinin in adult male rats fed an isoflavone supplemented diet. Exposure to soy in the diet suppressed phosphorylation of tau, a microtubule-associated protein in the brains cynomolgus and phosphorylation of tau has been implicated in neurodegeneration. Different studies suggest that dietary phytoestrogens can modulate proteins involved in calcium binding and inflammation in the CNS. Some reports indicate that phytoestrogens can influence some immune responses also. It is well established that sex hormones can have a substantial impact on the immune system and on the integrity of immune function. Estradiol is known to affect the development and organisation of lymphoid tissues and the activity of various cellular vectors of immune function. In many instances autoimmune diseases are more common in women than in men. Genistein has been shown to inhibit human T-cell proliferation, interleukin-2 (IL-2) production and IL-2 receptor expression *in vitro* and Daidzein even at lower concentrations can increase the proliferation of splenocytes after activation with either concanavalin A or lipopolysaccharide in a dose-dependent manner (Wang *et al*, 1997).

CONCLUSIONS

Phytoestrogens are biologically active when administered to animals and humans and have been shown to elicit their effects via a number of mechanisms: such as Interaction with estrogen receptors (ER) to modulate the expression of estrogen-responsive genes, inhibition of enzymes involved in estrogen biosynthesis and metabolism, inhibition of protein kinases and interaction with components of the cell cycle as well as proliferation, differentiation and apoptosis pathways, inhibition of topoisomerase, antioxidant reactions and played an important role in the mitigation of different types of cancers. In the present pattern of diet formulation and adopting western diet, it is becoming important that phytoestrogens derived biologically from the plants can pave the way to rescue population in general to design an appropriate disease free world.

REFERENCES

Adlercreutz H, Fotsis T, Heikkinen R, *et al.* 1982 Excretion of the lignans enterolactone and enterodiol and of equol in omnivorous and vegetarian postmenopausal women and in women with breast cancer. Lancet. **2:** 1295–1299.

Adlercreutz H, Mazur W. 1997 Phyto-oestrogens and western diseases. Ann Med. **29:** 95–120.

Alekel, D.L., Germain, A.S., Peterson, C.T., Hanson, K.B., Stewart,J.W., Toda, T., 2000. Isoflavone-rich soy protein isolate attenuates bone loss in the lumbar spine of perimenopausal women. Am. J.Clin. Nutr. **72:** 844–852

Artaud-Wild, S.M., Connor, S.L., Sexton, G., Connor,W.E. 1993 Differences in coronary mortality can be explained by differences in cholesterol and saturated fat intakes in 40 countries but not in France and Finland. A paradox.Circulation. **88:** 2771–2779.

Bennetts, H.W., Underwood, E.J., Shier, F.L., 1946. A specific breeding problem of sheep on subterranean clover pastures in Western Australia. Aust. Vet. J. **22:** 2–12.

Bingham, S.A., Atkinson, C., Liggins, J., Bluck, L., Coward, A.1998. Phyto-oestrogens: where are we now? Br J Nutr., **79:** 393-406.

Chen, Y.-M., Ho, S.C., Lam, S.S.H., Ho, S.S.S., Woo, J.L.F., 2003. Soy isoflavones have a favorable effect on bone loss in Chinese postmenopausal women with lower bone mass: a double-blind, randomized, controlled trial. J. Clin. Endocrinol. Metab. **88:** 4740–4747.

Clarke R, Hilakiui-Clarke L, Cho E, James RM, Leonessa F. 1996 Estrogens, phytoestrogens and breast cancer. In: American Institute for Cancer Research, ed. Dietary phytochemicals in cancer prevention and treatment. New York, London: Plenum Press; 63–85.

Davis, J.N., Kucuk, O., Sarkar, F.H., 2002. Expression of prostatespecific antigen is transcriptionally regulated by genistein in prostate cancer cells. Mol. Carcinog. **34:** 91–101.

Dixon, R.A., 2004. Phytoestrogens. Annu. Rev. Plant Biol. **55:** 225–261.

Dixon, R.A., Ferreira, D., 2002. Genistein. Phytochemistry **60:** 205–211.

Donn, A.S., Muir, C.S. 1985 Prostate cancer-some epidemiological features. Bull Cancer. **72:** 381–390.

Engelgau, M.M., Geiss, L.S., Saaddine, J.B., Boyle, J.P., Benjamin, S.M., Gregg, E.W., Tierney, E.F., Rios-Burrows, N., Mokdad, A.H., Ford, E.S., Imperatore, G., Narayan, K.M., 2004. The evolving diabetes burden in the United States. Ann. Intern. Med. **140:** 945–950.

Evans, B.A, Griffiths, K, Morton,M.S. 1995 Inhibition of 5-alpha reductase in genital skin fibroblasts and prostate tissue by dietary lignans and isoflavonoids. J Endocrinol. **147:** 295–302.

Franke, A.A., Custer, L.J., Cerna, C.M., Narala, K.K., 1994. Quantitation of phytoestrogens in legumes by HPLC. J. Agric. Food Chem. **42:** 1905–1913.

Fritz, W.A., Eltoum, I.-E., Cotroneo, M.S., Lamartiniere, C.A., 2002. Genistein alters growth but is not toxic to the rat prostate. J. Nutr. **132:** 3007–3011.

Fritz, W.A., Wang, J., Lamartiniere, C.A., 1998. Dietary genistein: perinatal mammary cancer prevention, bioavailability and toxicity testing in the rat. Carcinogenesis **19:** 2151–2158.

Glitsø, L.V., Mazur, W., Adlercreutz, H., Wa¨ha¨ la¨ , K., Ma¨ kela¨ , T., Sandstro¨ m, B., Wahala, K., 2000. Intestinal metabolism of rye lignans in pigs. Br. J. Nutr. **84:** 429–437.

Gooderham, M.J., Adlercreutz, H., Ojala, S.T., Wahala, K., Holub, B.J. 1996 A soy protein isolate rich in genistein and daidzein and its effects on plasma isoflavone concentrations, platelet aggregation, blood lipids and fatty acid composition of plasma phospholipid in normal men. J Nutr. **126:** 2000 –2006.

Gustafsson, J.-A., 2003. What pharmacologists can learn from recent advances in estrogen signalling. Trends Pharmacol. Sci. **24:** 479–485.

Hays, J., Ockene, J.K., Brunner, R.L., Kotchen, J.M., Manson, J.E., Patterson, R.E., Aragaki, A.K., Shumaker, S.A., Brzyski, R.G., LaCroix, A.Z., Granek, I.A., Valanis, B.G., Women's Health Initiative InvestigatorsEffects of estrogen plus progestin on healthrelated quality of life. N. Engl. J. Med. **348:** 1835–1837.

Hirose N, Inoue T, Nishihara K, *et al.* 1991 Inhibition of cholesterol absorption and synthesis in rats by sesamin. J Lipid Res. **32:** 629–638.

Humfrey, C.D.N. 1998.Phytoestrogens and human health effects: weighing up the current evidence. Nat.Toxins. **6:** 1-59.

Jonas, J., Plant, T., Gilon, P., Detimary, P., Nenquin, M., Henquin, J.,1995. Multiple effects and stimulation of insulin secretion by the tyrosine kinase inhibitor genistein in normal mouse islets. Br. J. Pharmacol. **114:** 872–880.

Kim, M.K., Chung, B.C., Yu, V.Y., Nam, J.H., Lee, H.C., Huh, K.B., Lim, S.K., 2002. Relationships of urinary phyto-oestrogen excretion to BMD in postmenopausal women. Clin. Endocrinol. (Oxf) **56:** 321–328.

Knuckles, B.E., DeFremery, D., Kohler, G.O., 1976. Coumestrol content of fractions obtained during wet processing of alfalfa. J.Agric. Food Chem. **24:** 1177–1180.

Kuiper, G.G., Lemmen, J.G., Carlsson, B., Corton, J.C., Safe, S.H., van der Saag, P.T., van der Burg, B., Gustafsson, J.A.1998. Interaction of estrogenic chemicals and phytoestrogens with estrogen receptor beta. Endocrinol. **139:** 4252-4263.

Kuiper, G.G., Lemmen, J.G., Carlsson, B., Corton, J.C., Safe, S.H.,van der Saag, P.T., van der Burg, B., Gustafsson, J.-A., 1998.Interaction of estrogenic chemicals and phytoestrogens with estrogen receptor b. Endocrinology **139:** 4252–4263

Lampe, J.W., Karr, S.C., Hutchins, A.M., Slavin, J.L,.1998. Urinary equol excretion with a soy challenge: influence of habitual diet. Proc Soc Exp Biol Med. **217:** 335-339.

Lephart, E.D., West, T.W., Webe,r K.S., Rhees, R.W., Setchell, K.D.R., Adlercreutz, H., Lund, T.D. 2002.Neurobehavioral effects of dietary soy phytoestrogens. Neurotox Teratol. **24:** 5-16.

Liu, J., Burdette, J.E., Xu, H., Gu, C., van Breemen, R.B., Bhat, K.P.,Booth, N., Constantinou, A.I., Pezzuto, J.M., Fong, H.H., Farnsworth,N.R., Bolton, J.L., 2001. Evaluation of estrogenic activity of plant extracts for the potential treatment of menopausal symptoms.J. Agric. Food Chem. **49:** 2472–2479.

Meagher LP, Beecher GR. Assessment of data on the lignan content of foods. J. Food Comp Anal. 2000, **13:** 935-947.

Meegan, M.J., Lloyd, D.G., 2003. Advances in the science of estrogen receptor modulation. Curr. Med. Chem. **10:** 181–210.

Morabito, N., Crisafulli, A., Vergara, C., Gaudio, A., Lasco, A., Frisina, N., D'Anna, R., Corrado, F., Pizzoleo, M.A., Cincotta, M.,Altavilla, D., Ientile, R., Squadrito, F., 2002. Effects of genistein and hormone-replacement therapy on bone loss in early postmenopausal women: a randomized double-blind placebo-controlled study. J. Bone Miner. Res. **17:** 1904–1912.

Morisset, A.S., Lemieux, S., Veilleux, A., Bergeron, J.., John Weisnagel, S., Tchernof A. 2009. Impact of a lignan-rich diet on adiposity and insulin sensitivity in post-menopausal women. *Br J Nutr.* **102(2):** 195-200

Morton, M.S., Chan, P.S., Cheng, C., Clacklock, N., Matos-Ferreria, A., Abranches-Monteiro, L., Correia, R., Lloyd, S., Griffiths, K. 1997. Lignans and isoflavonoids in plasma and prostatic fluid in men: samples from Portugal, Hong Kong and the United Kingdom. Prostate. **32:** 122-128.

Mousavi, Y., Adlercreutz, H. 1992 Enterolactone and estradiol inhibit each other's proliferative effect on MCF-7 breast cancer cells in culture. J Steroid Biochem Mol Biol. **41:** 615– 619.

Murkies, A.L., Lombard,C., Strauss, B.J., Wilcox, G., Burger, H.G., Morton, M.S. 1995 Dietary flour supplementation decreases post-menopausal hot flushes: effect of soy and wheat. Maturitas. **21:** 189 –195.

Nilsson, S., Gustafsson, J.A., 2002. Biological role of estrogen and estrogen receptors. Crit. Rev. Biochem. Mol. Biol. **37:** 1–28.

Nilsson, S., Gustafsson, J.A., 2002. Biological role of estrogen and estrogen receptors. Crit. Rev. Biochem. Mol. Biol. **37:** 1–28.

Norfleet, A.M., Thomas, M.L., Gametchu, B., Watson, C.S., 1999. Estrogen receptor a detected on the plasma membrane of aldehydefixed gh3/b6/f10rat pituitary tumor cells by enzyme-linked immunocytochemistry. Endocrinology **140:** 3805–3814.

Olshansky, S.J., Passaro, D.J., Hershow, R.C., Layden, J., Carnes, B.A., Brody, J., Hayflick, L., Butler, R.N., Allison, D.B., Ludwig, D.S., 2005. A potential decline in life expectancy in the United States in the 21st century. N. Engl. J. Med. **352:** 1138–1145.

Parker, L.P., Taylor, D.D., Kesterson, S., Gercel-Taylor, C.. 2009. Gene expression profiling in response to estradiol and genistein in ovarian cancer cells. *Cancer Genomics Proteomics.* **6(3):** 189-94

Parkin DM. 1989 Cancers of the breast, endometrium, and ovary: geographic correlations. Eur J Cancer Clin Oncol. **25:** 1917–1925.

Ramanathan, L., Gray, W.G., 2003. Identification and characterization of a phytoestrogen-specific gene from the MCF-7 human breast cancer cell. Toxicol. Appl. Pharmacol. **191:** 107–117.

Razandi, M., Pedram, A., Greene, G.L., Levin, E.R., 1999. Cell membrane and nuclear estrogen receptors (ERs) originate from a single transcript: studies of ERa and ERb expressed in Chinese hamster ovary cells. Mol. Endocrinol. **13:** 307–319.

Riggs, B.L., Hartmann, L.C., 2003. Selective estrogen-receptor modulators– mechanisms of action and application to clinical practice. N. Engl. J. Med. **348:** 618–629.

Rokhlin, O.W., Cohen, M.B. 1995 Differential sensitivity of human prostatic cancer cell lines to the effects of protein kinase and phosphatase inhibitors. Cancer Lett. **98:** 103–110.

Rossiter, R.C., Beck, A.B., 1966. Physiological and ecological studies on the estrogenic isoflavones in subterranean clover (*Trifolium subterraneum*) I. Effects of temperature. Aust. J. Agric. Res. **17:** 29–37.

Rowland, I., Wiseman, H., Sanders, T., Adlercreutz, H., Bowey, E. 2000. Interindividual variation in metabolism of isoflavonoids and lignans: the role of the gut microflora and habitual diet. Nutr Cancer. **36:** 27-32.

Rowland, I., Faughnan, M., Hoey, L., Wahala, K., Williamson, G.,Cassidy, A., 2003. Bioavailability of phyto-oestrogens. Br. J. Nutr. 89 (Suppl 1), S45–S58.

Sanghvi, A., Diven, W.F., Seltman, H., *et al.* 1985 Inhibition of rat liver cholesterol-7-alpha hydroxylase and acyl-coA; cholesterol acyl transferase activities by enterodiol and enterolactone. In: Kritchevsky D, Paoletti R, Holmes WL,eds. Drugs affecting lipid metabolism. New York: Plenum Press; 450.

Scanu, A.M. 1992 Lipoprotein(a): a genetic risk factor for premature coronary heart disease. JAMA. **267:** 3326 –3329.

Setchell, K.D., Lydeking-Olsen, E., 2003. Dietary phytoestrogens and their effect on bone: evidence from in vitro and in vivo, human observational, and dietary intervention studies. Am. J. Clin. Nutr. **78:** 593S–609.

Singletary, K.W., Frey, R.S., Li, J.Y., 2002. Differential effects of genistein on cell proliferation, cyclin B1, and p34cdc2 in transformed and nontransformed human breast cells. Pharm. Biol. **40:** 35–42.

So, F.V., Gunthrie, N., Chambers, A.F., Carroll, K.K. 1997 Inhibition of proliferation of estrogen receptor-positive MCF-7 human breast cancer cells by flavonoids in the presence and absence of excess estrogen. Cancer Lett.**112:** 127–133.

Tham, D.M., Gardner, C.D., Haskell, W.L. 1998. Potential health benefits of dietary phytoestrogens: a review of the clinical, epidemiological and mechanistic evidence. J Clin Endocrinol Metab. **83:** 2223-2235.

Virgili, F., Perozzi, G. 2008. From phytoestrogens to obesity and the metabolic syndrome: Health from food and food for health.Genes Nutr.**1(3-4):** 137-139.

Wang, W., Higuchi, C.M., Zhang, R. 1997. Individual and combinatorial effects of soy isoflavones on the *in vitro* potentiation of lymphocyte activation. Nutr Cancer. **29:** 29-34.

Wang, D., Gutkowska, J., Marcinkiewicz, M., Rachelska, G.,Jankowski, M., 2003. Genistein supplementation stimulates the oxytocin system in the aorta of ovariectomized rats. Cardiovasc. Res. **57:** 186–194.

Wang, G.S., Kuan, S., Francis, O.J., Ware, G.M., Carman, A.S.,1990. A simplified HPLC method for the determination of phytoestrogens in soybean and its processed products. J. Agric. Food Chem. **38:** 185–190.

Wei,H., Bowen, R., Cai, Q., Barnes, S., Wang, Y. 1995 Antioxidant and antipromotional effects of the soybean isoflavone genistein. Proc Soc Exp Biol Med. **208:** 124–130.

Wei, H., Wei, L., Frenkel, K., Bowen, R., Barnes, S. 1993 Inhibition of tumor promoter-induced hydrogen peroxide formation *in vitro* and *in vivo* by genistein. Nutr Cancer. **20:** 1–12.

Wu, A.H., Ziegler, R.G., Horn-Ross, P.L., *et al.* 1996 Tofu and risk of breast cancer. Cancer Epidemiol Biomarkers Prev. **5:** 901–906.

Xu, Y.J., Traystman, R.D., Hurn, P., Wang, M., 2003. Neuritelocalized estrogen receptor—a mediates rapid signaling by estrogen. J. Neurosci. Res. **74:** 1–11.

Yamaguchi, M. 2002. Isoflavone and bone metabolism: its cellular mechanism and preventive role in bone loss. J. Health Sci. **48:** 209–222.

Yanagihara, K., Ito, A., Toge, T., Numoto, M., 1993. Antiproliferative effects of isoflavones on human cancer cell lines established from the gastrointestinal tract. Cancer. Res. **53:** 5815–5821.

Zhao, R., Xiang, N., Domann, F.E., Zhong, W. 2009. Effects of selenite and genistein on G2/M cell cycle arrest and apoptosis in human prostate cancer cells. Nutr Cancer. **61(3):** 397-407.

Zhou, S., Turgeman, G., Harris Stephen, E., Leitman Dale, C., Komm Barry, S., Bodine Peter, V.N., Gazit, D., 2003. Estrogens activate bone morphogenetic protein-2 gene transcription in mouse mesenchymal stem cells. Mol. Endocrinol. **17:** 56–66.

Bioresources for Food Security and Rural Livelihood, Pages 127–139
Edited by : G.K. Kulkarni and B.N. Pandey

IDENTIFICATION OF A NOVEL LACTOFERRIN VARIANT IN MILK OF *BOS INDICUS* VECHUR BREED

S. Anisha, Anu Yamuna Joseph and C. Mohankumar

ABSTRACT

Lactoferrin gene has been studied in various breeds of cows, however the same has not been characterized in Vechur, an indigenous breed of Kerala. The gene was isolated and characterized in Vechur cows for the first time and the sequence analysed. The sequencing data revealed four nucleotide changes and a single amino acid substitution when compared with the already published sequences. A novel recognition site for KpnI enzyme was also generated which is exclusive to Vechur and can serve as a marker. More exhaustive studies on the structure and function of Lactoferrin protein in Vechur cows is recommended to assess the influence of this amino acid substitution on the functional efficiency of Lactoferrin when compared with the other breeds.

Keywords : Lactoferrin, Vechur, recognition site, marker, nucleotide, amino acid.

INTRODUCTION

Life science research in the past few decades has become more centered around two molecules: Genes and Proteins. More and more proteins are being subjected to investigations, to make the most out of their properties, which can be nutritional, therapeutic or otherwise. However, one major task to be accomplished before commencing investigations is the proper identification of beneficial protein molecules. One such mammalian protein that has proved to be having multifaceted properties is Lactoferrin (Lf).

Lf is a bioactive glycoprotein of the transferrin family found in secretions of a wide range of mammalian species including human, bovine, goat….(Adlerova, 2008). Owing to its multidimensional therapeutic properties and nutritional benefits, Lf continues to be the main attraction of researchers since its discovery in bovine milk in 1939.

The level of Lf in milk varies enormously between species with the highest concentration recorded in human milk (1.0- 5.0 mg/ml). However bovine milk which is the most common source of commercial Lf contains about one-tenth of the Lf found in other mammals with concentrations ranging between 0.02 and 0.2 mg/ml (O'Halloran *et al.* 2009).

Bovine Lf (BLf) gene is located on chromosome 22 covering approximately 35kb of genomic DNA and comprises 17 exons (Teng, 2002). The coding sequence of the functional BLf protein defines a single polypeptide of 708 amino acids (O'Halloran *et al. 2009).* The polypeptide is folded into two separate lobes each of which contains an iron-binding and a bicarbonate-binding site (Levay and Viljoen, 1995).

To date, single nucleotide polymorphisms (SNPs) in the promoter as well as intron and exon regions of BLf genes have been described in several breeds at the international level (Daly *et al.* 2006; Halloran *et al.* 2009). However, despite of the rich livestock diversity seen in India, not much has been done for the characterization of Lf gene including its polymorphism. Moreover, it is obvious from the literature that an absolute vacuum exists as far as Lf studies in the indigenous variety, Vechur is concerned. Vechur is a dwarf, indigenous breed of cow belonging to Kerala with good environmental adaptability to meet future challenges (Suprabha and Anilkumar, 2006).

Hence, the present study was conducted to detect and characterize Lf gene in Vechur cows.

MATERIALS AND METHODS

Collection of Samples

Milk samples (250ml) were collected from healthy Vechur cows from the Cattle farm at College for Veterinary and Animal Sciences, Kerala Agriculture University, Mannuthy, Trichur. The samples were collected into sterile bottles and immediately brought to the lab in ice. The sample was used for isolating RNA the same day. To serve as a control, samples were collected from hybrid cows (HF- Holstein Friesian Cross) also for RNA isolation.

Separation of Milk Cells

The sample was transferred to sterile centrifuge tubes and the cell pellet isolated by following the method of Talebkhan Garoussi *et al.* (2007) with slight modifications. The sample was centrifuged at 35[illegible] Xg for 15 minutes at 4°C and the supernatant whey and cream carefully removed without disturbing the cell pellet.

The cell pellet was washed in sterile, ice-cold PBS four times, each wash being followed by centrifugation at 600Xg for 10minutes at 4°C. The volume of PBS used for washing the cells was reduced each time, starting with 20ml.

RNA Isolation

Total RNA was extracted from the cells using TRI Solution (Genei, Bangalore). 3ml of the reagent was used for solubilizing the cell pellet. Subsequent steps were followed as per the manufacturer's instructions. The RNA was dissolved in DEPC-treated water and stored at -80°C in aliquots.

Reverse Transcription

The RNA was quantified at 260nm and 1μg was used for doing Reverse Transcription using 100ng of PolyT primer and 5 units of AMV-RT enzyme. The reaction was carried out at 42°C for 1 hr followed by denaturation of the enzyme at 70°C for 10 minutes.

Designing of Primers and PCR Amplification

For amplification of the Lactoferrin gene, primers were designed based on the published sequence data (Gene Accession No. L08604) using Fast PCR software. PCR was done in a volume of 25μl using 2.5μl of eDNA, 20 picomoles of either primer and 1 unit of Cesium Taq Polymerase enzyme. The cycling conditions were as: Initial denaturation at 94°C for 5min followed by 35 cycles of 94°C for 1 min, 58°C for 1 min, 72°C for 2min with a final extension at 72°C for 10min. The PCR product was checked on 1% agarose gel alongwith DNA ladder.

Cloning and Sequencing of Lactoferrin Gene

The expected band was excised from the gel and eluted using Gel extraction kit (Genei, Bangalore). The eluted product was ligated into a TA cloning vector (pTZ57R/T, Fermentas) using T4DNA ligase enzyme. The reaction was set in a volume of 10μl at 22°C overnight. 2.5μ1 of the ligation mix was used to transform E. *coli* JM109 cells using TA cloning Kit (Fermentas). White colonies representing cells transformed with recombinant vector were used to inoculate LB broth containing Ampicillin. The cultures were allowed to grow at 37°C overnight and used the next day to isolate the recombinant plasmid by Alkaline Lysis method. The plasmid was sent for sequencing of the Lactoferrin gene (Genei, Bangalore).

Restriction Enzyme Analysis

The Lf gene from both Vechur and the hybrid were subjected to Restriction enzyme digestion using Kpnl enzyme. The reaction was performed in a volume of 10μ1 at 37°C for 1 hour. The digested products were analysed on 1% agarose gel.

Sequence Comparison and Phylogenetic Analysis

The nucleotide as well as amino acid (aa) sequence of Lf gene from both the breeds were aligned with the published BLf sequence (Gene Ace. No. L08604) using CLUSTAL W Multiple Sequence Alignment tool (Thompson *et al.* 1994). A phylogenetic analysis was done for comparing the aa sequences of the BLf genes with the published Lf sequences of different mammals using the PHYLIP software package (Felsenstein, 1995). Bootstrap analysis was done with 1000 replicates to determine confidence values on the clades within trees. A phylogenetic tree was generated using UPGMA method and the distance analyses were done using the Kimura-2 parameter formula.

RESULTS

Blf Gene Detection

The RNA isolated from milk samples, both from Vechur and Hybrid cows, was found to be intact, showing two distinct bands of 288 and 188 rRNA (Fig.1) CDNA was synthesized using this RNA and Poly T primer. The BLf gene was amplified by PCR from cDNA of Vechur as well as hybrid cows using gene specific primers (Forward- 5' CAGCCA TGAAGCTCTTCGTCC 3' and Reverse- 5' TGCAGGCTTT ACCTCGTCAGGAA 3'). The amplicon showed an expected size of 2.141kb for both the set of samples (Fig. 2a and 2b) The optimum annealing temperature was observed to be 58°C in either case.1

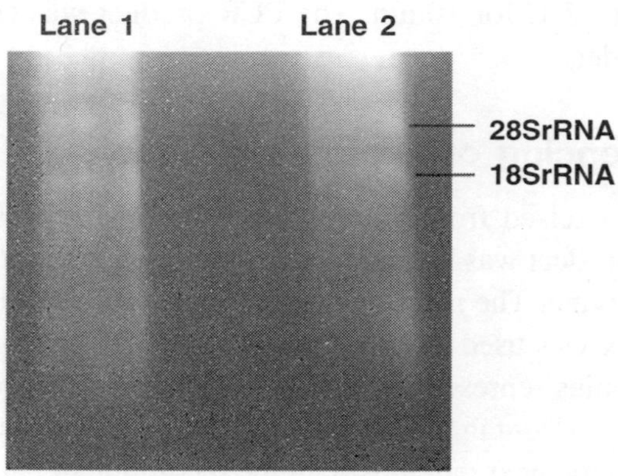

Fig. 1 : Milk RNA

Cloning and Sequencing of Blf Gene

The amplified BLf gene from Vechur (BLf-V) as well as hybrid (BLf-H) cows was cloned into pTZ57R/T vector and the gene sequenced. Both the gene sequences were deposited in Genbank, NCBI.

Sequence Comparison and Phylogenetic Analysis

The nudeotide sequences of Lf gene, BLf-V and BLf-H were compared with each other as well as with the other published sequence (Gene Acc. No. L08604). Blf-H showed 100% identity with the reference sequence, while it was 99% for BLf-V. BLf-V showed four nucleotide changes at positions 526, 979, 1281 and 1572 in the coding region respectively (Fig. 3). The nucleotide change from G to A at 1281 position in BLf-V creates an additional recognition site (5' GGTACJ\C 3') for Kpnl restriction enzyme. As observed for the nucleotide sequence, BLf-H showed no amino acid variation when compared with the published sequence. However, a single amino acid variation, from cysteine to arginine (at position 176) was observed for BLf-V (Fig. 4).

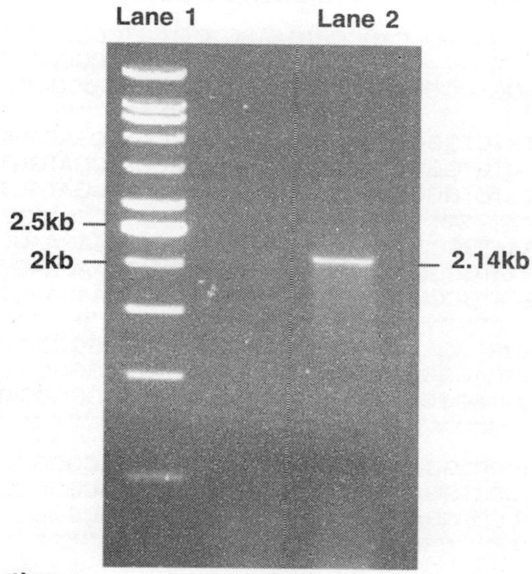

Fig. 2(a) : PCR amplification
Lane 1 : 1kb DNA Ladder
Lane 2 : PCR product of vechur

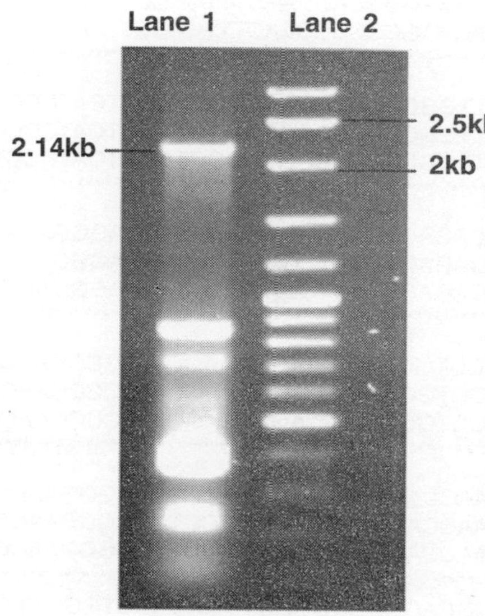

Fig. : 2 (b)
Lane 1 : PCR product of Hybrid
Lane 2 : 100 bp DNA Ladder

```
Vechur     ------------------------------------------------------------
FJ589071   ------------------------------------------------------------
L08604     GGGGGAGGAGGGAGGCTGGGGCGCTTATAGGACCACAGGGCGGGGCAAACCTCGTGAGGT

Vechur     ------------------------------------------------------------
FJ589071   ------------------------------------------------------------
L08604     CACCGAGCACTGGATAAAGGGACGCAGAACGAGCGCAGGTGGCAGACCTTCGTTCCGGAG

Vechur     --------------CAGCCATGAAGCTCTTCGTCCCCGCCCTGCTGTCCCTTGGAGCCCT
FJ589071   --------------CAGCCATGAAGCTCTTCGTCCCCGCCCTGCTGTCCCTTGGAGCCCT
L08604     TCGCCCCAGGACGCCAGCCATGAAGCTCTTCGTCCCCGCCCTGCTGTCCCTTGGAGCCCT
                         **********************************************

Vechur     TGGACTGTGTCTGGCTGCCCCGAGGAAAAACGTTCGATGGTGTACCATCTCCCAACCTGA
FJ589071   TGGACTGTGTCTGGCTGCCCCGAGGAAAAACGTTCGATGGTGTACCATCTCCCAACCTGA
L08604     TGGACTGTGTCTGGCTGCCCCGAGGAAAAACGTTCGATGGTGTACCATCTCCCAACCTGA
           ************************************************************

Vechur     GTGGTTCAAATGCCGCCGATGGCAGTGGAGGATGAAGAAGCTGGGTGCTCCCTCTATCAC
FJ589071   GTGGTTCAAATGCCGCCGATGGCAGTGGAGGATGAAGAAGCTGGGTGCTCCCTCTATCAC
L08604     GTGGTTCAAATGCCGCCGATGGCAGTGGAGGATGAAGAAGCTGGGTGCTCCCTCTATCAC
           ************************************************************

Vechur     CTGTGTGAGGAGGGCCTTTGCCTTGGAATGTATCCGGGCCATCGCGGAGAAAAAGGCGGA
FJ589071   CTGTGTGAGGAGGGCCTTTGCCTTGGAATGTATCCGGGCCATCGCGGAGAAAAAGGCGGA
L08604     CTGTGTGAGGAGGGCCTTTGCCTTGGAATGTATCCGGGCCATCGCGGAGAAAAAGGCGGA
           ************************************************************

Vechur     TGCTGTGACCCTGGATGGTGGCATGGTGTTTGAGGCGGGCCGGGACCCCTACAAACTGCG
FJ589071   TGCTGTGACCCTGGATGGTGGCATGGTGTTTGAGGCGGGCCGGGACCCCTACAAACTGCG
L08604     TGCTGTGACCCTGGATGGTGGCATGGTGTTTGAGGCGGGCCGGGACCCCTACAAACTGCG
           ************************************************************

Vechur     GCCAGTAGCAGCAGAGATCTATGGGACGAAAGAGTCTCCCCAAACCCACTATTATGCTGT
FJ589071   GCCAGTAGCAGCAGAGATCTATGGGACGAAAGAGTCTCCCCAAACCCACTATTATGCTGT
L08604     GCCAGTAGCAGCAGAGATCTATGGGACGAAAGAGTCTCCCCAAACCCACTATTATGCTGT
           ************************************************************

Vechur     GGCCGTCGTGAAGAAGGGCAGCAACTTTCAGCTGGACCAGCTGCAAGGCCGGAAGTCCTG
FJ589071   GGCCGTCGTGAAGAAGGGCAGCAACTTTCAGCTGGACCAGCTGCAAGGCCGGAAGTCCTG
L08604     GGCCGTCGTGAAGAAGGGCAGCAACTTTCAGCTGGACCAGCTGCAAGGCCGGAAGTCCTG
           ************************************************************

Vechur     CCATACGGGCCTTGGCAGGTCCGCTGGGTGGATCATCCCTATGGGAATCCTTCGCCCGTA
FJ589071   CCATACGGGCCTTGGCAGGTCCGCTGGGTGGATCATCCCTATGGGAATCCTTCGCCCGTA
L08604     CCATACGGGCCTTGGCAGGTCCGCTGGGTGGATCATCCCTATGGGAATCCTTCGCCCGTA
           ************************************************************

Vechur     CTTGAGCTGGACAGAGTCACTCGAGCCCCTCCAGGGAGCTGTGGCTAAATTCTTCTCTGC
FJ589071   CTTGAGCTGGACAGAGTCACTCGAGCCCCTCCAGGGAGCTGTGGCTAAATTCTTCTCTGC
L08604     CTTGAGCTGGACAGAGTCACTCGAGCCCCTCCAGGGAGCTGTGGCTAAATTCTTCTCTGC
           ************************************************************

Vechur     CAGCCGTGTTCCCTGCATTGATAGACAAGCATACCCCAACCTGTGTCAACTGTGCAAGGG
FJ589071   CAGCTGTGTTCCCTGCATTGATAGACAAGCATACCCCAACCTGTGTCAACTGTGCAAGGG
L08604     CAGCTGTGTTCCCTGCATTGATAGACAAGCATACCCCAACCTGTGTCAACTGTGCAAGGG
           **** *******************************************************

Vechur     GGAGGGGGAGAACCAGTGTGCCTGCTCCTCCCGGGAACCATACTTCGGTTATTCTGGTGC
FJ589071   GGAGGGGGAGAACCAGTGTGCCTGCTCCTCCCGGGAACCATACTTCGGTTATTCTGGTGC
L08604     GGAGGGGGAGAACCAGTGTGCCTGCTCCTCCCGGGAACCATACTTCGGTTATTCTGGTGC
           ************************************************************

Vechur     CTTCAAGTGTCTGCAGGACGGGGCTGGAGACGTGGCTTTTGTTAAAGAGACGACAGTGTT
FJ589071   CTTCAAGTGTCTGCAGGACGGGGCTGGAGACGTGGCTTTTGTTAAAGAGACGACAGTGTT
L08604     CTTCAAGTGTCTGCAGGACGGGGCTGGAGACGTGGCTTTTGTTAAAGAGACGACAGTGTT
           ************************************************************
```

```
Vechur     TGAGAACTTGCCAGAGAAGGCTGACAGGGACCAGTATGAGCTTCTCTGCCTGAACAACAG
FJ589071   TGAGAACTTGCCAGAGAAGGCTGACAGGGACCAGTATGAGCTTCTCTGCCTGAACAACAG
L08604     TGAGAACTTGCCAGAGAAGGCTGACAGGGACCAGTATGAGCTTCTCTGCCTGAACAACAG
           ************************************************************

Vechur     TCGGGCGCCAGTGGATGCGTTCAAGGAGTGCCACCTGGCCCAGGTCCCTTCTCATGCTGT
FJ589071   TCGGGCGCCAGTGGATGCGTTCAAGGAGTGCCACCTGGCCCAGGTCCCTTCTCATGCTGT
L08604     TCGGGCGCCAGTGGATGCGTTCAAGGAGTGCCACCTGGCCCAGGTCCCTTCTCATGCTGT
           ************************************************************

Vechur     CGTGGCCCGAAGTGTGGATGGCAAGGAAGACTTGATCTGGAAGCTTCTCAGCAAGGCGCA
FJ589071   CGTGGCCCGAAGTGTGGATGGCAAGGAAGACTTGATCTGGAAGCTTCTCAGCAAGGCGCA
L08604     CGTGGCCCGAAGTGTGGATGGCAAGGAAGACTTGATCTGGAAGCTTCTCAGCAAGGCGCA
           ************************************************************

Vechur     GGAGAAATTTGGAAAAAACAAGTCTCGGAGCTTCCAGCTCTTTGGCTCTCCACCCGGCCA
FJ589071   GGAGAAATTTGGAAAAAACAAGTCTCGGAGCTTCCAGCTCTTTGGCTCTCCACCCGGCCA
L08604     GGAGAAATTTGGAAAAAACAAGTCTCGGAGCTTCCAGCTCTTTGGCTCTCCACCCGGCCA
           ************************************************************

Vechur     GAGGGACCTGCTGTTCAAAGACTCTGCTCTTGGGTTTCTGAGGATCCCCTCGAAGGTAGA
FJ589071   GAGGGACCTGCTGTTCAAAGACTCTGCTCTTGGGTTTTTGAGGATCCCCTCGAAGGTAGA
L08604     GAGGGACCTGCTGTTCAAAGACTCTGCTCTTGGGTTTTTGAGGATCCCCTCGAAGGTAGA
           ************************************* **********************

Vechur     TTCGGCGCTGTACCTGGGCTCCCGCTACTTGACCACCTTGAAGAACCTCAGGGAAACTGC
FJ589071   TTCGGCGCTGTACCTGGGCTCCCGCTACTTGACCACCTTGAAGAACCTCAGGGAAACTGC
L08604     TTCGGCGCTGTACCTGGGCTCCCGCTACTTGACCACCTTGAAGAACCTCAGGGAAACTGC
           ************************************************************

Vechur     GGAGGAGGTGAAGGCGCGGTACACCAGGGTCGTGTGGTGTGCCGTGGGACCTGAGGAGCA
FJ589071   GGAGGAGGTGAAGGCGCGGTACACCAGGGTCGTGTGGTGTGCCGTGGGACCTGAGGAGCA
L08604     GGAGGAGGTGAAGGCGCGGTACACCAGGGTCGTGTGGTGTGCCGTGGGACCTGAGGAGCA
           ************************************************************

Vechur     GAAGAAGTGCCAGCAGTGGAGCCAGCAGAGCGGCCAGAACGTGACCTGTGCCACGGCGTC
FJ589071   GAAGAAGTGCCAGCAGTGGAGCCAGCAGAGCGGCCAGAACGTGACCTGTGCCACGGCGTC
L08604     GAAGAAGTGCCAGCAGTGGAGCCAGCAGAGCGGCCAGAACGTGACCTGTGCCACGGCGTC
           ************************************************************

Vechur     CACCACTGACGACTGCATCGTCCTGGTGCTGAAAGGGGAAGCAGATGCCCTGAACTTGGA
FJ589071   CACCACTGACGACTGCATCGTCCTGGTGCTGAAAGGGGAAGCAGATGCCCTGAACTTGGA
L08604     CACCACTGACGACTGCATCGTCCTGGTGCTGAAAGGGGAAGCAGATGCCCTGAACTTGGA
           ************************************************************

Vechur     TGGAGGATATATCTACACTGCGGGCAAGTGTGGCCTGGTACCTGTCCTGGCAGAGAACCG
FJ589071   TGGAGGATATATCTACACTGCGGGCAAGTGTGGCCTGGTGCCTGTCCTGGCAGAGAACCG
L08604     TGGAGGATATATCTACACTGCGGGCAAGTGTGGCCTGGTGCCTGTCCTGGCAGAGAACCG
           *************************************** ********************

Vechur     GAAATCCTCCAAACACAGTAGCCTAGATTGTGTGCTGAGACCAACGGAAGGGTACCTTGC
FJ589071   GAAATCCTCCAAACACAGTAGCCTAGATTGTGTGCTGAGACCAACGGAAGGGTACCTTGC
L08604     GAAATCCTCCAAACACAGTAGCCTAGATTGTGTGCTGAGACCAACGGAAGGGTACCTTGC
           ************************************************************

Vechur     CGTGGCAGTTGTCAAGAAAGCAAATGAGGGGCTCACATGGAATTCTCTGAAAGACAAGAA
FJ589071   CGTGGCAGTTGTCAAGAAAGCAAATGAGGGGCTCACATGGAATTCTCTGAAAGACAAGAA
L08604     CGTGGCAGTTGTCAAGAAAGCAAATGAGGGGCTCACATGGAATTCTCTGAAAGACAAGAA
           ************************************************************

Vechur     GTCGTGCCACACCGCCGTGGACAGGACTGCAGGCTGGAACATCCCCATGGGCCTGATCGT
FJ589071   GTCGTGCCACACCGCCGTGGACAGGACTGCAGGCTGGAACATCCCCATGGGCCTGATCGT
L08604     GTCGTGCCACACCGCCGTGGACAGGACTGCAGGCTGGAACATCCCCATGGGCCTGATCGT
           ************************************************************
```

```
Vechur      CAACCAGACAGGCTCCTGCGCATTTGATGAATTCTTTAGTCAGAGCTGTGCCCCTGGGGC
FJ589071    CAACCAGACAGGCTCCTGCGCATTTGATGAATTCTTTAGTCAGAGCTGTGCCCCTGGGGC
L08604      CAACCAGACAGGCTCCTGCGCATTTGATGAATTCTTTAGTCAGAGCTGTGCCCCTGGGGC
            ************************************************************

Vechur      TGACCCGAAATCCAGACTCTGTGCCTTGTGCGCTGGCGATGACCAGGGCCTGGACAAGTG
FJ589071    TGACCCGAAATCCAGACTCTGTGCCTTGTGTGCTGGCGATGACCAGGGCCTGGACAAGTG
L08604      TGACCCGAAATCCAGACTCTGTGCCTTGTGTGCTGGCGATGACCAGGGCCTGGACAAGTG
            ****************************** *****************************

Vechur      TGTGCCCAACTCTAAGGAGAAGTACTATGGCTATACCGGGGCTTTCAGGTGCCTGGCTGA
FJ589071    TGTGCCCAACTCTAAGGAGAAGTACTATGGCTATACCGGGGCTTTCAGGTGCCTGGCTGA
L08604      TGTGCCCAACTCTAAGGAGAAGTACTATGGCTATACCGGGGCTTTCAGGTGCCTGGCTGA
            ************************************************************

Vechur      GGACGTTGGGGACGTTGCCTTTGTGAAAAACGACACAGTCTGGGAGAACACGAATGGAGA
FJ589071    GGACGTTGGGGACGTTGCCTTTGTGAAAAACGACACAGTCTGGGAGAACACGAATGGAGA
L08604      GGACGTTGGGGACGTTGCCTTTGTGAAAAACGACACAGTCTGGGAGAACACGAATGGAGA
            ************************************************************

Vechur      GAGCACTGCAGACTGGGCTAAGAACTTGAATCGTGAGGACTTCAGGTTGCTCTGCCTCGA
FJ589071    GAGCACTGCAGACTGGGCTAAGAACTTGAATCGTGAGGACTTCAGGTTGCTCTGCCTCGA
L08604      GAGCACTGCAGACTGGGCTAAGAACTTGAATCGTGAGGACTTCAGGTTGCTCTGCCTCGA
            ************************************************************

Vechur      TGGCACCAGGAAGCCTGTGACGGAGGCTCAGAGCTGCCACCTGGCGGTGGCCCCGAATCA
FJ589071    TGGCACCAGGAAGCCTGTGACGGAGGCTCAGAGCTGCCACCTGGCGGTGGCCCCGAATCA
L08604      TGGCACCAGGAAGCCTGTGACGGAGGCTCAGAGCTGCCACCTGGCGGTGGCCCCGAATCA
            ************************************************************

Vechur      CGCTGTGGTGTCTCGGAGCGATAGGGCAGCACACGTGAAACAGGTGCTGCTCCACCAGCA
FJ589071    CGCTGTGGTGTCTCGGAGCGATAGGGCAGCACACGTGAAACAGGTGCTGCTCCACCAGCA
L08604      CGCTGTGGTGTCTCGGAGCGATAGGGCAGCACACGTGAAACAGGTGCTGCTCCACCAGCA
            ************************************************************

Vechur      GGCTCTGTTTGGGAAAAATGGAAAAAACTGCCCGGACAAGTTTTGTTTGTTCAAATCTGA
FJ589071    GGCTCTGTTTGGGAAAAATGGAAAAAACTGCCCGGACAAGTTTTGTTTGTTCAAATCTGA
L08604      GGCTCTGTTTGGGAAAAATGGAAAAAACTGCCCGGACAAGTTTTGTTTGTTCAAATCTGA
            ************************************************************

Vechur      AACCAAAAACCTTCTGTTCAATGACAACACTGAGTGTCTGGCCAAACTTGGAGGCAGACC
FJ589071    AACCAAAAACCTTCTGTTCAATGACAACACTGAGTGTCTGGCCAAACTTGGAGGCAGACC
L08604      AACCAAAAACCTTCTGTTCAATGACAACACTGAGTGTCTGGCCAAACTTGGAGGCAGACC
            ************************************************************

Vechur      AACGTATGAAGAATATTTGGGGACAGAGTATGTCACGGCCATTGCCAACCTGAAAAAATG
FJ589071    AACGTATGAAGAATATTTGGGGACAGAGTATGTCACGGCCATTGCCAACCTGAAAAAATG
L08604      AACGTATGAAGAATATTTGGGGACAGAGTATGTCACGGCCATTGCCAACCTGAAAAAATG
            ************************************************************

Vechur      CTCAACCTCCCCGCTTCTGGAAGCCTGCGCCTTCCTGACGAGGTAAAGCCTGCAA-----
FJ589071    CTCAACCTCCCCGCTTCTGGAAGCCTGCGCCTTCCTGACGAGGTAAAGCCTGCAA-----
L08604      CTCAACCTCCCCGCTTCTGGAAGCCTGCGCCTTCCTGACGAGGTAAAGCCTGCAAAGAAG
            *******************************************************
Vechur      ------------------------------------------------------------
FJ589071    ------------------------------------------------------------
L08604      CTAGCCTGCCTCCCTGGGCCTCAGCTCCTCCCTGCTCTCAGCCCCAATCTCCAGGCGCGA
Vechur      ------------------------------------------------------------
FJ589071    ------------------------------------------------------------
L08604      GGGACCTTCCTCTCCCTTCCTGAAGTCGGATTTTTGCCAAGCTCATCAGTATTTACAATT
Vechur      ---------------------------------------------------------
FJ589071    ---------------------------------------------------------
L08604      CCCTGCTGTCATTTTAGCAAGAAATAAAATTAGAAATGCTGTTGATTTTCATTCCCT
```

Fig. 3: Comparison of Nucleotide Sequence of Lf gene from Vechur, Hybrid and Reference cow

```
L08604      MKLFVPALLSLGALGLCLAAPRKNVRWCTISQPEWFKCRRWQWRMKKLGAPSITCVRRAF
FJ589071    MKLFVPALLSLGALGLCLAAPRKNVRWCTISQPEWFKCRRWQWRMKKLGAPSITCVRRAF
Vechur      MKLFVPALLSLGALGLCLAAPRKNVRWCTISQPEWFKCRRWQWRMKKLGAPSITCVRRAF
            ************************************************************

L08604      ALECIRAIAEKKADAVTLDGGMVFEAGRDPYKLRPVAAEIYGTKESPQTHYYAVAVVKKG
FJ589071    ALECIRAIAEKKADAVTLDGGMVFEAGRDPYKLRPVAAEIYGTKESPQTHYYAVAVVKKG
Vechur      ALECIRAIAEKKADAVTLDGGMVFEAGRDPYKLRPVAAEIYGTKESPQTHYYAVAVVKKG
            ************************************************************

L08604      SNFQLDQLQGRKSCHTGLGRSAGWIIPMGILRPYLSWTESLEPLQGAVAKFFSASCVPCI
FJ589071    SNFQLDQLQGRKSCHTGLGRSAGWIIPMGILRPYLSWTESLEPLQGAVAKFFSASCVPCI
Vechur      SNFQLDQLQGRKSCHTGLGRSAGWIIPMGILRPYLSWTESLEPLQGAVAKFFSASRVPCI
            ******************************************************* ****

L08604      DRQAYPNLCQLCKGEGENQCACSSREPYFGYSGAFKCLQDGAGDVAFVKETTVFENLPEK
FJ589071    DRQAYPNLCQLCKGEGENQCACSSREPYFGYSGAFKCLQDGAGDVAFVKETTVFENLPEK
Vechur      DRQAYPNLCQLCKGEGENQCACSSREPYFGYSGAFKCLQDGAGDVAFVKETTVFENLPEK
            ************************************************************

L08604      ADRDQYELLCLNNSRAPVDAFKECHLAQVPSHAVVARSVDGKEDLIWKLLSKAQEKFGKN
FJ589071    ADRDQYELLCLNNSRAPVDAFKECHLAQVPSHAVVARSVDGKEDLIWKLLSKAQEKFGKN
Vechur      ADRDQYELLCLNNSRAPVDAFKECHLAQVPSHAVVARSVDGKEDLIWKLLSKAQEKFGKN
            ************************************************************

L08604      KSRSFQLFGSPPGQRDLLFKDSALGFLRIPSKVDSALYLGSRYLTTLKNLRETAEEVKAR
FJ589071    KSRSFQLFGSPPGQRDLLFKDSALGFLRIPSKVDSALYLGSRYLTTLKNLRETAEEVKAR
Vechur      KSRSFQLFGSPPGQRDLLFKDSALGFLRIPSKVDSALYLGSRYLTTLKNLRETAEEVKAR
            ************************************************************

L08604      YTRVVWCAVGPEEQKKCQQWSQQSGQNVTCATASTTDDCIVLVLKGEADALNLDGGYIYT
FJ589071    YTRVVWCAVGPEEQKKCQQWSQQSGQNVTCATASTTDDCIVLVLKGEADALNLDGGYIYT
Vechur      YTRVVWCAVGPEEQKKCQQWSQQSGQNVTCATASTTDDCIVLVLKGEADALNLDGGYIYT
            ************************************************************

L08604      AGKCGLVPVLAENRKSSKHSSLDCVLRPTEGYLAVAVVKKANEGLTWNSLKDKKSCHTAV
FJ589071    AGKCGLVPVLAENRKSSKHSSLDCVLRPTEGYLAVAVVKKANEGLTWNSLKDKKSCHTAV
Vechur      AGKCGLVPVLAENRKSSKHSSLDCVLRPTEGYLAVAVVKKANEGLTWNSLKDKKSCHTAV
            ************************************************************

L08604      DRTAGWNIPMGLIVNQTGSCAFDEFFSQSCAPGADPKSRLCALCAGDDQGLDKCVPNSKE
FJ589071    DRTAGWNIPMGLIVNQTGSCAFDEFFSQSCAPGADPKSRLCALCAGDDQGLDKCVPNSKE
Vechur      DRTAGWNIPMGLIVNQTGSCAFDEFFSQSCAPGADPKSRLCALCAGDDQGLDKCVPNSKE
            ************************************************************

L08604      KYYGYTGAFRCLAEDVGDVAFVKNDTVWENTNGESTADWAKNLNREDFRLLCLDGTRKPV
FJ589071    KYYGYTGAFRCLAEDVGDVAFVKNDTVWENTNGESTADWAKNLNREDFRLLCLDGTRKPV
Vechur      KYYGYTGAFRCLAEDVGDVAFVKNDTVWENTNGESTADWAKNLNREDFRLLCLDGTRKPV
            ************************************************************

L08604      TEAQSCHLAVAPNHAVVSRSDRAAHVKQVLLHQQALFGKNGKNCPDKFCLFKSETKNLLF
FJ589071    TEAQSCHLAVAPNHAVVSRSDRAAHVKQVLLHQQALFGKNGKNCPDKFCLFKSETKNLLF
Vechur      TEAQSCHLAVAPNHAVVSRSDRAAHVKQVLLHQQALFGKNGKNCPDKFCLFKSETKNLLF
            ************************************************************

L08604      NDNTECLAKLGGRPTYEEYLGTEYVTAIANLKKCSTSPLLEACAFLTR
FJ589071    NDNTECLAKLGGRPTYEEYLGTEYVTAIANLKKCSTSPLLEACAFLTR
Vechur      NDNTECLAKLGGRPTYEEYLGTEYVTAIANLKKCSTSPLLEACAFLTR
            ************************************************
```

Fig. 4: Comparison of Amino acid Sequence of Lf protein from Vechur, Hybrid and Reference Cow

Phylogenetic analyses were done for both the Lf aa sequences, BLf-Vand BLf-V. A phylogenetic tree comparing Lf aa sequences from different mammalian genera (cow, buffalo, goat, sheep, yak, human, horse, camel, rat, mouse) was constructed (Fig. 5) using the UPGMA method in the PHYLIP software package (Version 3.68). The phylogenetic analysis of Lf from different species revealed three major groups: all the members of the Bovidae family belonged to one group, rat and mouse formed another group while Human, horse and camel belonged to the third group.

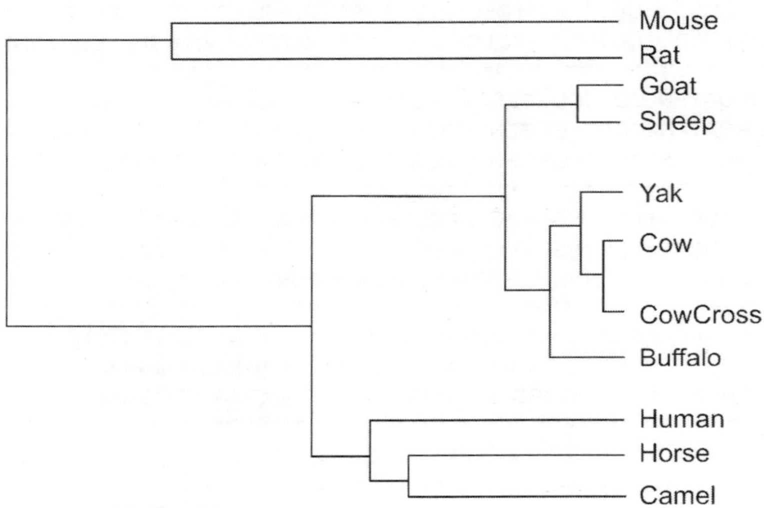

Fig. 5: Phylogenetic Tree

Restriction Enzyme Analysis

The restriction analysis of BLf-V and BLf-H genes with Kpnl enzyme yielded an expected restriction pattern (Fig. 6) Blf-H gene bears a single recognition site for Kpnl enzyme and thus was expected to show two bands of 1496 and 645 bp. However, BLf-V possesses two recognition sites for the same enzyme and the expected bands were of size 1422, 74 and 645 bp respectively.

DISCUSSION

BLf has remained a subject of exhaustive research since its discovery and it has been applied to commercial food products for the last 25 years. The protein is well-known for its therapeutic and nutritional merits which include antimicrobial properties, immunomodulatory role and cancer-preventive effects. Characterization of the Lf gene in *various* mammals has *revealed* variations of different sorts among genera, species and varieties. These variations when reflected in the amino acid sequences may affect the properties of the mature protein, depending upon the position and the nature of amino acid (Li *et al*, 2004; Velliyagounder et ai, 2003, Lee *et al*, 1997).

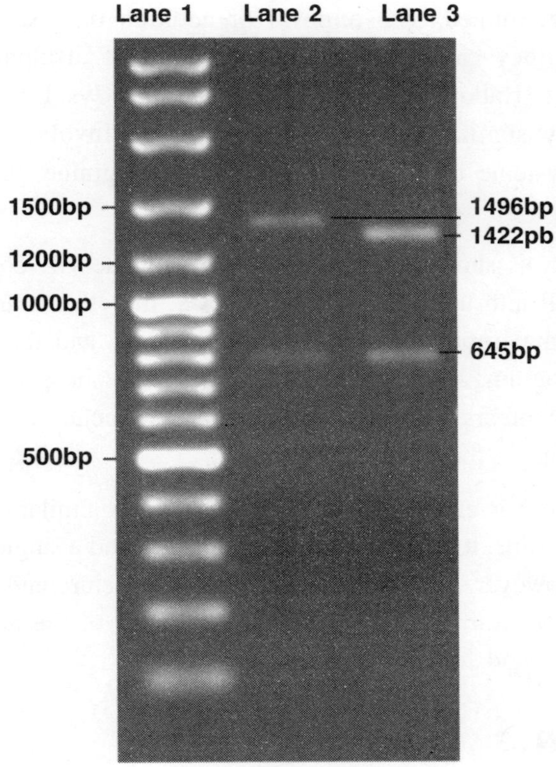

Fig. : 6
Lane 1 : 100 bp DNA Ladder
Lane 2 : RE digest of Lf from hybrid cow
Lane 3 : RE digest of LF from Vechur cow

Determination of the Lf gene and aa sequences from cows (breeds) hitherto not subjected to characterization is expected to reveal novel and useful variations. As far as the Indian indigenous breed Vechur is concerned, no studies have been done on its Lf gene. In the present study, for the first time, Lf gene was isolated and sequenced from the Vechur cow. The comparison of BLf-V nucleotide sequence with BLf-H and other published sequences revealed four nucleotide changes which are all exclusive to BLf-V, thus indicating its polymorphic nature. Only one of the Single Nucleotide Polymorphisms (SNPs) was found to be associated with an aa change involving a cysteine to arginine substitution (Cys176Arg). The aa change was observed in exon 5 of the gene which comes within the N-terminal domain of Lf. Polymorphisms have been studied in Lf in different species (O'Halloran *et al*, 2009; Teng and Gladwell, 2006). Polymorphisms become significant if they are associated with aa substitutions that can potentially affect the structure or function of the protein. Conservative aa substitutions are believed not to affect protein structure, but the influence on the substrate recognition properties needs to be determined. The removal or addition of cysteine residues are structurally significant as they can influence the formation of disulphide bridges that stabilize protein structures. (Halloran et al, 2009). Hence, the Cys 176 Arg substitution seems to be a potentially significant change. Moreover, it involves replacement of the negatively charged Cysteine by the positively charged Arginine, thus affecting the overall charge of Lf.

The phylogenetic analysis showed that members of Bovidae viz. cow, buffalo, yak, Vechur, goat and sheep fell into the same clade. However, the other mammalian species like human, horse and camel formed a separate group. Mouse and rat belonged to yet another group in the cladogram. Among the Bovidae goat and sheep formed a separate branch while the other members belonged to one branch. Vechur cow showed close similarity with cow and yak in the dendrogram.

From the present study it is clear that though BLf-V shows similarity with the other BLf sequences, at the same time, it exhibits nucleotide changes and a single aa substitution, making it polymorphic. However, an indepth study into the structure and properties of Lf protein from Vechur cows is warranted to ascertain the effects of the nucleotide and aa variations in the expression and functioning of Lf protein.

ACKNOWLEDGEMENTS

Our sincere thanks are due to Dr. G P C Nayar, Chairman, SCMS Group for providing the Lab facilities for the work. We also thank Dr. Chandrashekharan Nair and Dr. Raghunandan of College of Veterinary and Animal Sciences, Thrissur, Kerala for supplying milk samples for the work.

REFERENCES

Adlerova L, Bartoskova A, Faldyna M (2008). Lactoferrin: a review. Veterinami Medicina, **53(9):** 457- 468.

Daly M, Ross P, Giblin L (2006) Polymorph isms within the lactoferrin gene promoter invarious cattle breeds. Anim. Biotechnol. **17:** 33-42.

Felsenstein J (1995). PHYLlP: Phylogeny inference package. Version 3.57c. University of Washington, Seattle.

Garoussi M T, Bassami M R, Afshari S E (2007). Detection of bovine viral diarrhea virus using a nested RT -PCR assay in bulk milk samples of dairy cattle herds in suburb of Mashhad-Iran. Iranian Journal of Biotechnology, **5(1):** 52-55.

Lee T H, Shimazaki K, Yu S L, Nam, M 8, Kim S J, Lee K K, Yu D U (1997). Polymorphic sequence of Korean ladoferrin exhibiting greater antibacterial activity. Anim. Genet. **28:** 367-369.

Levay P F, Viljoen M (1995). Lactoferrin: a general review. Haematologica **80:** 252-267.

Li G H, Zhang Y, Sun D X, Li N (2004). Study on the polymorphism of bovine lactoferrin gene and its relationship with mastitis. Anim. Biotechnol. **15:** 67- 76.

O'Halioran F, Bahar B, Buckley F, O'Sullivan 0, Sweeney T, Giblin L (2009). Characterisation of single nucleotide polymorphisms identified in the bovine lactoferrin gene sequences across a range of dairy cow breeds. Biochimie, **91:** 68-75.

Supraha P, Anilkumar K (2006). DNA fingerprinting and genetic distance estimation among native cattle ecotypes by random primers. Proceedings of the 12th Baltic Animal Breeding Conference. 185-195.

Teng T Christina (2002). Lactoferrin gene expression and regulation: an overview. Biochem. Cell Biol. **80:** 7-16.

Teng T Christina, Gladwell W (2006). Single nucleotide polymorphisms (SNPs) in human lactoferrin gene. Biochem. Cell BioI. **84:** 381-384.

Thompson JD, Higgins DG, Gibson TJ (1994). CLUSTAL W: improving of the sensitivity of progressive multiple sequence alignment through sequence rating, position specific gap penalties and weight matrix choice. Nucleic Acids Res. **22:** 4673-4680.

Velliyagounder K, Kaplan JB, Furgang D, Legarda D, Diamond G, Parkin R E, Fine D H (2003). One of two human lactoferrin variants exhibits increased antibacterial and transcriptional activation activities and is associated with localized juvenile periodontitis. Infect. Immun. **71:** 6141-6147.

Bioresources for Food Security and Rural Livelihood, Pages 141–143
Edited by : G.K. Kulkarni and B.N. Pandey

OESTROGEN AND PROGESTERONE LEVELS DURING SUPEROVULATORY TREATMENT IN FRIESWAL COWS

S.K. Maurya and A.K. Mathur

ABSTRACT

Ten normal adult healthy cyclic Frieswal cows were considered for this study. These cows were maintained under standard management condition of Military farm and were fed additional mineral mixture @ 50g/day/animal through out the experimental period. Their regular cyclicity were confirmed after rectal palpations at 11 days intervals. A double dose of prostaglandin F_2 alpha (PG) was utilized for synchronization of oestrus in these animals. These animals were divided into two groups of 5 animals each super ovulation was induced in the first group by utilizing 280 mg of NIH-FSH P_1 in 8 divided doses given in 4 days. In the second group PMSG was utilized @ 1500 IU/animal as a single i/m injection. All the ten animals were given injection of PGF_2 alpha (PG) for induction of oestrus and were subsequently inseminated with frozen thawed semen. The blood samples were collected for the estimation of oestradial 17B and progesterone at the end of natural oestrus prior to superovulatory treatment, just before PG injection and at the end of induced oestrus.

The mean oestradiol 17β concentration estimated just before the start of super ovulation treatment were 163 ± 37.20 and 140 ± 47.54 pg/ml where as the concentration of progesterone were 4.95 ± 1.53 and 4.45 ± 1.48ng/ml in I and IInd superovulatory treatment group respectively. It was observed that after 60h of start of supervoulatory treatment the levels of estradiol 17β were 109 ± 40.70 and 90 ± 30.78 pg/ml in group I & II, these values at the end of induced oestrus were 148.0 ± 54.4 and 85.0 ± 29.88 pg/ml showing a lower trend as compared to levels obtained from same animals at the end of natural oestrus just before the start of experiment (i.e. 210 ± 136.64 and 93 ± 44.71 pg/ml in I & IInd group). Similarly, mean progesterone concentration after 60h of start of superovualtory treatment were 4.28 ± 1.19 and 4.19 ± 0.99 ng/ml in group I & II which were 2.91 ± 0.83 and 3.65 ± 1.58 ng/ml at the end of induced oestrus. Here also levels were showing a lower trend on compared to mean values obtained just before the start of experiment (i.e. 4.8 ± 2.12 and 4.9 ± 2.01 ng/ml).

Keywords : Superovulation, Progesterone, Oestrogen, Prostaglandin, FSH-P.

INTRODUCTION

Embryo transfer technology has been practiced world wide for several years. The variable and unpredictable superovulatory response of doner cows remains the major limiting factor for successful embryo transfer in cattle (Adame *et al* 1992). The important factor that affects the superovulatory response and quality of embryos are endocrine events that occur during the superovulatory treatment (Kewon *et al* 1987 and Roberge *et al* 1995). In view of this the study was under taken to know the circulatory levels of oestrogen and progesterone at different stages of superovulation in Frieswal animals.

Ten adult elite Frieswal cows were maintained under standard management conditions as per the Military Farm schedule. They were fed additional ration and mineral mixture (Agrimin forte, Agrivet farm care division, Glaxosmithkine Pharmaceuticals Ltd.) @ 50g per animal per day through out the experimental period.

Oestrus was regularly observed twice daily in the morning and evening followed by confirmation of oestrus by rectal palpation. These animals were divided into two groups of 5 cows each and were superovulated by FSH-P (Follotropin – V Bioniche Animal health Canada) and PMSG (Folligon, Intervet India Pvt. Ltd.). In the first group of 5 cows, 280 mg of FSH – P was injected twice daily in descending doses schedule for 4 days. In the second group PMSG was used in a single dose of 1500 IU as single i/m injection. All the cows were given injection of prostaglandin on 3rd day of FSH/PMSG injection. The animals exhibited oestrus were inseminated with frozen-thawed semen. Samples of jugular venous blood were collected at the end of natural oestrus, prior to superovulatory treatment just before prostaglandin injection and at the end of induced oestrus from all the 10 cows for estimation of oestrogen (oestradial 17 – P) and progesterone. Estimation of oestrogen and progesterone was done by enzyme linked immunosorbaut assay (ELISA) technique using kit pathozyme (Omega Diagnostic Ltd. Scotland UK).

It was observed that the mean concentration of hormone oestradiol 17-β in FSH superovulated doner animals (group-I) at the end of natural oestrus was 210 ± 136.64 pg ml. After day 10 of the cycle i.e. just before the start of superovulatory treatment the levels were 163 ± 3 7 .20 pg/ml and after 60h of superovulatory treatment (i.e. just before PG injection) were 109 ± 40.70 pg/ml. These levels were 148 ± 54.44 pg/ml at the end of induced oestrus. In the second group (superovulated with eCG) the respective concentration of oestradiol 17-β during the four different stages of sample collection were 93 ± 44.71, 140 ± 47.04; 90 ± 30.78 and 85 ± 29.88 pg/ml. Observations on oestradiol concentration of individual animal before PG injection, revealed a trend suggesting increased concentration at this stage probably gives an indication of number of developing estrogenized follicle which could have turned out into a superovulatory response (with 175 pg/ml-14 CL and 200 pg/ml-l0 CL) however we observed a decline in the mean estradiol concentration (163 ± 37.20 to 109 ± 40.70 pg/ml and 140 ± 47.54 to 90 ± 30.78 pg/ml in I & II group respectively) which does not support our aforesaid presumption, however, increase in concentration of estradiol before induction of oestrus with PG have been correlated with more number of follicle formation by other workers. (Kweon *et al*, 1987, Agarwal *et al*, 1993).

The mean concentration of progesterone at the end of natural oestrus was 4.8 ± 2.12 and 4.9 ± 2.01 ng/ml in group I & II animals. These values were 4.95 ± 1.53 and 4.45 ± 1.48 ng/ml in two group just before the start of superovulatory treatment. The concentration of progesterone after 60h post FSH and PMSG treatment were 4.28 ± 1.19 and 4. 19 ± 0.99 ng/ml. However the values at the end of induced oestrus were 2.9 ± 0.83 and 3.65 ± 1.58 ng/ml in group I & II respectively. Perusal of overall mean concentration (irrespective of stage of sample collection) of progesterone in two groups broadly indicates that there was a low superovulatory response with higher concentration of progesterone Observations on the progesterone concentration of individual cows in our experiment revealed no definite trend, the mean concentration of progesterone at the end of natural oestrus were 4.8 ± 2.12 and 4.9 ± 2.01 for two groups where as others have reported lower concentrations (Kharche *et al* 1997; Prasad *et al* 2003). The higher values of progesterone at the end of induced oestrus (2.91 ± 0.83 and 3.65 ± 1.58 ng/ml) suggests some unovulated follicle undergoing lutenization. This could also be a reason for lower recovery of embryos despite good superovulation response.

ACKNOWLEDGEMENT

Authors are thankful to Director Frieswal project for providing necessary facilities to conduct experiment at Military Farm, Meerut.

REFERENCES

Agarwal, S.K.; Taneja, V.K.; Shanker U, Yadav M.C.; Sanwal, P.C. and Varshney V.P. (1993). Superovulation, embryo recovery and endocrine response in crossbred cattle treated with PMSG and FSH-P. Indian Journal of dairy Science **46:** 101.

Kewon, O.K., Kanagwa, H; Takahashi, Y., Miyamato, A., Masnki, J. Wmezu M; Kagabu, S; Iwazumi, Y and Aoyayi Y (1987). Plasma enocrine profile and total cholesterol levels in superovulated cows. Theriogenology **27:** 841-57.

Kharche, S.D., Ansari, M.R., Varshney, V.P., Majumdar, A.C and Taneja, V.K. (1997). Endocrine profile in superovulated crossbred cattle. Indian Journal of Animal Reproduction. **18(2):** 101-103.

Prasad Shiv, Maurya S.N. and Varshney, V.P. (2003). Serum progesterone 2. and estradiol 17-β concentration in relation to superovulatory response in embryo doner (sahiwal) and conception rate in embryo recipient (crossbred) cattle. Indian Journal of Animal Science **73(4):** 353-356.

Roberge S, Rieger D; and Rawlings N.C (1995). Preovulatory LH and FSH 4. 5. and Steriod hoemone profile insuperovulated and unstimulated Holstein heifers. Theriogenology **44:** 59-70.

Bioresources for Food Security and Rural Livelihood, Pages 145–163
Edited by : G.K. Kulkarni and B.N. Pandey

BIORESOURCES: CHALLENGES OF WETLAND RESOURCES IN NORTHEAST INDIA

D.K. Sharma and Chittaranjan Baruah

ABSTRACT

Bio resources include living resources, biodiversity and bio-information. Bioresources availability is highly diversified and it depends on the region's agro-climatic conditions. India's northeast region is a transitional zone at the conjunction of the Himalaya and Indo-Burma biodiversity hotspots. The gene pools of the region are invaluable genetic resources and can be put to use for the economic growth. The region hosts many potential bioresources such as Medicinal Plants, Orchids, Microbial Bio-resources, Aquatic Genetic Resources and Insect bio-resources etc. The north – eastern states of India harbor many spectacular and biologically rich wetlands. With increasing human pressure and prevalent underlying causes these wetlands are under increasing pressure. The threats, opportunities and challenges of wetland resources for sustainable livelihood with the rich indigenous traditional knowledge system are relevant. On the other hand the availability of food has posed as challenge to address the hunger. Hunger is an unbalanced energy intake, vitamins and minerals account for more than half of the worlds deficiency disease burden (FAO). The bioresources have not either been properly utilized or remains unutilized to arrest hunger and malnutrition with sustainable livelihood. The essential elements like Ca, P, I, Fe, I Mg etc. are found in adequate quantity in these bioresources to maintain adequate nutritional requirements. Over exploitation of bioresources has been explained citing the decline trend of turtle diversity. Being biodiversity hot spots the region has enough resources, however, demand full characterization and effective utilization.

Keywords : Food scarcity; *gene pool;* hunger; livelihood; over exploitation

INTRODUCTION

The "God Almighty" created nature and its inhabitant to live together in harmony and so the Bio resources. By definition, bio resources include living resources, biodiversity and bio information as such the forests and their products, plants, animals and other organisms. Minerals such as coal and petroleum are also included in this category because of their origin from decayed organic matter. Bioresources availability is highly diversified and it depends on the region's agro-climatic conditions. Inventory of these resources is required for describing the quality, quantity, change, productivity of bio resources. The distinct tribes in the region have rich indigenous traditional knowledge system on the uses of components of biodiversity for their daily sustenance like food, fodder; shelter and healthcare are also the source of bioresources.

Over the years, people of the NE region of India have been surviving with nature and have developed location-specific traditional ecological knowledge systems tuned to culture and nature. The survival strategies of these communities were of subsistence in mode, without greed of destroying biodiversity and other resources. However, the inquisitiveness of mankind to explore the unknown, has led us to an era of a technically advanced age called "Kalyug" meaning the age of machine and tools, have weakened the dynamics of sustainable conservation of biodiversity and other natural resources.

CHALLENGES AND OPPORTUNITIES

The technological advancements have adversely affected the nature and as such the basic natural inputs to human body-air, water, plants (bioresources) are highly depleted and so the global food crisis. The large mass of the total populace are from rural area and have been confronted with hunger and poverty. However, sustainable medium and long term approach are needed to fight hunger, disease and poverty, but the need of the hour, to be specific enough is the Food sovereignty- A real challenge in fact, which really needs a long term solution depending on our own resources. Thus the conservation and management of our bioresources are the principal challenge of the hour.

Hunger is an unbalanced energy intake, vitamins and minerals account for more than half of the worlds deficiency disease burden (FAO). The United Nations Secretary General Ban Ki-moon categorically warns of a rapidly escalating crisis of food availability around the globe reaching emergency proportions. One eight of the world's people do not have enough to eat (Pinstrup- Anderson and Cheng, 2007). The so called nutrition transition in which diets change from basic grain and tubers to meat, dairy products and processed foods high in sugar and fat, is the full force in developing countries bring with it a dual nutritional problem of deficiency and hunger in some households and obesity and related disease in others. Globalization and improved communication as well as enough food production in the globe, have not been able to meet the challenge of the developing and underdeveloped nations in the food front despite of huge bioresources. The main reason of hunger and nutritional deficiencies persist is poverty, sustained by poor family planning and reproductive health care along with the presence of a large group of environmental refugee.

Poverty is widespread in rural areas of India, which is caused by marginalization, deprivation and exclusion of certain section of communities, and characterized by the food deficit and malnutrition, inaccessibility to health and education, inability to participate local governance, and vulnerability to insecurities. Rural livelihoods are also prone to dispossession phenomenon. The unequal distribution of food, overexploitation of bioresources, and lack of awareness for the need of conservation in addition to the transitory hunger phase led to the hunger problem.

The reasons for food scarcity etc.

- Food production not keeping pace with population increase.
- Challenges of food habit- whereby people consume more of the three key staples- rice, maize and wheat as well as increased consumption of live stock and dairy products and hence more grain conversion to feed.
- Increase in fossil fuel prices-quick turn to biofuel crops including food biofuel crops.
- Climate change- too much water (flood) and or too little (drought) and low (freeze) and high temperature- enhances crop failure- grain shortage.
- Further, market where we see huge capital flow to speculative commodity index

The crucial components are the rural development and agriculture since many of the Indians in general and the northeast Indians in particular, poor live in rural areas. The effective and sustainable bioresources have not been either properly utilized or remains unutilized.

India's Northeastern Region consists of eight states – Arunachal Pradesh, Assam, Manipur, Meghalaya, Mizoram, Nagaland, Sikkim, and Tripura. It occupies 7.7% of India's total geographical area and with a population of 39 million (2001 census; **Table 1**). The region is supporting 50% of the flora (ca. 8000 species), of which 31.58% (ca. 2526 species) is endemic. It is a transitional zone between the Indian, Indo-Malaysian and Indo-Chinese regions. The region is rich in natural resources, especially water and forests, and there is a feeling that it could potentially be one of the wealthiest bioresources zone of India. However, several indicators reveal a different picture: in spite of some progress in a few of the northeastern states (for example Mizoram), overall growth rates over the past years have remained low, poverty incidence (especially in Assam) is high, there are still a number of areas subject to continued violence, and there is an abundance of reports documenting natural resource degradation, depleting the very assets that are usually highlighted as offering the greatest potential for growth and development in the Northeastern Region. Thus, in recent years, the region has missed out on the economic growth acceleration witnessed in much of India.

Table 1 : Population Density in northeastern states (2001 census)

State	Population (in million)	Land Area (in sq. km)	Population Density (per sq. km)
Arunachal	1.09	83743	13
Assam	26.65	78438	340
Manipur	2.29	22327	103
Meghalaya	2.31	22429	103
Mizoram	0.89	21087	42
Nagaland	1.99	16579	120
Sikkim	0.54	7096	76
Tripura	3.19	10486	313
Northeast	38.98 (88% rural)	262185	149
All India	1028.61 (75% rural)	3287263	313

The Northeastern region of India is home to 13,500 plant species (2.3% of the global agencies) of which 7000 are endemics and 2,185 vertebrates (1.9 % of the global species) of which 528 are endemics. Avian fauna is represented by 1,170 species with 140 as endemics. Mammalian diversity constitutes with 329 species with 73 endemics. The species richness in terms of numbers of reptiles and amphibians are 484 and 202 with 201 and 114 endemics respectively. It may be noted that only in Manipur the known biodiversity includes 4000 angiospermic plants, 430 medicinal plants 34 species of edible fungi, 500 orchids, 55 species of bamboo, 40 endemic rice cultivars, 160 fish species, 21 migratory aquatic birds and multitude of low and high altitude butterflies. It is one of the rainiest regions of the world being fed with two monsoons. The highest rainfall area is also present in this region, the town called Mawsynram, about 10 km away from earlier highest rainfall area (Chirapunjee) in the state of Meghalaya. The tropical climate of the regions with high rain fall and plenty of sun-light coupled with unique bio-geographical positioning of this region is responsible for rich biodiversity in terms of floral and faunal elements (**Table 2**). Recognizing the importance of this region as one of the hot spots, majority of the biodiversity rich areas of the region has been placed inside the protected area network system comprising mainly of the National Parks and Sanctuary. However, the total area under the protected area network is only about 6.5%, which ideally should be around 10% as per the policy of the government.

Table 2 : Biodiversity in Northeast India (Ghosh & Tiwari, 1984)

Plant/Animal Group		No. of Species
Plants		
	Flowering plants	7,500
	Orchids	700
	Citrus	64
	Bamboos	58
	Conifers	28
	Mosses	500
	Ferns	700
	Lichen	728
Animals		
	Butterflies	183
	Molluscs	50
	Fishes	236
	Amphibians	64
	Reptiles	137
	Birds	541
	Mammals	160

GENE POOLS OF NORTHEAST REGION OF INDIA

The gene pools of the N.E. Region are invaluable genetic resources and can be put to use for the economic growth of this region in particular and to the nation in general. Biotechnological interventions are needed for conservation and sustainable uses of the bio-resources.

Medicinal Plants

Realizing the great role of traditional medicine in global health-care program especially in the developing countries, the World Health Organization (WHO) gave (1976) a formal recognition to traditional medicine, and emphasized for the global resources into medicinal herbs. The Belem Declaration (1988) the Chiang Mai Declaration (1988) and the Bangalore Declaration (1998) also drew the attention of the United Nations, its agencies and Member States, other international agencies and their members and NGOs to the vital importance of medicinal health care, and the significant economic value of the medicinal plants. The North-Eastern India is blessed with nature's bounty as regards the wealth of industrially potential plants. From early periods of human civilization in this region, many of these plants have been using as different species of plants for treating different types of ailments. Modernization of the traditional knowledge and use of the medicinal plants sustainably through biotechnological interventions will ensure the economic prosperity of the region.

Orchids

Orchids are accepted to be the world's most exotic and fascinating flowers. The unmatched ornamental value of Orchids accounts for multi-million dollar cut flower industries in several countries like Europe, America, Japan, etc. Several South-East countries, viz., Thailand, Singapore, Malaysia, and Philippines have also resorted to orchid growing. North - Eastern Region with its diverse climatic condition and topography is identified as the most potential areas for orchid cultivation. Out of the 1,200 species, fewer than 160 genera are found in India. The region grows more than 650 species of which Manipur grows about 500 of which 472 have been identified. The need of the hour is to bring awareness, popularization, conservation, propagation and commercialization of orchids so as to convert the existing national orchid wealth into rewarding resources.

Horticultural Bio-resources

Of the horticultural plants, gene pools of Citrus (oranges and lemons), Musa (banana), Prunus (plums and other pomaoeous fruits) etc. probably constitute the major centre of gene diversity. Besides "Queen" which is a world class pineapple in terms of sweetness and high vitamin C content is widely cultivated as cash crop by the weaker sections (SCs/ STs).

Microbial Bioresources

The climatic condition of N.E. region favours rich sources of microbial species. The rich microbial resources of the region have tremendous future scope for human survival and sustainable livelihood management. Although, less than 10 % of the total microbial community is currently known, the contribution of this versatile group in the agricultural, industrial and pharmaceutical sector is very impressive. The total number of bioactive fungal products is approximately 8600, representing 38 % of all microbial products. More than 10,000 bioactive compounds have been reported from the filamentous *Actinomycetalles species*. A variety of blue-green algae found in N.E. Region can successfully be used as 'bio-fertilizers' the application of which improves crop quality and yield. The mushroom flora is highly rich in protein and the edible species are highly tasty (with good flavor), nutritive and of commercial importance. In this region, many edible species (34 in Manipur) have been recorded. Bioprospecting these rich, soil micro-organisms may lead to the discovery of novel antibodies and other industrial products.

Aquatic Genetic Resources

A variety of endemic fish species are found in the Wet land ecosystem of northeast India. Several migratory fishes are found along the Chindwin-Irrawaddy and Barak river systems as well as the Brahmaputra basin. As many as 140 fish species (only in Manipur) have been found, including a number of small ornamented fishes in the hill streams and 149 fish species of Nagaland (Ao *et al*., 2008). In addition, many edible snails and mussels, edible red algae, Lemanea australis (Nungsam) are also available in this region to challenge the nutritional deficiency. There is an ample scope for commercial culture of freshwater mussels and prawn and also hill stream fishes, many of which are highly prized.

Seribio-technology and Insect Bio-resources

The insect Bio-resources of N.E. Region comprise of (a) mulberry silkworm, (b) Mug silk worm), (c) Oak tasar and Eri silkworm, (d) Pollinators - honey bees, (e) predators and parasitoids of pests, (f) bio-control agents, (g) wild butterflies of aesthetic value and (h) edible insects, which are of valued bioresources for the rural livelihood of the rural masses either directly or indirectly.

WETLANDS – THE RICH BIORESOURCES FOR LIVELIHOOD

"A wetland can be biologically far more productive than a woodland or grassland of comparable size".

Wetland is a generic term for water bodies of various types, and includes diverse hydrological entities, namely, lakes, marshes, swamps, estuaries, tidal flats, river flood plains and mangroves. The Ramsar convention defined wetlands as "areas of marsh, peat land or water, whether natural or artificial, permanent or temporary, with water that is static or flowing, fresh, brackish or salt, including areas of marine water the depth of which at low tide does not exceed six meters". Wetlands all over the world are threatened directly due to reclamation for development activities, reduction in function due to pollution, water demand, change in hydrologic regime, over exploitation of wetland resources (Shine and de Klemm 1999).

> 'India's freshwater resources comprise the single most important class of natural endowments enabling its economy and its human settlement patterns. The fresh water resources comprise the river systems, groundwater and wetlands. Each of these has a unique role, and characteristic linkages to other environmental entities—— Wetlands, natural and manmade, freshwater or brackish, provide numerous ecological services. They provide habitat to aquatic flora and fauna, as well as numerous species of birds, including migratory species'— A holistic view of Wetlands is necessary which looks at each identified Wetlands in terms of its causal linkages with other natural entities, human needs, and its own attributes.
>
> ***National Environment Policy, 2006***

The North – eastern states of India harbor some of the spectacular and biologically rich wetlands in India. Some of these wetlands have been explored, but most of them are still unknown. With increasing human pressure and prevalent underlying causes these wetlands are under increasing pressure. With increasing infrastructure development and other pressures in this region, there is a danger of losing this natural heritage, the species it supports, the ecosystem services it provides as well as the spiritual and cultural ties the local population has with these wetlands.

Benefits of Wetlands

Wetlands offer several substantive benefits. Unfortunately, they are often not fully understood. They, in fact support life system, extend suitable habitat to flora and fauna, wide variety of microbes. They are effective in flood control, waste water treatment, reducing sediment loads and recharging of aquifers. They are valuable for their educational and scientific interest (especially their high diversity or Species richness) and their recreational benefits are significant.

A Brief Account on Wetlands in Northeast India

Assam

Assam is dominated by two river plains -the Brahmaputra plains (56,480 sq. kms) drained by the river Brahmaputra and its 43 tributaries and the Barak plains (6962 sq.kms) drained by the river Barak and its tributaries. The wetlands of this region are in form of beels, swamps and marshes hosting a rich array of flora and fauna. Two species of endemic aquatic plants, two critically endangered, seven endangered and 26 vulnerable species of fishes, five near threatened, six vulnerable species of freshwater turtles. 133 species of birds have been recorded from the 3556 major wetlands in Assam. SACON study prioritized 75 wetlands, 25 of these being the potential candidates for designation as Ramsar Sites. One of them the Deepor Beel has already been designated as a Ramsar site.

Fig. 1 : The Chandubi beel (A tectonic lake of Assam) (See colour view also)

Fig. 2 : The Deepor beel Ramsar site (See colour view also)

Meghalaya

Meghalaya has 98 wetlands of the size 2.25 ha and above, covering a total area of 22.21 sq km. There are four Endangered and 14 vulnerable species of fishes, eight species of freshwater turtles, two of which are Endangered, and two Near threatened species of birds. About forty seven species of wetland birds have been recorded. Three wetlands have been prioritized as potential Ramsar sites, namely, Birbah Lake, Umiam Lake and

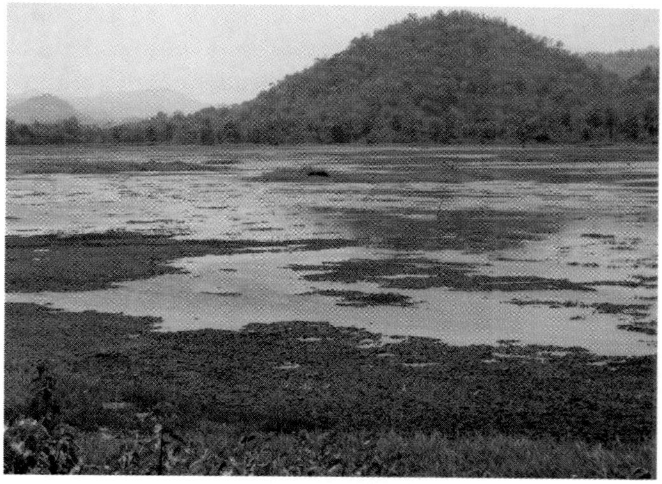

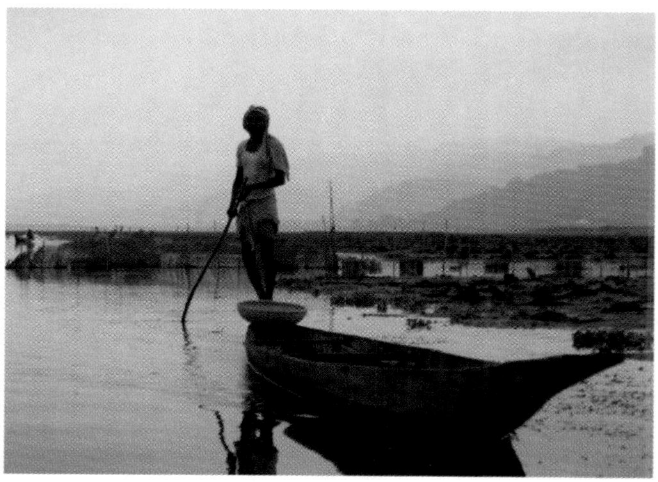

Tasek lake complex. Some other important wetlands on priority list are Chidampong and Bolgisim, Kopilli reservoir, Umtru lake and Kyrdemkullai lake.

Arunachal Pradesh

Arunachal Pradesh has 806 wetlands of the size 2.5 ha and above with an area of about 560 sq

Km). Arunachal has three Endangered and 14 Vulnerable fish species, two each of endangered and vulnerable and five near threatened species of freshwater turtles, and seven threatened and one near threatened species of birds. SACON study surveyed 18 wetlands and prioritized eight of these for designation as Ramsar sites. WWF-India National Consultation on high altitude wetlands, prioritized following wetlands from Tawang district: PT Tso, Oriangdukpu, Sangetsar and Paradise Lake.

Nagaland

Wetlands cover around 9.16 sq km area in this narrow strip of rugged mountainous State. Four Endangered and 18 vulnerable species of fishes, one each of an endangered and near threatened species of freshwater turtles and one threatened species of bird reported. Three wetlands namely, Shilloi lake, Doyang reservoir and Dzudu lake have been prioritized by SACON.

Manipur

Manipur is home to the Paradise flower-Shiroy lily (Lilium macklinae), which is not found elsewhere in the world and has the only habitat for Sangai, the Dancing deer in the famous wetland Keibul Lamjao national park. The State has 155 wetlands of 2.5 ha and above in size covering a total area of 529 sq km Manipur has seven endangered and 15 vulnerable species of fishes, one endangered , two vulnerable and three near threatened species of freshwater turtles, and three threatened and two near threatened species of birds. Six wetlands were prioritized out of which Loktak lake is already a Ramsar site. The others are Pumlen pat, Ikop pat, Kharung pat, Zeilad lake complex and Waithou lake.

Mizoram

The State does not have large wetlands, the total area of wetlands of the size 2.25 ha and above is only 1.5 sq km. Mizoram has four endangered and 11 Vulnerable species of fishes, two endangered, three near-threatened and one vulnerable species of turtles and one each of threatened and near-threatened species of birds. Three wetlands namely, Palak dil, Mampui dil and Tam dil have been prioritized for designation as wetlands of national and international importance.

Tripura

408 wetlands of 2.25 ha and above covering an area of 98.95 ha have been recorded. Tripura has two Endangered and 12 vulnerable species of fishes, and one near-threatened species of bird. Dumbur lake and Rudrasagar lake have been prioritized of which Rudrasagar is already designated a Ramsar site. Trishna wetlands, Sattar Mia's Haor, Batapura lake and College Tilla lake have been recommended for designation as community reserves.

Sikkim

Sikkim has 160 wetlands of the size 2.25 ha and above, covering 19.85 sq km. These support four endangered and fifteen vulnerable species of fishes, three species of freshwater turtles- one of which is Vulnerable and one near threatened, and two threatened species of birds. SACON study has prioritized four wetlands namely Thepley Tso, Khechiperi, Batangcho and Tsomgo lake.

THREATS

Threats to wetland ecosystems comprise the increasing biotic and abiotic pressures and perils.

Biotic

- Uncontrolled siltation and weed infestation
- Uncontrolled discharge of waste water, industrial effluents, surface run-off, etc. resulting in proliferation of aquatic weeds, which adversely affect the flora and fauna
- Tree felling for fuel wood and wood products causes soil loss affecting rainfall pattern, loss of various aquatic species due to water-level fluctuation
- Habitat destruction leading to loss of fish and decrease in number of migratory birds.

Abiotic

- Encroachment resulting in shrinkage of area.
- Anthropogenic pressures resulting in habitat destruction and loss of biodiversity.
- Uncontrolled dredging resulting in successional changes.
- Hydrological intervention resulting in loss of aquifers.

Agriculture

In the whole NE region agriculture system is predominantly traditional. The overall geographical land-to-man ratio for the NE region (0.81 Hectares/Person) is much higher than the National average (0.39 Hectares/Person). This could be attributed to the fact that the areas covered by hills in this region are more in comparison to the national average.

About 80 % of the farmers in the NE region belongs to the small (< 1.44 ha) and marginal (< 0.40 ha) category. Over the years the average land holding is reducing owing to the increase in population in a rapid pace. As a result many wetlands have been converted to agriculture sector.

Aquaculture and Fisheries

Assam has the largest surface water area among the NE states and produces the largest quantity of fishes in the region. In 1990, the land fish production in the NE region was about 90 thousand tones, or over 6 % of the production in the country. The total freshwater resource of the region is estimated at 15,728 km of river and canal; almost 89000 hectares of reservoirs; about 150000 hectares of streams and lakes and almost 85000 hectares of tanks and ponds which could extend better dividend for poor rural people.

Those nutritional resources available some of the fish bioresources have been depicted in the **Table 3**. The macro elements like Ca, P, Fe, I and Mg mixed with micro elements found in adequate quantity in these bioresources help other to maintain adequate nutritional condition for the body. Further, these bioresources are the good source of vitamins too.

Traditionally used Non-piscian Food Resources

Non-piscian fishery resources including snails, crabs, prawns shrimps, lobster, frog, etc. are regarded as important resourcs along with both fresh and marine fishes. The amount of fish produce in India could not meet the required demand corresponding to the number of population in our country. Freshwater molluscs are common in quiet water pool, ponds, lakes and in flowing water like perennial rivers, irrigation canals etc. (Anand and Prasad, 1919; Blandford *et al*., 1980). The farming of snails and crabs has not been systematically developed in northeast India. The various groups of tribals simply harvest them from the natural or wild water through traditional technique. Also, there is lack of well-organized market system for such products. An attempt has made to present the food values of some non-piscian food in the **Table 4.**

Food values of snails and crabs

Edible snails are eating by various tribal communities of Northeast India. The major edible snail species of this region are- *Pila globosa, Pila theobaldi, Pila virens, Bettamya bengelensis, Brotia costula.*

Proximate composition shows that edible varieties of snails have got high nutrient value. Considering the protein, carbohydrate and lipids, the high amount of polyunsaturated fat showed the high quality of snails and crab meat for nutrition. Protein of snails and crabs is 1st grade in nature showing its easy digestibility after confirming the pepsin and pepsin-trypsin digestion reaction. Liberation of essential amino acid showed that the protein is of high quality.

Table 3 : Essential macro and micro elements in Muscle of some small fishes (All values are in µg/kg dry weight) ND= Not Detected (Source: Chakraborty, 2008). Analyzed by Atomic Absorption Spectrophotometre

Sample	Co	Cu	Fe	Mn	Se	Zn	Cr
Anabas testudineus	16.83± 2.22	10.83± 4.61	33.03± 23.48	90.24± 10.47	5.35± 0.96	34.54± 4.11	ND
Ompok bimaculatus	10.78± 1.36	ND	160.54± 48.56	82.10± 33.47	1.54± 0.40	132.58± 11.52	ND
Puntius siphore	ND	ND	158.65± 31.35	83.15± 22.26	17.35± 7.73	152.77± 6.05	ND
Labeo pangusia	16.85± 26.91	ND	87.76± 6.84	41.62± 10.30	15.26± 2.84	56.46± 12.68	ND
Labeo bata	21.46± 4.56	7.61± 2.13	150.65± 23.27	34.36± 5.68	11.30± 3.46	29.98± 2.18	ND
Puntius ticto	26.51± 7.90	ND	137.52± 35.31	162.56± 29.23	5.39± 1.01	78.37± 8.35	ND
Clarius batrachus	34.55± 7.31	ND	160.00± 20.55	51.61± 7.71	9.71± 1.65	55.80± 3.43	ND
Labeo derio	22.49± 5.73	ND	122.24± 20.67	84.97± 39.83	ND	38.27± 4.34	ND
Cyprinus carpio	3.05± 0.90	2.65± 1.58	11.75± 1.88	4.17± 2.56	12.63± 8.62	52.56± 23.91	ND
Amblyphery-ngodon mola	6.64± 0.41	2.68± 0.38	111.64± 18.01	9.32± 0.53	ND	211.50± 18.43	ND
Botia derio	7.68± 0.42	3.70± 0.41	39.32± 5.34	12.27± 2.58	ND	174.97± 21.97	1.48± 0.84
Pseudeutropius atherinoides	9.79± 2.27	4.16± 1.35	88.30± 26.35	13.34± 2.94	11.89± 2.03	205.73± 27.86	2.01± 0.02
Puntius chola	8.58± 3.54	3.27± 0.81	80.41± 31.29	7.15± 2.17	ND	123.44± 14.49	0.60± 0.42
Ailia colia	10.35± 2.06	2.71± 0.42	28.96± 4.75	10.57± 1.69	ND	182.05± 3.56	2.18± 1.32
Danio aequipinatus	5.00± 1.23	4.85± 0.43	51.06± 6.51	7.88± 0.42	ND	215.17± 6.76	2.33± 1.31
Lepidocephalus guntia	10.45± 3.38	4.68± 0.41	91.02± 7.39	17.11± 2.19	ND	261.33± 27.22	1.71± 0.42
Salmostomata bacaila	15.34± 0.89	1.00± 0.01	32.42± 0.86	4.84± 1.75	ND	123.74± 1.5	0.80± 0.45
Mystus tyngara	ND	0.53± 0.42	3.51± 0.40	2.17± 0.79	13.88± 1.08	33.65± 5.80	ND

Table 4 : Certain biochemical parameters of different Prawn, *Macrobrachium spp.* (Source: Goswami, 2009)

Name of Species	Calcium	Crude Fibre	Crude Fat	Crude Protein	Dry Matters µg/g	Carotenoid (µg/g) Carapace	Carotenoid (µg/g) Muscles
M. dayanum	4.29± 0.88	1.54± 0.48	0.69± 0.03	51.52± 1.79	19.14± 2.09	19.36± 1.79	1.97± 0.34
M. assamensis	5.08± 1.5	1.61± 0.55	0.79± 0.03	53.87± 2.01	20.39± 2.72	22.16± 1.65	5.15± 0.85
M. birmanicum choprae	4.98± 0.95	1.42± 0.47	1.09± 0.12	63.87± 2.92	29.18± 2.07	65.07± 1.52	4.41± 0.62
M. malcolmsonii	4.29± 0.65	1.69± 0.48	0.95± 0.09	61.11± 2.12	25.31± 2.09	58.45± 2.03	3.98± 0.63
M. memoni	3.91± 0.74	1.65± 0.57	0.796± 0.04	55.13± 1.93	22.614± 2.49	42.27± 2.06	2.95± 0.56
M. lamarrie	3.23± 0.76	1.37± 0.47	0.68± 0.04	37.08± 1.39	16.11± 1.92	14.35± 1.23	4.61± 0.51
M. altifrons	3.45± 0.96	1.39± 0.47	0.71± 0.05	37.54± 0.95	17.32± 2.26	16.25± 1.93	2.93± 0.68
M. tiwari	3.52± 0.87	1.36± 0.45	0.68± 0.05	36.74± 0.63	17.59± 2.48	17.576± 1.51	3.21± 1.16

OVER EXPLOITATION OF BIORESOURCES: AN EXAMPLE ON THE TURTLE DIVERSITY

Chelonians are the oldest survived reptilian group in this planet. They have enormous role as one of the best scavengers in the earth, however, very timid and docile in its behavior, with a prolonged life span compared to other animals. They are the precious creatures of an aquatic body. The northeastern region is regarded as one of the major centre of turtle diversity (Das, 1990). However, the region has been poorly explored scientifically with regard to surveys, conservation and monitoring of turtle fauna. The turtle and tortoise diversity of northeast India is the highest in the country and 23 of 29 species are found in this region (Das, 1995; Baruah and Sharma, 2009). However, the lack of recent field information has created a gap in conservation and management of turtle species in the region.

IUCN/SSC Tortoise and Freshwater Turtle Specialist Group (IUCN, **2007),** in conjunction with the **Turtle Conservation Fund (TCF),** have released lists of the most endangered species of tortoises and freshwater turtles of the world. Out of 23 species found in northeast India, seven species namely **, *Chitra*** *indica, Manouria emys, Nilssonia nigricans, Batagur dhongoka, Cuora mouhotii, Indotestudo elongata and Pangshura sylhetensis* are listed as ***"Asia's Most Endangered Tortoises and Freshwater Turtle –2007"* (Table 6) .** The soft-shell turtle *Chitra indica* needs special attention as it is on the new list of the world's 25 most **Endangered Tortoises and Freshwater Turtles included in the year 2007.**

Table 5 : Proximate composition of edible snails and crab.

Biochemical parameters (gm %)	*Pila globosa* (gm %)	*Bellamya bengelensia* (gm %)	*Brotia costula* (gm %)	*L. corrionus* (gm %)	*L. wood mansoni* (gm %)	Serteriana *spinigera* (gm %)
Protein	47.50	38.06	28.25	35.37	17.5	19.0
Carbohydrate	6.13	17.25	8.48	15.37	3.2	3.5
Fat	0.84	1.30	1.25	2.80	0.22	0.25
Ash.	23.69	28.04	45.70	26.64	1.2	1.25
Moisture	16.48	11.35	11.82	16.32	79	76
Glycogen	15.2	7.5	10.2	10.5	ND	ND

(Source: Sharma et al., 1998**)**

Table 6 : Endangered turtle species found in different states of the region (2007 IUCN Red List for Testudines).

Sl No.	Family	Common Name	Taxon	AP	AS	NL	MN	MZ	ML	TR	IUCN Status
1.	*Geoemydidae*	Assam roofed turtle	*Pangshura sylhetensis*	+	+	–	+	+	+	–	EN
2.		Keeled box turtle	*Cuora mouhotii*	+	+	–	+	+	+	–	EN
3.		Three-striped roof turtle	*Batagur dhongoka*	+	+	–	–	–	–	–	EN
4.	Trionychidae	Black soft-shell turtle	*Nilssonia nigricans*	+	+	–	+	–	+	–	EW
5.		Narrow-headed softshell turtle	*Chitra indica*	+	+	–	–	–	–	–	EN
6.	Testudinidae	Elongated tortoise	*Indotestudo elongata*	–	+	–	–	+	+	–	EN
7.		Asian brown tortoise	*Manouria emys*	+	+	+	+	+	+	–	EN

AP= Arunachal Pradesh; AS= Assam; NL= Nagaland; MN= Manipur; MZ= Mizoram; ML= Meghalaya; TR= Tripura.EW = Extinct in the Wild; CR = Critically Endangered; EN = Endangered.

Habitat loss and over exploitation for meat is considered persistent threat to chelonians of the region. Religious superstition is another threat particularly for *Nilssonia hurum.* The killing of adult female freshwater turtles and incidental catch of breeding adults has created pressure on the freshwater turtle population. Immediate adoption of conservation measures is essential for the conservation of endangered turtles of this region. As a part of the conservation of genetic diversity, the following studies are relevant: (i) survey of

Fig. 3 : Endangered Turtles of northeast India: Family-Geoemydidae: *Pangshura sylhetensis (a), Batagur dhongoka (b), Cuora mouhotii(c);* Family-Trionychidae: *Nilssonia nigricans (d), Chitra indica (e);* Family-Testudinidae: *Manouria emys(f), Indutestudo elogata(g).* (See colour view also)

the different turtle species, (ii) reproductive cycle of endangered species, (iii) molecular phylogeny & DNA bar-coding to know the evolutionary lineages and identification, (iv) *in-situ* hatching (v) breeding in captivity, and (vi) generate awareness against killing and need of conservation through community participation.

Study on chelonians of northeast India is very scanty, however, a few field surveys has been carried out for conservation management of turtle species of northeast India by the Centre for Wildlife Research and Conservation Action (CWRCA), Department of Zoology, Gauhati University, Assam since January 2001. Under the Turtle Survival Alliance (freshwater turtle conservation task force of IUCN) India program, the centre instituted a modest "Chelonian Conservation Research Program" (CCRP) to initiate the endangered turtle conservation research and inculcate the resultant knowledge among the students and youths of the area. The work plan for the turtle conservation work in northeast India is being designed by the centre. This is to be noted here that most of the proposed field activities for endangered turtle conservation programme of the centre are outlined in the "**Conservation Action Plan for Endangered Turtles and Tortoises of India**" drafted in 2006 during Lucknow meeting supported by TSA and Conservation International and being attempted for most of the endangered turtle species. Comparative multi-gene phylogeography of fresh water turtles of the region is needed to determine appropriate conservation units as well as identifying priority areas for effective conservation management of freshwater turtles within the Indian subcontinent.

Opportunities of the Wetland for Livelihood

Among the most easily noticeable services offered by wetlands of the state is fishing, an activity that keeps alive scores of communities that possess no other livelihood opportunity. The fish they harvest help to fulfill their economic needs, but also provide nutritional needs of many who might live far away from the wetlands. In areas, where irrigation is yet to make a difference, it is water from the wetlands that is used for agriculture and sanitation. Wetlands since the time of their emergence have been the natural refuge of a vast range of organisms, some of which are in fact endemic to the region. They are home to a fantastic assembly of aquatic life, which include many that await scientific scrutiny. Habitat for a variety of water fowl, wetlands attract migratory birds in great numbers every winter. The peculiar physical condition of wetlands in Assam has proved to be conducive for the one-horned rhinoceros, sometimes referred to as a prehistoric species. Thus this precious bioresources has to be conserved.

Future challenges

Protection of wetlands may soon become vital to save endangered fish species and their ecosystems, particularly in the Northeast, where they form the main source of livelihood. Recent scientific studies suggest that wetlands serve the nutrient recycling process of elements such as nitrogen and phosphorus. They play another role of maintaining the carbon cycle in fine balance. Yet, nothing substantial has happened in conserving the invaluable landscapes. On the contrary, with expansion of human habitation and infrastructure, wetlands have become an easy prey, especially those which are close to towns and cities.

It is indeed surprising that the issue of conserving and managing the region's wetlands have been unable to gain attention even though their importance is so palpable. Apparently we have noticed the dots, but yet refrained from connecting those to make a strong and sincere pitch. For some time a few of us have been calling for a wetland policy involving all the probable stakeholders. A well laid out wetland policy could be a red light for all those involved in the process of destroying a living heritage of the region. It could establish precautionary principles, which act as psychological deterrence and thwart at least some illegal activities. Perhaps more significantly, a clear-cut policy would guide the efforts of many departments, where even the most well intentioned officers are unaware how their efforts could cause irreparable damage to the ponds, lakes, and swampy areas.

But it should be a policy that is pragmatic and can be implemented by a single competent authority. An authority that would comprise qualified people, who can be held responsible for any gain or loss. Multiple agencies leave the scope for blame games among officials that benefits none, a practice not infrequent in the region s public affairs. Here it might be further beneficial to bring in local communities and academia to monitor and review the work, a step to ensure transparency and probity into an area that ultimately affects their lives.

LIVELIHOODS AND FOOD SECURITY PROGRAM

Livelihood and food security must go hand in hand for sustainable development. And the capacity building is a major tool without which no conservation activity is possible. We need to have good infrastructure, trained people, and case studies to teach values and functions of wetlands in an integrated and multi-disciplinary manner. At the same time without addressing the problem of livelihood the conservation action is irrelevant. Thus Impart training on various components of wetland conservation including wise use, catchment area treatment, weed control, hydrological aspects, research methodology, preparation of management action plans and community participation are the much needed component. Recent studies show that wetlands serve the nutrient recycling process.

MEETING THE CHALLENGES AHEAD

Need for Technology transfer

Appropriate technologies have to be developed for and adopted in rural areas for:

a. income/employment generation

b. reduction of drudgery

c. upgrading habitat and social infrastructure

d. otherwise improving quality of life

In the present context, dominant requirement is for technologies for income/employment generation in the non-farm sector, i.e., for rural industrialization.

I. Value-addition to rural resources

II. Utilization and up-gradation of local skills

III. Sustainable farm and off-farm livelihoods

CONCLUSIONS

Bio-resources and their sustainable use for socio-economic growth of the N.E. Region is the present need of the hour. Government/relevant authorities must take up steps to generate technological packages leading to the development of bio-resources products and processes applicable to enhance the productivity in agriculture, horticulture and forestry, aquaculture etc. Being biodiversity hot spots the region has enough resources, however, demand full characterization and effective utilization.

ACKNOWLEDGEMENTS

The authors are thankful to DBT, Govt of India; University Grant Commision; Dr. Kartik Shanker, Centre for Ecological Sciences , IISc, Bangalore; Dr. Shailendra Singh, Scientist, TSA/Madras Crocodile Bank Trust and Dr. Peter Praschag, Am Katzelbach 98; 84 Graz, Austria.

REFERENCES

Ao, S., Dey, S.C. and Sarmah, S.K. 2008. Fish and Fisheries of Nagaland. Govt. of Nagaland publication.

Anand, N. and Prasad, B. 1919. Fauna of certain small streams in the Bombay Presidency. Rec. Ind. Mus. **16:** 139

Baruah, C. and Sharma, D. K. 2009. Software for exploring the turtle and tortoise diversity of northeast India. *Turtle and Tortoise Newsletter*, **13:** 9-10.

Blandford, W.T., Godwin, T.S. and Austin, H.H. 1908. Fauna of British India. Series I, Cited by Society Publication.

Chakraborty, R. 2008. Surface microstructureal features and elemental status of some common and traditional edibles of northeast India with reference to human health, Ph.D Thesis, Gauhati University

Das, I. 1990. Distributional records for Chelonians from North Eastern India. Journal of Bombay Natural History Society, 87, 91-97.

Das, I. 1995. Turtles and Tortoises of India. Oxford University Press, pp 179.

Ghosh, A. K. and Tiwari, K. K. 1984. Faunal Resources of Northeast India. In Resource Potential of Northeast India, Vol. II (Living Resources), Tripathi, R. S. (Ed.). Meghalaya Science Society, Shillong, Meghalaya.

Goswami, A.K. 2009. A study on prawns of Assam and their fisheries, Ph. D. Thesis, Gauhati University

IUCN. 2007. Turtles in Trouble: Asia's Most Endangered Tortoises and Freshwater Turtle. IUCN SSC.

Per Pinstrup-Anderson and Fuzhi Cheng. September 2007 . Still Hungry: One-eighth of the world's people do not have enough to eat, *cover story , Scientific American.*

Sharma, D.K., Goswami, U.C. and Dutta, A. 1998. Studies on non-piscian resources such as crabs & snails as food by the plain tribes of Assam, AIRTSC (STE) project Report.

Shine, C. and C. de Klemm. 1999. Wetlands, water and the law- using law to advance wetland conservation and wise use. IUCN Environmental Policy and Law Paper, No. 38. IUCN – The World Conservation Union.

Bioresources for Food Security and Rural Livelihood, Pages 165–175
Edited by : G.K. Kulkarni and B.N. Pandey

ROLE OF FISHERY RESOURCES FOR FOOD SECURITY AND RURAL LIVELIHOOD IN 21ST CENTURY WITH SPECIAL REFERENCE TO RAJASTHAN

Prahlad Dube

ABSTRACT

Rajasthan possesses a large area of inland water bodies which offer potential for development of both intensive and extensive system of culture based fisheries. From the available fresh water resources in the state. 3.30 lakh ha of inland water sheets in the form of reservoir (1.2 lakh ha.) tanks, and ponds (1.8 lakh ha.) and rivers (0.30 lakh ha.) have been identified for capture cum culture fishery management. Besides, there exists 0.04 lakh ha brackish water bodies and perennial flowing system, 214 Km. Indira Gandhi Feeder Canal and about 500 Km under I.G.N.P. in North West Rajasthan.

The beginning of the new millennium century the next millennium looks not only to secure enough food for the existing population but also to provide a guarantee to sustained food availability for future generation. Though scientific aquaculture has substituted the traditional practices in a big way, it cannot be fully relied upon to meet the demands of future generation. This is due to the fact that continued over fishing, heavy intensification of farms in localized areas, degradation of aquatic environment etc. will put insurmountable strain on the fishery sector which is capable enough for a sustainable food supply.

Understanding the livelihood systems of the poor is crucial to effective poverty reduction.

Livelihoods of the poor can never be understood in any one-track logic - be it economic, social, technical, cultural or political. In the present paper, the changing face of rural livelihood, new challenges and probable solutions in the context of Rajasthan are discussed.

Keywords : Fishery resources, rural livelihood, Rajasthan, food security, climate change.

INTRODUCTION

The changing faces of rural livelihoods in India : India is witnessing a series of changes since early nineties. Recently, the sensex crossed 20,000 points and simultaneously

India ranked in 94th out of 118 countries in the Global Hunger Index — behind Ethiopia. Unprecedented numbers of farmer suicides, big corporate houses entering into retail business, land allocation for Special Economic Zones (SEZ), boom in information technology and IT enabled services, zero growth rates in employment, are among many such events that needs introspection.

Understanding the livelihood systems of the poor is crucial to effective poverty reduction.

Livelihoods of the poor can never be understood in any one-track logic - be it economic, social, technical, cultural or political. The livelihood systems are made up of very diverse elements which - taken together - constitute the physical, economic, social and cultural universe wherein the families live (Hogger, 2006). Thus, the livelihood system is more than just a set of physico-economic preconditions for continued existence. It also encompasses psycho-social dimensions of experience of living. The livelihoods approach puts households of the poor as its central focus. It takes holistic consideration of things that the poor might be vulnerable to; assets and resources that help them thrive and survive, policies and institutions that impact their livelihoods, how the poor respond to threats and opportunities and what sort of outcomes the poor aspire to. Agriculture and allied activities support livelihoods of nearly 70 percent of India's rural population.

In recent years, land based livelihoods of small and marginal farmers are increasingly becoming unsustainable, since their land has not been able to support the family's food requirements and fodder for their cattle. As a result, rural households are forced to look at alternative means for supplementing their livelihoods.

Globalisation and Vulnerability

Rural livelihoods are affected by the process of Globalisation. Special Economic Zones (SEZ) are created across India to promotion exports. The Commerce and Industries Minister claims that exports will ultimately grow five times, GDP will increase by 2 percent, and 30 lakh jobs will be generated by SEZs across India. The government also claims that SEZs will attract global manufacturing through Foreign Direct Investment (FDI), enable the transfer of modern technology and create incentives for infrastructure.

To date, the central government has approves 237 SEZs in 19 states (occupying 86,107 hectares) out of which 63 of these SEZs have already been notified. Around 23 SEZs are operational, 18 in the IT sector. Total amount of land to be acquired across India is estimated to be around 150,000 hectares. This land – predominantly agricultural and typically multi-cropped - is capable of producing close to 1 m tonnes of food grains. If SEZs are seen to be successful in the future and more cultivated land is acquired, they will endanger the country's food security.

Rural Livelihoods and Climate Change : Threats and Opportunities

Climate change is increasingly important as a policy issue for both the international community and for India. It is likely to impact on rural India in two ways :

- Changes in temperature, rainfall regime, more variable weather and increased frequency and severity of natural disasters will affect the livelihood strategies of rural people;
- The market for carbon offsets may create new opportunities for resource flows to poor rural communities.

Climate change poses a new challenge to rural livelihoods in India. Programmes need to identify the increased risks resulting from climate change and assess their significance for design and implementation. Our rural livelihoods programmes emphasize sustainable management of natural resources and diversification of livelihood opportunities and are operating in areas where seasonal and inter-annual variability are key determinants of production and income.

International recognition of climate change as a global problem has resulted in the creation of a market for carbon offsets, through which carbon credits can be created, either by avoided emissions of greenhouse gases (GHGs) or by sequestering carbon in vegetation or soils. To date almost all of the benefits of the carbon market have been captured by medium and large scale industry and very few have flowed to the poor. However as custodians of natural resources that can be used to either reduce or increase carbon dioxide emissions, the rural poor could potentially earn significant revenue from the sale of carbon credits provided the current barriers to the carbon market can be addressed.

There is urgent need for designing a Climate Change Innovation Programme with the Ministry of Environment and Forests that will focus on helping the poor adapt to climate change and benefit from any opportunities offered by the carbon market.

SMALL-SCALE FISHERIES : ASSESSING THEIR CONTRIBUTION TO RURAL LIVELIHOODS

The total fish production in the country has grown nearly six- fold over the last three decades, from 0.76 MT in 1950-51 to 6.1 MT in 2002-2003. The share of fisheries in total agricultural GDP has increased impressively during the last five years, from 0.84% in 1950-1951 to 4.19% in 1999-2000 (Anjani Kumar, 2003). Fisheries have not so good reputation nowadays in the international community. FAO recently estimated that worldwide 18 percent of the fish stocks or species groups are overexploited, while 10 percent have become significantly depleted, or are recovering from depletion (FAO, 2002). For the World Environment Day (5 June 2004), the United Nations Environment Programme (UNEP) chose the slogan "Wanted! Seas and Oceans, Dead or Alive?" and recently the World Bank posted on its front webpage an article by Ian Johnson 2 entitled "One World, One Ocean. It's time to save it" (World Bank, 2004). The responsibility of fisheries (both large- and small-scale) in this situation is undisputable and it is likely that it will affect and influence the way development agencies and governments will consider initiatives related to the support of fisheries in the near future.

"Small-scale fisheries can be broadly characterized as a dynamic and evolving sector employing labour intensive harvesting, processing and distribution technologies to exploit marine and inland water fishery resources. The activities of this sub-sector, conducted full-time or part-time, or just seasonally, are often targeted on supplying fish and fishery products to local and domestic markets, and for subsistence consumption. Export oriented production, however, has increased in many small-scale fisheries during the last one to two decades because of greater market integration and globalisation. While typically men are engaged in fishing and women in fish processing and marketing, women are also known to engage in near shore harvesting activities and men are known to engage in fish marketing and distribution. Other ancillary activities such as net-making, boat-building, engine repair and maintenance, etc. can provide additional fishery-related employment and income opportunities in marine and inland fishing communities. Small-scale fisheries operate at widely differing organizational levels ranging from self-employed single operators through informal micro-enterprises to formal sector businesses. This sub-sector, therefore, is not homogenous within and across countries and regions and attention to this fact is warranted when formulating strategies and policies for enhancing its contribution to food security and poverty alleviation." (FAO, 2004)

Food Security

Definition

Food security was defined during the 1996 World Food Summit as:

> *"a condition when all people, at all times, have physical and economic access to sufficient, safe and nutritious food to meet their dietary needs and food preferences for an active and healthy life".*

Contribution of small-scale fisheries to environmental conservation

The relation between small-scale fisheries and environment revolves around the question on whether small-scale fisheries contribute to environmental conservation or, on the contrary, participate to the current general decline affecting fisheries resource in the World. On this question, it is striking to notice that totally antagonistic clichés co-exist in the literature. On one hand a large number of academics and practitioners are claiming that small-scale fisheries are more "environment-friendly" than larger-scale fisheries. If one refers to indicators such as these used by Tompson in his Table (e.g. the fish caught per tonne of fuel consumed or the estimated by-catch rate), it seems indeed justified to claim that small-scale fisheries are less detrimental than industrial fisheries. This fact probably explains why the perception about the "eco-friendly" dimension of small-scale fisheries is so frequently re-asserted in the literature. This perception is further reinforced by the equally widely accepted view that formal and/or informal community-based management approaches such as those observed in many small-scale fisheries in developing countries also contribute to more effective fisheries management regulation enforcement, thus strengthening the capacity of the stakeholders to reduce the risk of overexploitation of the resource.

Yet at the same time it is widely recognized that small-scale fishers can, and do, make use of detrimental fishing techniques (e.g. dynamite, reef bleaching, etc.) to increase, or sometimes simply maintain, their capture levels in a general context of increasing fishing pressure. From a different –but related- perspective, it should also be recognized that over-capacity is becoming a concerning issue in an increasing number of small-scale fisheries around the world. Those facts, which implicitly highlight the possibility of resource overexploitation by small-scale fisheries, clearly go against the position promoted by people who emphasize the eco-friendly dimension of small-scale fisheries. The conclusion may be that no generalization is possible regarding the contribution of small-scale fisheries to environmental issues, and therefore that sweeping generalizations which assert the existence of poverty-environment nexus or on the contrary which advocate the environment-friendly dimension of small-scale fisheries are inappropriate at such a level of generalization and may in fact be misleading for appropriate supportive interventions at local levels. In fact it is probably correct to say that to every single case-study which tends to suggest/ demonstrate that small-scale fisheries are less detrimental than larger-scale fisheries, a counter-example could be brought up that contradicts these conclusions and shows that small-scale fisheries can also lead to resource-base degradation and overexploitation.

The key factor is therefore not the scale of the fishery (small versus large; artisanal versus industrial) but the effectiveness of the management setting to ensure the sustainability of the fisheries. This effectiveness has more to do with the institutional arrangements of the fisheries and the collective capacity of the stakeholders to control individual fishing effort and investment, rather

FISHERY RESOURCES OF RAJASTHAN

Rajasthan possesses a large area of inland water bodies which offer potential for development of both intensive and extensive system of culture based fisheries. From the available fresh water resources in the state. 3.30 lakh ha of inland water sheets in the form of reservoir (1.2 lakh ha.) tanks, and ponds (1.8 lakh ha.) and rivers (0.30 lakh ha.) have been identified for capture cum culture fishery management. Besides, there exists 0.04 lakh ha brackish water bodies and perennial flowing system, 214 KM. Indira Gandhi Feeder Canal and about 500 KM under I.G.N.P. in north west Rajasthan.

Classification of Inland Fisheries Resources

The river system of the country comprises 14 major rivers (catchments >20000 km2), 44 medium rivers (catchments 2000 to 20000 km2) and innumerable small rivers and desert streams. Different river systems of the country, having a combined length of 29 000 km, provide one of the richest fish genetic resources in the world (Government of India, 2001).

Reservoirs constitute the single largest inland fishery resource, both in terms of resource size and productive potential. A detailed study made by FAO in 1995 has estimated a total of 19 370 reservoirs in the boundary with total area of 3.15 million ha. As per this study, the distribution of small medium and large reservoirs is as follows:

Table 1 : The distribution of small medium and large reservoirs in Rajasthan.

	Small	Medium	Large	Total
Number	19134	19134	56	19370
Area (ha)	1485557	527541	1140268	3153366

Inland water resources of the country are classified as rivers and canals; reservoirs; tanks and ponds; beels, oxbow lakes, and brackish water. The state-wise distribution of water resources may differ both in quality and quantity.

Fisheries, though an important source of livelihood in Rajasthan, does not find a place in the water policy or management decisions. The study shows that water management if integrated with other sector needs such as fisheries, can benefit a number of poorer groups and at the same time increase water use efficiency.

The study showed that fisheries in Rajasthan supported the livelihoods of a significant section of the poor and small scale fishermen. Despite the dependency, the Water Resources Department does not consider fisheries as one of the priorities while allocating water. The State Water Policy does mention that water should be ensured for various sectors and livelihoods, but does not specify how it should be implemented. The development initiatives by the Fisheries Department are not integrated with the activities of the water resources department or the Agriculture Department. Overall, there appears to be declining trend in fisheries production. There is also a change in the institutional mechanisms that is not in the interests of the small scale fishermen. The government should reconsider to issue fishing rights to small scale and traditional fishermen rather than open tenders, customize training programs and implement government welfare schemes to benefit the poorest amongst the small scale fishers.

Current status of Fisheries in Rajasthan

Fishing communities were spread across the Rajasthan, concentrated around major water bodies. The number of households within these communities varied from 4 to 100 families. Depending on the location, fish catch and season; these families migrated as labourers to different parts of the state.

A majority of the fishermen were small scale and small scale fishing was mostly a household enterprise with the women and children often sharing the burden. Women were engaged mostly in sorting and selling fish. The role of the women largely depended on the socio-economic conditions of the households.

Overall, the conditions and quality of life for small scale fishermen in general and women in particular were poor across different fishing groups. This included long working days, poor wages as compared to men and in addition the burden of household maintenance. Most of the fishermen around major water bodies are large scale, who operated through fishing camps. They hired small fishers and labourers who migrated from other regions to work at the camps. Annual migrations were a common pattern in the fisheries activities in Rajasthan. Often this caused hardships to families and their social life, as it involved living in temporary settlements where facilities are poor. Migration to other places was more a constraint than advantage to the fishermen.

Decline in Fisheries

The study revealed that fisheries in the Rajasthan have declined by almost half from 2000 to 2005.. This can be attributed to: inadequate and poor quality fingerlings, lack of integrated efforts by the Department of Fisheries, illegal fishing, pollution from industries and lack of integrated water resources management. During the study it was observed that efforts to develop fisheries were not integrated with activities of other sectors, for example water or irrigation. As a result, decisions made by the Water Resources Department did not take into consideration the needs of the fisheries sector while almost half from 2000 to 2005. This is a major drawback for fisheries in the Rajasthan as irrigation and drinking water gets the priority. The government should take note of the fact that fisheries is a non-consumptive use and efforts to promote fisheries in Rajasthan are inadequate and much more is needed to be done.

Constraints In Increasing Production

- Uncertain and irregular monsoon.
- Draining/ pumping out of maximum stored water for Irrigation and Drinking purposes.
- Shortage of quality fish seed.
- Lack of traditional fishermen community.
- Lack of awareness among rural masses.
- Lack of technical know how in rural sector for fish culture.

PROBLEMS FACED BY FISH FARMERS

- Poaching
- Water availability to maintain a minimum level of water in the ponds for the fish to survive during the lean months
- Capital*
- Poisoning of ponds
- Market access

- Labour
- Diseases to fish-stock
- Resistance from villagers.

*Banks do not Lend to Fish Farmers Due to the Following Reasons

1. High risk nature of the fishing business which can wipe out entire investments during natural calamities like floods and droughts or due to poaching.
2. Insecure property rights in case of leased out ponds due to the multi-use and multi-ownership of ponds. Also lease periods of less than 3 years do not encourage banks to offer loans.
3. No collateral on basis of which poor fish farmers could obtain loans.
4. Dispersed nature of the economy which leads to high transaction costs for banks to lend to the fish farmers.
5. Cumbersome banking procedures which do not encourage fish farmers to take loans.

The Main Objects of the Fisheries and Aquaculture Development

Programme identified for Eleventh Five Year Plan in Rajasthan are :

- Maximization of Biological Productivity with optimum utilization of available resources.
- Enhancement of the quality fish seed production.
- Generation of employment opportunities.
- Production of all commercially important fish species available in Rajasthan.
- Improvement in socio-economical condition of Fishermen.
- Conservation of aquatic biodiversity and ecosystem.
- Intensive training and demonstration for propagation of advance technologies in aquaculture practices.
- Establishment of fish marketing yard, cold chain and retail outlets.
- Promotion of nutritional fish feed production units.
- Strengthening of the fisheries research and education net work.
- Strengthening the data base information net work.

Development of Inland Fisheries and Aquaculture

Under Centrally Sponsored Scheme, 15 Fish Farmer's Development Agencies (F.F.D.A.) are functional in the state with the object:

(a) Increase of Fish Production through intensive fish culture, particularly in rural area for utilization of small water bodies.

(b) Generation of rural employment potential.

(c) Additional source of income for local bodies.

During the annual plan 2009-10 about 1500 fish farmers are proposed to be trained in aquaculture techniques and an additional water area of about 1000 ha is to be brought under aquaculture. An amount of Rs. 5.65 lacs is proposed as spillover liability for following activities (Table 2).

Table 2 : Proposed amount as spillover liability for various activities.

S.No	Item	Amount (Rs. In lacs)	Remarks
1.	Grant in aid to FFDA for Training and subsidy	4.00	25% CSS matching
2.	Subsidy for Nets & Boats	1.50	25% CSS matching
3.	Subsidy for Pan culture	0.15	25% CSS matching
	Total	5.65	

Fisheries Training and Extension

For updating the knowledge of advance aquaculture techniques among departmental personnel, creating awareness regarding adoption of advance practices of aquaculture among fish farmers, for fish culture and capture activity. An amount of Rs. 0.50 lacs is being proposed for Annual Plan 2009-10, for spillover liability for following activities (Table 3).

Table 3 : Proposed amount of Rs. 0.50 lacs for annual plan 2009-2010 for spillover liability for various activities.

S.No.	Particulars	Amount (Rs. In lacs)	Remarks
1.	Training to progressive farmers	0.30	CSS matching 20%
2.	National/State level workshop	0.20	CSS matching 20%

Group Accident Insurance Scheme for Active Fishermen

Active fisher folk which are engaged with fisheries activities are proposed to be insured through this scheme. For this purpose a premium of Rs. 14/- per fisher folk has to be paid to FISHCOPPED, New Delhi. The 50% of the premium is paid by Government of India directly to FISHCOPPED, and 50% has to be contributed by state government. A provision of Rs. 0.50 lacs has been proposed for the year 2009-10 as committed liabilities. About 6000 fishers will be benefited under this scheme.

National Scheme of Welfare of Fishermen

During Tenth Five Year Plan, the National Cooperative Development Corporation (NCDC) proposed new schemes

(i) Beel Fishery Development ;

(ii) Reservoir Fishery Development ;

(iii) Motorisation of Traditional Craft ;

(iv) Intermediate Fishing Craft ;

(v) Fisheries Training / Extention;

(vi) Fish-finder/positioning Equipment ; and

(vii) Introduction of new technology (total cost of Rs. 530 crores.).

Under central sponsored scheme (50:50) " National scheme of Welfare for Fishermen" an amount of Rs. 13.35 lacs is proposed as state share for the year 2009-10 as state marketing share for following activities.

1. **Development of Model Fishermen Village :** 8.38 For development of model fishermen village in Tribal sub plan area of the state a provision of Rs. 10.00 lacs is proposed as state share during the year 2009-10, for construction of fishermen houses, hand pumps and community hall.
2. **Saving cum Relief :** 8.39 Under this scheme financial assistance is provided to fishermen during lean fishing season. Fishermen has to contribute Rs. 50.00 per month for nine month of fishing period the amount so collected along with the equal contribution of government (Central and State Government sharing 50:50 basis) will be paid to the beneficiaries in three equal installment of Rs. 300.00 during three months of close season. 8.40 An amount of Rs. 3.35 lacs is proposed as matching share for the year 2009-10 for this purpose.

CONCLUSIONS

From the above discussion it is clear that Globalization and climate change has posed new challenges before the rural population of India. Rajasthan being a state provided with relatively more adverse conditions to face, has to deal these changing situations more efficiently. Rural livelihood in Rajasthan is changing its face and agriculture and livestock keeping only proves to be insufficient. This creates food problem: insufficiency and malnutrition (quantity and quality of food). Fishery sector offers some solutions to this problem. Where management has been ineffective the greater efficiency of fishing methods and aquaculture production has sometimes led to over fishing and environmental degradation. This points to the need to develop more effective fisheries management frame works, together with safer and more environment-friendly methods of production.

REFERENCES

Anjani, K., Joshi, P.K. and Pratap, S.B. (2003), Fisheries Sector in India: An Overview of Performance, Policies and Programmes. In: Anjani, K., Pradeep, K.K. & Joshi, P.K.

(Eds.), A Profile of People, Technologies and Policies in Fisheries Sector in India. pp. 1-16.

Anonymous 1996a. *Handbook on Fisheries Statistics.* Ministry of Agriculture (Department of Agriculture and Co-operation, Fisheries Division), Government of India. Economic Survey, (2000-2001).

FAO. (2005), Aquaculture production, 2003. Yearbook of Fishery Statistics - Vol. 96/2. Food and Agriculture organization of the United Nations, Rome, Italy. (Also available at Fishstat Plus, Version 2.3)

Government of India 2000-2001, Economic Survey, Ministry of Finance, Government of India.

Hogger, Ruedi (2004). Understanding Livelihood Systems as Complex Wholes In "*In Search of Sustainable Livelihoods*: *Managing Resources and Change*", Ruedi Baumgartnet and Ruedi Hogger (Editors), pp. 94-125, Sage Publications, New Delhi.

Indu, Rajnarayan and Ghosh, Santanu, (2005), Inland Fishery in a Traditionally Vegetarian State: A Gujarat Story, IWMI-TATA Programme, Vallabh Vidyanagar, (Unpublished)

Katiha, Pradeep K., (2000) Freshwater Aquaculture in India: Status, Potential and Constraints. *In* Krishnan, M. and Birthal, P.S. (eds) *Aquacultural Development of India: Problems and Prospects*.

Report of the Working Group on Fisheries for the Tenth Five Year Plan, Government of India, June 2001 pp. 45.

Sinha, M and Pradeep K Katiha (2002), "Management of Inland Fisheries Resources under Different Property Regimes" in Dinesh Marothia (ed.) *Institutionalising Common Pool Resources,* Concept Publishing Company, New Delhi.

Bioresources for Food Security and Rural Livelihood, Pages 177–187
Edited by : G.K. Kulkarni and B.N. Pandey

PRAWN FISHERY RESOURCES OF INDIA FOR FOOD SECURITY AS WELL AS FOR RURAL EMPLOYMENT

K.V. Jayachandran and B. Indira

ABSTRACT

Prawns comprise about 2500 species throughout the world. They belong to highly confusing taxonomic groups. They are commercially important organisms fetching huge foreign exchange to the country. Prawns as a group comes under a few taxa -Penaeidae, Sergestidae, Solenoceridae, Pandalidae, Palaemonidae and so on. Of the above except the last one all others are marine prawns whereas the last one has marine and freshwater inhabitants. The present paper discusses on the prawns of the families, . Penaeidae and Palaemonidae of India.

The resources of the family Penaeidae consist of 50 valid species and that of subfamily Palaemoninae under Palaemonidae includes 60 species under the genus *Macrobrachium.* There are more than 25 species under the former family is commercially important whereas the latter with 20 species. Judicious utilization of these resources is of prime importance. The paper describes about the resources of India, present production trends, ways of utilization at present for the sustainable utilization of the resources. Utilization of lesser sized prawn species for value addition and also ornamental ones for aquarium purposes are the thrust areas for rural employment generation. The paper presents full details of the above facts.

Keyword : Prawn resources, Penaeidae, Palaemonidae, Production & Utilization trends.

INTRODUCTION

Prawns and shrimps comprise about 2500 species throughout the world They belong to complex taxonomic groups. Extensive studies on the taxonomy/biodiversity of penaeid prawns of the world have been carried out by Perez Farfante & Kensley (1997- for review) and that of palaemonid prawns by Jayachandran (2001 - for review) Our knowledge on Indian penaeid prawns are mainly from the fundamental research work of Alcock, Rai, Panikkar, Chopra, Nataraj Kubo, Barnard, Srivastava, Chacko, Hall, Menon, Da11, Kunju, Ramamurthy, George, George *et al,* George & Rao. George & Muthu, Thomas, Muthu, Silas & Muthu and these works have been reveiwed by Kurien & Sebastian (1986),

Suseelan (2003) and Jayachandran (2008). A total of 26 genera and 225 species have been reported under the family so far. Around 90 species are reported to be commercially important. Details are given in the Table 1.

Table 1 : Details of genera and species under the family Penaeidae Rafmesque, 1815

Sl. No	Genera	Number of of reported	Numberspedes of species importance
1.	*Artemesia* Bate, 1888	1	1
2.	*Atypopenaeus* Alcock, 1905	5	2
3.	*Farfanupenaeus* Burukovsky. 1997	8	8
4.	*F ermeropenaeus* Perez Farfante. 1969	5	4
5.	*Funchalia* 1. Y. Johnson, 1868	5	-
6.	*Heteropenaeus* De Man, 1896	1	-
7.	*Litopenaeus* Perez Farfante. 1969	5	5
8.	*Macropetasma* Stebbing, 1914	1	1
9.	*Marsupenaeus* Tirmizi, 1971	1	1
10.	*Megokris* Perez Farfante and Kensley. 1997	4	2
11.	*Melicertus* Rafmesqu-Schmaltz, 1814	7	6
12.	*Metapenaeopsis* Bouvier. 1905	73	17
13.	*Metapenaeus* Wood-Mason, 1891	26	23
14.	*Miyadiella* Kubo, 1949	2	-
15.	*Parapenaeopsis* Alcock. 1901	21	15
16.	*Parapenaeus* Smith, 1885	13	7
17.	*Pelagopenaeus* Perez Farfante and Kensley. 1997	1	-
18.	*Penaeopsis* Bate. 1881	5	2
19.	*Penaeus* Fabricius. 1798	3	3
20.	*Protrachypene* Burkenroad, 1934	1	1
21.	*Rimapenaeus* Perez Farfante and Kensley. 1997	6	-
22.	*Tanypenaeus* Perez Farfante. 1972	1	-
23.	*Trachypenaeopsis* Burkenroad, 1934	3	1
24.	*Trachypenaeus* Alcock. 1901	12	12
25.	*Trachysalambria* Burkenroad, 1934	9	1
26.	*Xiphopenaeus* Smith, 1869	2	2

From the Indian region, 50 species have been reported so far and are listed below :

1. *Atypopenaeus sunodactylus* (Stimpson, 1860)
2. *Fenneropenaeus indicus* (H. Milne Edward~ 1837) 3. *Fenneropenaeus merguiensis* (De Man. 1888)

4. *Fenneropenaeus penicillatus* (Alcock, 1905)
5. *Fermeropenaeus silasi* Muthu & Motoh, 1979
6. *Marsupenaeus japonicus* (Bate. 1888)
7. *Megokris granulosus* (Haswell. 1879)
8. *Megokris pescadoreensis* (Schmitt. 1931)
9. *Megokris sedili* (Hall, 1961)
10. *Melicertus canaliculatus* (Olivier. 1811)
11. *Melicertus latisulcatus* (Kishinouye. 1896)
12. *Metapenaeopsis andamanensis* (Wood-mason, 1891) 13. *Metepenaeopsis barbata* (De Haan. 1844)
14. *Metapenaeopsis borradaili* (DeMan, 1911)
15. *Metapenaeopsis coniger(Wood-Mason,* 1891)
16. *Metapenaeopsis gallensis* «Pearson, 1905)
17. *Metapenaeopsis hilarula* (De Man. 1911)
18. *Metapenaeopsis mogiensis* (Rathbun, 1902)
19. *Metapenaeopsis philipp;;* (Bate. 1881)
20. *Metapenaeopsis stridulans* (Alcock. 1905)
21. *Metapenaeus affinis* (H. Milne Edwards. 1837)
22. *Metapenaeus alcocki* George & Rao. 1968
23. *Metapenaeus brevicomis* (H. Milne Edwards, 1837) 24. *Metapenaeus dobson;* (Miers, 1878)
25. *Metapenaeus ensis* (De Haan. 1844)
26. *Metapenaeus krishnatrii* Silas & Muthu, 1976
27. *Metapenaeus kutchensis* George. George & Rao. 1963 28. *Metapenaeus lysianassa* (De Man. 1888)
29. *Metapenaeus monoceros* (Fabricius. 1798)
30. *Metapenaeus moyebi* (Kishinouye. 1896)
31. *Metapenaeus stebbingi* Nobili. 1904
32. *Parapenaeopsis acclivirostris* Alcock. 1905
33. *Parapenaeopsis comuta* (Kishinouye. 1900)
34. *Parapenaeopsis coromandelica* Alcock, 1906
35. *Parapenaeopsis hardwickii* (Miers. 1878)
36. *Parapenaeopsis indica* Muthu, 1972
37. *Parapenaeopsis moxillipedo* Alcock. 1905
38. *Parapenaeopsis nana* Alcock. 1905

39. *Parapenaeopsis sculptilis* (Heller. 1862)
40. *Parapenaeopsis styli/era* (H. Milne Edwards. 1837) 41. *Parapenaeopsis tenella* (Bate. 1888)
42. *Parapenaeopsis uncta* Alcock. 1905
43. *Parapenaeusfissurus* (Bate. 1881)
44. *Parapenaeus investigatoris* Alcock & Anderson. 1899 45. *Parapenaeus longipes* Alcock, 1905
46. *Penaeopsisjerryi* Perez Farfante, 1979
47. *Penaeus monodon* Fabricius. 1798
48. *Penaeus semisulcatus* De Haan, 1844
49. *Trachypenaeopsis minicoyensis* Thomas, 1972
50. *Trachysalambria curvirostris* (Stimpson. 1860)

Our knowledge on the Indian palaemonid decapod crustaceans are mainly from the research work of Anantha Raman *et al.* Bate, atopra & Tiwari, Dutt & Ravindranath, Holthuis. Jalihal *et al.* Jayachandran, Jayachandran & Joseph, Jayachandran & Raji. Jayachandran *et al.* Kemp. Nobili. Sankolli & Shenoy. Tiwari, Tiwari & Pillai and reviewed by Jayachandran (2001.2009). These studies have recorded 75 species of inland prawns of which 60 species belongs to the genus *Macrobrachium.* A list of these species are given below :

1. *Macrobrachium aemulum* (Nobili. 1906)
2. *Macrobrachium altifrons altifrons* (Henderson, 1893)
3. *Macrobrachium altifrons ranjhai* Tiwari, 1963
4. *Macrobrachium andamanicum* (Tiwari, 1952)
5. *Macrobrachium assamense assamense* (Tiwari. 1955)
6. *Macrobrachium assamense peninsularae* (Tiwari. 1955)
7. *Macrobrachium australe* (Guerin-Meneville. 1838)
8. *Macrobrachium banjarae* (Tiwari. 1958)
9. *Macrobrachium birmanicum* (Schenkel. 1902)
10. *Macrobrachium bombayense* Almelkar & Sankolli, 2007
11. *Macrobrachium canarae* (Tiwari, 1958)
12. *Macrobrachium cavemicola* (Kemp, 1924)
13. *Macrobrachium dayanum* (Henderson, 1893)
14. *Macrobrachium divakarani* Jayachandran, 2001
15. *Macrobrachium elatum* Jayachandran.1989
16. *Macrobrachium equidens* (Dana, 1852)

17. *Macrobrachium gangenticum* Bate, 1868
18. *Macrobrachium gurudeve* Jayachandran & Raji, 2004
19. *Macrobrachium hendersodayanum* (Tiwari, 1952)
20. *Macrobrachium hendersoni hendersoni* (De Man, 1906)
21. *Macrobrachium hendersoni cacharense* (Tiwari, 1952)
23. *Macrobrachium hendersoni platyrostre* (Tiwari, 1952)
24. *Macrobrachium honnaense* Thampy, Jayachandran &: Arunachalam, 2007
25. *Macrobrachium idae* (Heller, 1862)
26: *Macrobrachium idella idella* (Hilgendorf, 1898)27227.
27. *Macrobrachium idella georgii* Jayachandran &: 27 Joseph, 1985
28. *Macrobrachium indicum* Jayachandran &: Joseph, 1986
29. *Macrobrachiumjavanicum* (Heller, 1862)
30. *Macrobrachium jayasreei* Jayachandran & Joseph, 1985
31. *Macrobrachiumjohnsoni* Ravindranath, 1919
32. *Macrobrachiumjosephi Jayachandran, 2001*
33. *Macrobrachium kempi* (Tiwari, 1952)
34. *Macrobrachium Idstnense* (fiwari, 1952)
35. *Macrobrachium ladkamii* Almelkar & Sankolli, 2007
36. *Macrobrachium kulsiense* Jayachandran, Lal Mohan & Raji, 2007
37. *Macrobrachium lamjuramani* Jayachandran & Joseph, 1985
38. *Macrobrachium lamarrei lamarrei*(H. Milne Edwards, 1837)
39. *Macrobrachium lamarrei lamarroides* (Tiwari, 1952)
40. *Macrobrachium lar* Fabricius, 1798
41. *Macrobrachium latimanus* (von Martens, 1868)
42. *Macrobrachium malco/msonii* (H. Milne Edwards, 1844)
43. *Macrobrachium manipurense* (Tiwari, 1952)
44. *Macrobrachium mirabile* (Kemp, 1917)
45. *Macrobrachium naso* (Kemp, 1918)
46. *Macrobrachium nobilii* (Henderson &: Matthai, 1910)
47. *Macrobrachium novaehollandiae* (De Man, 1908)
48. *Macrobrachium omatus* Jayachandran &: Raji, 2004
49. *Macrobrachium peguense* (Tiwari, 1952)
50. *Macrobrachium rogersi* (Tiwari, 1952)
51. *Macrobrachium rosenbergii* (De Man,1879)
52. *Macrobrachium rude* (Heller, 1862)

53. *Macrobrachium sankolli* Jalilial &: Shenoy, 1988
54. *Macrobrachium scabriculum* (Heller, 1862)
55. *Macrobrachium siwalikense* (Tiwari. 1952)
56. *Macrobrachium sulcatus* (Henderson & Matthai, 1910)
57. *Macrobrachium tiwarii* Jalihal, Sankolli & Shenoy, 1988
58. *Macrobrachium unicamatakae* Jalihal, Sankolli &: Shenoy. 1988
59. *Macrobrachium veliense* Jayachandran & Joseph, 1985
60. *Macrobrachium villosimanus* (Tiwari, 1947)

PRESENT UTILIZATION OF THE RESOURCES

At present the resources are utilized in the following ways :

1. **Species of capture importance :** A number of species are reported to be commercially important in different states of India. These prawns are either marketed locally or exported. *F. indicus, F. merguiensis, F. penicillatus, P. monodon, P. semisulcatus,* M. *dobson;, M. monoceros, M qffinis, M kutchensis, M. brevicomis, P. styliferus, P. hardwickii, P. sculptilis, P. uncta, P. longipes, M andamanensis,* M. *stridulans, T. curvirostis, M. rosenbergii, M. malcolmsonii, M gangeticum. M ideUa iclella,* M. *iclella georgii,* M. *divakarani, M equidens,* M. *sulcatus, M dayanum,* M. *lamarrei lamarrei,* M. *lamarrei lamarroides, M mirabile,* M. *scabriculum,* M. *rude,* , *M. villosimanus* are important ones belong to such groups. Catch data for most of the states and water bodies are lacking. But fragmentary data available for *M. rosenbergii* is given below.

Year	Landings (kg)
1997	186685
1999	330095
2000	390444
2001	490748
2002	489502
2003	232929
2004	266068

Source : Jayachandran, 2009.

It is disappointing to note that during the peak breeding season (June to December) nearly 40 to 60 % of the catch constituted berried females. This single factor is the major threat to the wild population of the species in the lake. Personal observation is that the population size of the lake is dwindling alarmingly and the present catch revival reports were mainly due to the artificial stocking of hatchery produced seeds by Government agencies. Conservation measures are discussed elsewhere.

Penaeids contribute to about 7.5% of all India marine fish production in 2005. It roughly consists of 57% of edible crustaceans. All Indian penaeid prawn landings varied between 1,21,673 t (1985) to 2,24.620 t (1994) with an annual average landings of 1,78,534 t during the period, 1985-20059 (Radhakrishnan *et al.,* 2007).

2. **Species of Aquaculture importance :** Though the species diversity of the genus is rich, only a few species are at present utilized for aquaculture production. The species of importance include: *M. rosenbergii, M. malcolmsonii, M. gangeticum,* F. *indicus, P. monodon, P. semisulcatus, L. vannamei.* Mass larval production attempts were successful to an extent With regard to these speecies. There are different tyoes of equaculture practices going on in Kerala, namely, monoculture, polyculture, integrated culture of freshwater prawns. Of these special mention has to be made about Rice cum Prawn culture ('Oru Nellum oru Meenum'- in Malayalam or One rice-fish programme). This programme was developed by Kumarakom Research Station, Kerala Agricultural University (padmakumar, 2006). In this system the rice and fish are grown alternatively or in sequence. Joseph (2003) also revealed that yield and returns (per ha) in rice-prawn rotational farming system is highly profitable (cost-benefit ratio is 1.46).

   The advantages of this system are :

   a. It enriches the soil, thus increases the rice production

   b. It helps to control insect pests and aquatic weeds

   c. Large areas of paddy fields utilizing for such fish culture do not demand any major modifications in its natural physiography

   d. Such fields a more suitable for rice farming

   e. Such fields are ecological harmonious

   f. More productive and profitable than popular crop rotations.

   g. It is a strategic technological intervention to protect our vanishing wetlands and sustain rice production

   h. Generates employment opportunities

3. **Species having aquaculture potential, but unutilized :** M. *josephi, M. villosimanus,* M. *latimanus* are potential species which can be utilized for aquaculture, because they grow to bigger size. Practically very little information is available on various aspects of biology and management of these species.

4. **Species with ornamental value :** Recently Jayachandran (2006 a;b; 2007) and Jayachandran *et al* (2006) have made pioneering attempt to introduce 8 species of ornamental prawns of the genus *Macrobrachium* to the aquarium. This is the first step in this direction. The species of ornamental value include - M. *canarae, M. latimanus,* M. *ornatus,* M. *gurudeve,* M *rosenbergii,* M. *kulsiense,* M *assamense,* M. *rogersi* and M. *sulcatus.* This is a novel method of utilization of biodiversity. But judicious exploitation from the wild is necessary for which regulatory

measures are to be urgently taken up to protect them from over exploitation. This programme will certainly improve the livelihood security of rural people. Not much work has been carried out in these lines.

5. **A new concept in Aquaculture :** Cultivation of lesser species along with *M. rosenbergii,* M. *malcolmsonii,* M. *gangeticum* etc. proved beneficial. The lesser species will act as forage. This is a novel approach of organic farming. This is being, practiced under the supervision of the author. The advantages of the present practice

Sl. No	State	Natural Nos	Area (ha)	Nos	Artificial Area (ha)
1	Andhra Pradesh	219	1,00,457	19,020	4,25,892
2	Arunachal Pradesh	2	20,200	NA	NA
3	Assam	1394	86,355	NA	NA
4	Bihar	62	2,24,788	33	48,607
5	Goa	3	12,360	NA	NA
6	Gujarat	22	3,94,627	57	1,29,660
7	Haryana	14	2,691	4	1,079
8	Himatlon Pradesh	5	702	3	19,165
9	Jammu and Kashmlr	18	7,227	NA	21,880
10	Karnataka	10	3,320	22,758	5,39,195
11	Kerala	32	24,329	2,121	2,10,579
12	Madhya Pradesh	8	324	53	1,87,818
13	Maharashtra	49	21,675	1,004	2,79,025
14	Manlpur	5	26,600	NA	NA
15	Meghalaya	2	NA	NA	NA
16	Mlzoram	3	36	1	1
17	Nagaland	2	210	NA	NA
18	Orlssa	20	1,3,7,022	36	1,48,454
19	Punjab	33	17,085	6	5,391
20	Rajasthan	9	14,027	85	1,00,217
21	Slkklm	42	1,107	2	3
22	Tamll Nadu	31	58,068	20,030	2,01,132
23	Trlpura	3	575	1	4,833
24	Uttar Pradesh	125	12,832	28	2,12,470
25	West Bengal	54	2,91,963	9	52,564
26	Chandlgarh	-	-	1	170
27	Pondlcherry	3	1,533	2	1,131
	Total	**2170**	**14,60,113**	**65,254**	**25,89,266**

source: http://ces.iisc.emet.in/energy/TR86/intro.html

are improved health of cuhivating species, reduction of disease problem in species under cultivation and drastic reduction of feed inputs and maintenance of good water quality.

6. **Vahle addition :** Many species mentioned above can be utilized for production of value added products. KAU has developed a few products like: Prawn Pickle, prawn cutlet, prawn stick (pavunny et at, 2007). The other products that can be produced include - flavour extract and chitin and chitosan production from shell waste.

7. **Rural employment generation :** Value addition of lesser species and utilization ornamental prawns are highly beneficial for generating employment for rural people. Considerable work has to be carried out in this regard

Wet land potential of India : India has a vast potential area of wet lands. The available reports indicate that the total inland area of wetlands comprising 1.2 million hectares of brackish water; 2.38 million hectares of fresh water ponds and tanks; 3.15 million hectares of reservoirs; besides about 1,91,000 krn of rivers, rivulets, streams and canals. Wet lands provide innumerable beneficial functions including sustaining life processes (fauna and flora), water storage (domestic, agricultural and industrial usage), protection from storms and floods, recharge of ground water, water purification, storehouse for nutrients, erosion control, stabilization of local climate, help maintaining ecological balance so on and so forth. Since it is an important ecological system of food production, the availability of major wetlands of India (state wise) would be of great relevance and hence provided below :

Road ahead : There must be major policy formulation for proper utilization of the inland wetlands for food production and security. The following points may be valuable for increasing food production and this may partially help to tide over the anticipated food scarcity.

- Present annual production is 2.6 million tomes and projected annual production is 5.0 million tonnes.
- Increase in the coverage of areas of ponds and tanks for aquaculture practices, from 10 lakh hectares to 15 lakh hectares
- Increasing productivity of existing water bodies, from 2.4 tonnes/ha/yr to 3 tonnes/ha/yr
- Diversification and intensification of culture practices
- Integrated fish farming practices are to be advocated
- Ornamental prawn and fish culture for employment generation and foreign exchange earning for rural folk
- Provision of feed mills
- Establishment of Aqua-shops as single-window outlets for all aquaculture inputs
- Establishment of rural cold chairs for shell fish and fish marketing

- Sanitary markets for fish and shellfish (models in selected towns)
- Strengthen systems to up-grade the existing baseline data on wetlands across the country
- Proper transfer of technologies to stakeholders and provide appropriate training for sustainable food production
- Formulate region-wise development plans
- Effective management of ecological components of ecosystems
- Conservation of aquatic fauna so that sustainable productive systems can be developed
- Catch ban period are to be suggested. so that breeding of aquatic fauna can be assured
- Effective integrated management and networking for fishery development initiatives with different line departments
- The sharing of information state! regional level to achieve sustainability
- Understanding the impact of weather changes on inland aquatic organisms
- Diversified use of aquatic fauna for the beneficial use of man kind .
- Suggest alternative occupation for wetland dependent people during catch ban period
- Management of ecosystems on peoples participatory mechanism

REFERENCES

Jayachandran, K.V., 2008. Palaemonid Prawns - Biodiversity, Taxonomy, Biology and Management - Science Publishers, Inc., Enfield (NH), USA: Plymouth. UK. (International Edition, 2001); Oxford & IBH Publishing Co. Pvt. Ltd., New DeDti (Special Indian Edition, 2001), 624 P + 12 prelims and 4 pages colour photographs.

Jayachandran, K.V., 2008. Freshwater prawns of ornamental value from erala. *Training Manual on Ornamental Fish Culture* 96-98 + a colour plate.

Jayachandran, K.V., 2006b. hy not prawns and shrimps in aquariua? *Proceedings. Aqua ornamentals - 2006:* 19-22. Jayachandran, K.v.. 2007. Ornamental Freshwater prawns In : *Ornamental Fish Breeding, Farming and Trade* urup. B.M.. Boopendranath, M. R.. Ravindran. K., Saira Banu and Nair, A G. (eds.)), Dept. of Fisheries, Govt. of Kerala, India: 99-106.

Jayachandrn K.V., 2008 Biodiversity of marine prawns of the family Penaeidae Refinesqu,1815 of Indian Waters.***Glimpses of Biodiversity -Rajiv Gandhi Chair Spl. Pub*****., 7:** 207-220. (Indaia)

Jayachandran, K. V., 2009. Indian palaemoind decapod crustaceans: taxonomic status, research challenges and conservation needs. ***National Bureau of Fish Genetic Resources*** (Abstract). (India)

Jayachandran, K.V., A. V. Raji & Tessa Thomas, 2006 Prawns and shrimps of ornamental value from Kerala *Proc. Recent Trends in Mariculture:*77-79 (+1colour plate).

Joseph, K.J. 2003. *Coastal economy of Kerala* : a *profile* keraka Agricultural University, Therissur, p 40

Kurien, C.V.& Sebastian, V. O. 1986. *Prawns and prown fisheries of Indisa*. Hindustan Publoshing Corporation, New Delhi, (3rd ed.) 295p.

Padmakumar, K.G. 2006. Rice Fish Integration -a win-win farming model for low lands. In Rice-*Fish integration through Organic Farming*. Agrotech Publishing. Academy, Udaipur: 86-100.

Pavunny, O., S. Krishnahumar, P.M. Sherief, D. D. Nambudiri & S. M. Joseph 2007 . Development of value added products from undersized freshwater prawns In: Advances in biology, aquaculture and marketing. *Proceedings. Freshwater Prawns,* **2003:** 648-658.

Perez-Farfante, I. & Kensley, B. 1997. Penaeoid and Sergestiod shrimps and prawns of the world Keys and diagnoses for the families and genera. *Memoires du Museum Naturelle,* **175:** 1-233.

Redhakrishnan, E.V., Mary K. Manisseri & Nandakumar, G., 2007 Status of Research on Crustacean resources In : Mohan Koseph Modayil and N.G.K. Pillai (eds.) *Stahus and Perspectives in Marine Fisheries Research in India*, CMFRI, Kochi : 135-172.

Suseelan, C. 2003 Taxonomy of penaeidean shrimps of India coast. In : *Texonomy and biodiversity of fishes and Crustaceans, ICAR training manual* (Dr.K. V. Jayachandra & Dr. J. Rajasekharan Nair, eds) : 129-138

Bioresources for Food Security and Rural Livelihood, Pages 189–204
Edited by : G.K. Kulkarni and B.N. Pandey

SMALLER FRESH WATER PRAWNS: THEIR AQUACULTURE POTENTIAL AND SUITABILITY AS GOOD LABORATORY MODEL

Sanjive Shukla and U.D. Sharma

ABSTRACT

Two fresh water prawns, *Macrobrachium lamarrei* and *Macrobrachium dayanum* are abundantly available in river Gomti and other water reservoirs in and around Lucknow (U.P.) India throughout the year. These are smaller prawns having average length and weight, 48.6+5.5 mm & 1.107+0.26 gm (*M. lamarrei*) and 56.4+4.2 mm & 3.26+0.68 gm (*M. dayanum*). Fecundity of both the species ranges from 55-250 (*M. lamarrei*) and 38-142 (*M. dayanum*) having large sized eggs of about 1.2±0.05 mm & 2.56±0.06 mm respectively. These prawns are continuous breeders and have many advantages in terms of their culture over larger fresh water prawn species, like abbreviated development fully completing in fresh water, short larval period, high percentage of survival upto post larval stage, availability of seed throughout the year, resistance to certain diseases, compatibility for polyculture in small natural water reservoirs and cost effectiveness. Their smaller size, hardy nature, easy maintenance in laboratory, being type animal of syllabi of various Indian universities and researches performed in different fields like physiology, receptor biology, molecular biology, reproductive biology, neuro-endocrinology, toxicology, environmental monitoring *etc*. makes them better suited model for fundamental and applied research. Aquaculture potential of these smaller fresh water prawns and their role in improving the nutritional status and economy of poor people of Uttar Pradesh has also been discussed.

Keywords : Fresh water prawn, *Macrobrachium lamarrei*, *Macrobrachium dayanum*, Aquaculture.

INTRODUCTION

Crustaceans are major constituents of aquatic ecosystem that provides a variety of ecological and economic services. They have been included in human diet since antiquity. Prawns, shrimps, crabs and lobsters are on top of foreign exchange earners among fisheries products (Jhingaran 1991, Kurian and Sebastian 1993). Tremendous increase in human population as well as habitat destruction due to anthropogenic interference led a pressure on natural fisheries resources therefore farming of these valuables came into existence and now attaining form of industry all over world. Farming of fish & other aquatic organisms like Crustaceans, Molluscs and aquatic plants has been defined as "aquaculture" by Food and Agricultural Organization (FAO). In India mid-eighties was the period when shrimp production gained importance, flourished in 1990 and tremendously increased during 1995-96. The shrimp exports contributed about 71% of total seafood export ranging to about Rs. 50040 million (Rajagopal 1995). The shrimp production in India registered 14% increase in 2005-06 in terms of value of Rs. 3194.38 crore.

Among aquaculture, freshwater prawn aquaculture is a recent one and has many advantages over marine like being less expensive, can be performed in smaller ponds, rivers and lakes, less vulnerable to diseases etc. To fulfill day by day increasing demand all over the world, aquaculture is the only solution. Palaemonid prawns are very important in fresh water aquaculture and contribute about 50% of crustacean aquaculture. Prawn culture has developed at a very fast pace in present decade in many countries and 30000 nit of *Macrobrachium* was produced through culture practices in 2001 (New 2003). It has now taken shape of a worldwide industry with annual expansion rate of over 29% during 1992-2001 and global expansion rate 48% (New 2003). India produced 453 mt of fresh water prawns in 1997 and among Palaemonid prawn producing countries holds 8th position (Jaychandran 2001). According FAO (2001) reports Indian production of "Scampy" ranged about 24230 mt which is estimated to become just double about 50000 mt in 2010 (New 2003).

Prawns belong to the family Palaemonidae of Decapod crustaceans which bears eggs in their pleopods. Fresh water prawns used so for in aquaculture and aquaculture experiments belong to Genus *Macrobrachium*, which is the largest one. About 200 species of the genus have been described throughout the world and are distributed in tropical and sub-tropical zones (Holthuis 1980). Among these 40 species are well reported from India (Tiwari 1955, Jaychandran 2001).

Freshwater prawns have an important place in fresh water aquatic food chain and are used as food, particularly a human delicacy, and contains rich amount of proteins, free amino acids, vitamin A&D etc. Due to less fat contents they are advised to diabetic people as replacement of meat and are also used for various medicinal purposes in Unani & Ayurvedic systems of medicine (Proudfit & Robinson 1955, Martin 1966, Singh 1977, Khan *et al.* 1994).

Among fresh water prawns, *Macrobrachium rosenbergii, M. malcolmsonii, M. gangeticum, M. villosimonis, M. josephi, M.idella, M. rude, M. equidens, M. lanchesteri, M. kistenensis, M.idae, M. scabriculum, M. mirabile, M.nobilii, M. latimanus, M. lamarrei and M. dayanum etc.* are found in India and are commercially important species (Holthuis 1980, Jhingaran 1991, Kanoujia 2002). According to Holthuis (1980) 84 species of *Macrobrachium* are consumed as food and hence economically important. About half of these species have been examined for their potential suitability to aquaculture (Brown 1991, Jaychandran & Joseph 1992).

Out of above said species, first three i.e. *M. rosenbergii, M.malcolmsonii* and *M. gangenticum* are thought to be suitable for aquaculture in India. Most of the efforts and technological developments have been centered around aquaculture of these species due to their considerable size & fast growing rate only (Rao & tripathi 1993, Prasad & Kanoujia 2006). Culture practices of these larger prawn species have so many problems particularly in regions other than coastal areas, must be taken into consideration.

Disadvantages with Aquaculture of Larger Freshwater Prawn Species

No doubt larger prawn species like *M. rosenbergii* is highly productive and demanded in market worldwide from India and earn foreign exchange. Some part of demand is also fulfilled by culture of *M.malcolmsonii* and *M. gangeticum*, through are in beginning phase and constituting only 10% of total Palaemonid production (Kutty 2003, Kanoujia 2006, Upadhayay *et al* 2006). Culture of these prawn species have some disadvantages for regions other than coastal areas like M.P., Punjab, Bihar, Haryana & Uttar Pradesh, having enormous capabilities for freshwater prawn aquaculture in future. Some of disadvantages are :-

1. All larger prawn species require brackish water for their initial larval development therefore directly and indirectly connected to sea via migration.
2. Most of the hatcheries to obtain seed are set near coastal areas therefore seed and transportation cost makes their culture expensive.
3. Due to long larval period they are more vulnerable to environmental stress therefore success rate upto post larval period is very less despite high fecundity.
4. Prawn processing facilities and marketing units are not properly set throughout India which makes prawns culture difficult.
5. Culture of larger prawns is not successfully performed in natural water reservoirs therefore use of productive land is required which is not eco-friendly.
6. Due to high international demand these prawns are beyond reach common man of India therefore have very limited role in improving nutritional status and economy of poor people in India, particularly states like U.P.

To sustain local population in terms of nutrition and economy other alternatives are to be explored. Smaller freshwater prawns which are native and readily consumed by local people are the best candidates in this direction.

Culture of Smaller Freshwater Prawns

Concept to culture smaller and native species of fresh water prawns started in China with *Macrobrachium nipponense* which opened new gates for aquaculture diversification (Kutty 2003). Dobkin (1967) investigated and suggested the culture of *M. jelskii,*

M. borellii, M. quelehi, M. potunus, M. brasilience, M. iherigi from USA and recommended search for species with similar abbreviated development from other parts of the world for their culture.

In Africa, culture trails are going on with *M. carcinus, M. amazonicum, M. acaranthus*, and *M. vollenhovenii. Cambaroides japonicus,* a small sized fresh water crayfish, has been successfully adapted for culture in Japan and artificial incubation system for their eggs has been developed (Nakata *et al.* 2004, Nakata & Goshima 2004). Culture trails of *M. lamarrei* have been successfully done in Pilippines, Malaysia & Bangladesh; of *M.lanchesteri* in Malaysia, Thialand & Vietnam; and of *M. rude* in East Africa and Bangladesh (New and Valenti 2000).

India has rich heritage of indigenous small freshwater prawn species distributed in almost all rivers and water reservoirs throughout the country. Most them are readily consumed by local population and have high potential for aquaculture (Jhingaran 1991, Kurian & Sebastian 1993). Common species of smaller freshwater prawns found in India are *Macrobrachium rude, M. josephi, M.idella, M.equidens, M.lanchesteri, M.idella, M. scrabiculum, M. mirabile, M. nobilli, M. lamarrei and M. dayanum etc*. These prawn species are easily available in natural reservoirs throughout the year and have good market price. Most them are smaller in size, having abbreviated development and can be easily cultured. These *Macrobrachium* species have many advantages in terms of their culture over larger freshwater prawn species :-

1. They have abbreviated development, fully completing in freshwater therefore seed can be obtained locally and hatcheries can be set in all parts of country.
2. Availability of seed in local rivers throughout the year.
3. Despite low fecundity, they have short larval period and high survival rate which make them better suited for culture.
4. Resistant to diseases than larger prawn species.
5. Culture can be easily performed in natural water reservoirs which is eco friendly in nature.
6. Due to smaller size they can be sun dried & stored for longer periods.
7. Polyculture of these prawns with some common fishes is possible.

All above features indicate high potential of smaller freshwater prawns for aquaculture in future, which is a "untapped" fisheries resource allover the world, particularly in India.

Studies on *Macrobrachium Lamarrei* (H. Milne Edward) and *Macrobrachium Dayanum* (Henderson)

The biology of freshwater prawns, *M. lamarrei* and *M. dayanum* has been thoroughly studied in Prawn Research Centre, Department of Zoology, University of Lucknow, Lucknow (U.P.) India (N-25°51′59″ E-81°56′17″). *Macrobrachium dayanum*

(Henderson) was reported first time from Lucknow region by Sharma *et al.* (1997). *M. lamarrei* and *M. dayanum* are abundantly available in river Gomti and other water reservoirs in and around Lucknow (U.P.) India. Both the prawn species are potential animals for fresh water aquaculture in future.

River Gomti, a tributary of river Ganga, is a medium sized river originates from swampy forested area of Pilibhit district near Himalayan foothills. It travels about 950 km. in Uttar Pradesh passes through Lucknow, the capital of U.P., and joins river Ganga near Varanasi.

Macrobrachium lamarrei (H. Milne Edwards)

It is a smaller, translucent fresh water prawn with size range 30-55 mm in length (Mean=48.6+5.5mm) and 1.107+0.26gm in weight. Light green coloured chromatophores along with some brown once are distributed through out exoskeleton. The maximum recorded size of this species was 65mm. No marked difference in size of males and females was observed. Berried females are abundantly available from March-Sept. bearing egg in pleopods. Fecundity ranges from 55-250. Eggs are elliptical measuring 1.2-1.5 mm on longer axis which are dark green in colour after spawning turns brown and finally colourless with two externally visible eye spots before hatching. This prawn species has abbreviated development and completes its development in fresh water. This is highly active species and shows frequent, active movements and responses to various stimuli in laboratory.

***Macrobrachium dayanum* (Henderson) :-** Reproductive biology of *M. dayanum* has been studied in detail from this lab (Tripathi 2007). It is a robust prawn than *M. lamarrei*, having size range 35-65 mm in length (mean=56.4+4.2mm) and average weight 3.26+0.68 gm. Males are slightly bigger (Maximum recorded length=71mm) then females (Maximum recorded length=65mm). These have dark colour patterns due to presence of blue and brown chromatophores. II chelate legs are more robust than *M. lamarrei* and covered with dense setae in case of males. *M. dayanum* is more aggressive in behaviour than *M. lamarrei*.

M. dayanum is continuous breeder, berried females are regularly found throughout the year except in months of Nov., Dec., Jan. & Feb. Berried females carry egg in pleopods. Fecundity ranges from 38-142. Eggs are elliptical, slightly bigger (2.56+0.06mm) than *M. lamarrei*, dark green in colour which intern changes brown and colourless just before hatching, Fecundity of *M. dayanum* in different months (August 2004-July 2005) has been summarized in Table 1 and Fig. 1. Fecundity was found higher in months like March, April and May. Minimum mean fecundity (38.2+4.5) was recorded in September in prawn size 47.9+4.9mm and Weight 1.01+0.04 gm. while maximum fecundity (142.9+4.9) was recorded in month of April with prawn size 64.90+00.69mm and weight 2.896+0.97 gm.

Table 1 : Relation between length, weight & fecundity of *M. dayanum* in different months of the year. Data is Mean ± S.E., N=10 Data followed by different letters is significant at P<0.001

S. No.	Months	Length of Animal (Cm)	Weight of Animal (Gm)	Fecundity (mean ± Se)	Relative Fecundity
1.	January	4.560 ± 0.125 (a)	0.785 ± 0.053 (a)	0 (a)	0 (a)
2.	February	4.79 ± 0.212 (a)	1.126 ± 0.168 (a)	0 (a)	0 (a)
3.	March	6.280 ± 0.107 (b)	2.962 ± 0.169 (c)	118 ± 9.824 (c)	40.334 ± 2.675 (b)
4.	April	6.490 ± 0.069 (b)	2.896 ± 0.097 (c)	142.90 ± 4.89 (c)	50.125±2.404 (b)
5.	May	6.410 ± 0.142 (b)	3.033 ± 0.199 (c)	129.60 ± 8.184 (c)	44.063±2.550 (b)
6.	June	4.67 ± 0.318 (a)	1.676 ± 0.124 (b)	77.40 ± 6.275 (c)	46.600±3.966 (b)
7.	July	5.147± 0.195 (b)	1.457 ± 0.171 (b)	68.70 ± 11.278 (b)	55.475 ± 13.812 (b)
8.	August	4.930 ± 0.148 (a)	1.227 ± 0.134 (b)	57.100 ± 6.613 (b)	42.248±3.931 (b)
9.	September	4.790 ± 0.086 (a)	1.010 ± 0.047 (b)	38.2 ± 4.505 (b)	38.766 ± 4.621 (b)
10.	October	5.150 ± 0.104 (a)	0.97 ± 0.083 (b)	57 ± 3.740 (b)	64.533±6.067 (b)
11.	November	4.88 ± 0.235 (a)	1.164 ± 0.118 (a)	0 (a)	0(a)
12.	December	5.23 ± 0.138 (a)	1.341 ± 0.107 (a)	0 (a)	0(a)
	F-Value	11.26	78.78	55.16	16.44
	P-Value	<0.000	<0.000	<0.000	<0.000

The relationship of length & weight with fecundity have been summarized in Figs. 2 & 3. A linear relationship between length & fecundity is expressed by equation $y=62.01\times-269.83$ ($R^2=0.6899$). The relationship was found highly significant ($r=0.665$, $P<0.001$), similarly the weight and fecundity relationship was also found linear with equation $y=55.36\times-33.23$($R^2=0.7723$) and was found highly significant ($r=0.759$, $P<0.001$).

M. dayanum is more aggressive and canabalistic behaviour is more pronounced, usually attacking weak & moulting animals. This prawn species is not very active and resides mainly at the bottom of aquaria in lab. Feeding frequency is less but consumes more food at a time than *M. lamarrei*. Grooming behaviour is more frequent than

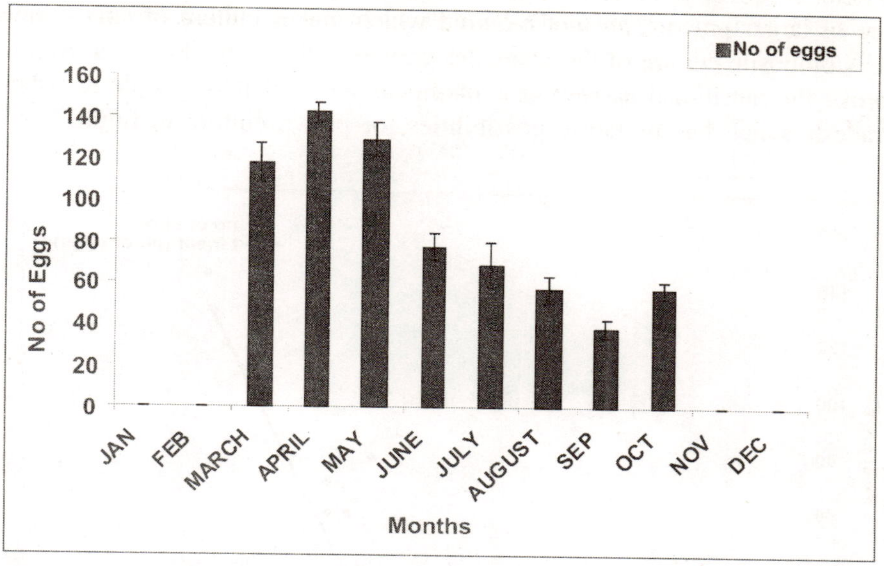

Fig. 1 : Seasonal variation in fecundity of *M. dayanum*.

M. lamarrei. Colour change capability is lesser than *M. lamarrei* which frequently changes its colour to back ground colouration. Both the species of fresh water prawns have been successfully reared in lab upto post larval stage.

These two prawn species viz. *M. lamarrei* and *M. dayanum* are potential animals for fresh water aquaculture which has not been exploited till date. These prawns are good source of proteins for the poor people and are economically important. Sun dried prawns cast Rs. 100-250/Kg. in Lucknow depending on season.

Biology of *M. lamarrei*, *M. dayanum* and other smaller prawn species has also been studied in other parts of India (Katre 1977, Katre & Pandian 1972, Bhattacharjee & Das Gupta 1989, Manna & Raut 1991, Biswas & Qureshi 1994, Tewari & Quershi 1997, 1998, charlse 1999, Mariappan 2003) as well as from other countries (Yaqoob 1980, Praveen *et al*. 2002, Nakata & Goshima 2004, Nakata *et al*. 2004, Zarari *et al*. 2003, Saifullah 2005, sharma & subba 2005, Buiyan *et al* 2007). These studies are indicative of potential for culture of these smaller but economically important prawns.

Scattered attempts to culture these smaller prawns have been done in India as well as outside and reported possible production of *M. lamarrei* as 50.83-108kg/ha and 106.5kg/ha (Tewari *et al* 1997, Chandrasekharan *et al* 1997, Singh & Qureshi 1997. With in-vitro hatching techniques (Mathavan & Murugadass 1988, Nakata *et al*. 2004) it is evident that hatching and survival rate of *M. lamarrei* in higher than larger freshwater prawn species. Abbreviated development of *M. lamarrei* & *M. dayanum* has been studied in detail by various workers (Barrakpoor 1973 & 74, Rajyalakshmi 1961, Sharma & Subba 2005, Jalihal & Sankoli 1975) which indicate that they have shorter larval span against longer larval span of larger fresh water prawn species. *Artemia* naupli, a costly & essential food for larvae of larger prawns, are not required which makes culture of small prawns cost effective. Composite culture of these smaller prawns with fishes like Catla, Rohu & Nain can improve the nutritional as well as economical status of poor people & fishermen of Uttar Pradesh which has immense possibilities for prawn culture in future.

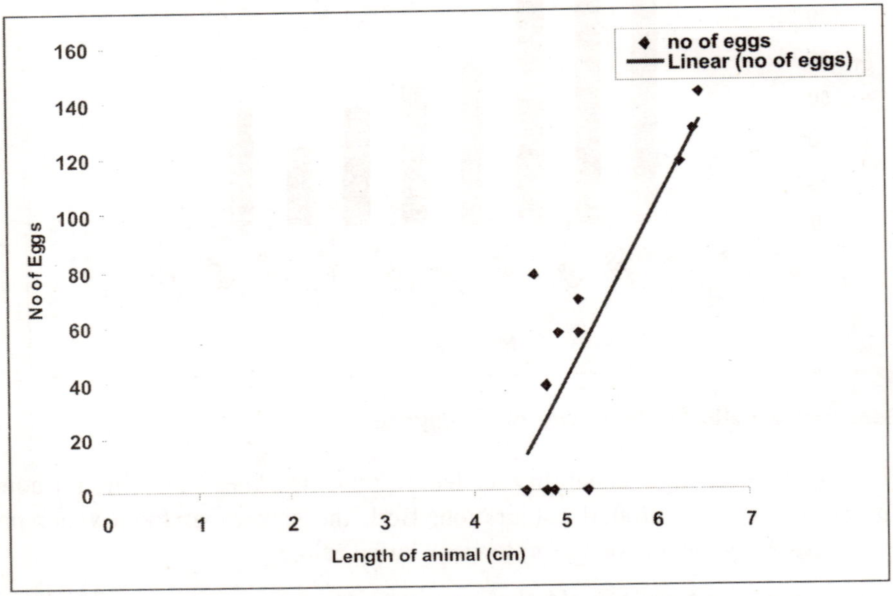

Fig. 2 : Relation between length and fecundity of *M. dayanum*.

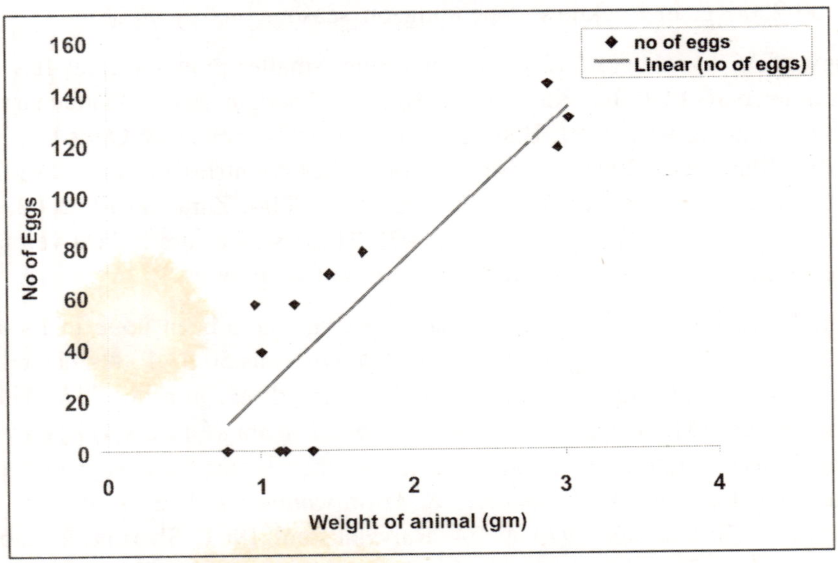

Fig. 3 : Relation between weight and fecundity of *M.dayanum*.

Smaller Fresh Water Prawns as Good Laboratory Model

Smaller fresh water prawns, *M. lamarrei* and *M. dayanum* have been proved as better laboratory model for fundamental and applied research like toxicology & environmental monitoring (Shukla *et al.* 2009). Prawns are widely studied crustaceans and have been included as type animal in syllabi of different Indian universities. They are proffered in comparison to other animals due to following reasons :-

1. Due to smaller size & hardy nature they can be easily maintained in aquaria in lab.
2. Due to omnivorous nature, maintenance cost is very less.
3. They are easily available throughout the year.
4. Due to abbreviated development, animals of known age can be grown in laboratory.
5. Due to high sensitivity to environmental fluctuations & xenobiotics they can serve as better bio-indicators in environmental monitoring.
6. Due to translucent colour some parameters like heart beat rate, scaphognathite oscillations, colour change, gut movement etc. can be observed with necked eyes without sacrificing the animal.
7. New researches on them can directly be of use of undergraduate & post graduate students.
8. Results of studies on them can be used to improve commercial aquaculture of fresh water prawns.

Important studies carried out on *M. lamarrei* & *M. dayanum* in various disciplines are:- (1) Taxonomy & disease (Shukla 1988, Sharma *etal* 1997). (2) Digestive physiology & Ultra Structure (Shukla & Sharma 1990, Sharma & Shukla 1992) (3) Receptor biology (Sharma & Shukla 1990) (4) Heart & Haemocytes (Lodhi *et al.* 2004, Lodhi *et al.* 2008, 2009) (5) Reproductive Biology (Tripathi 2007) (6) Neuroendocrine system (Tewari 2008) (7) Toxicological studies- Effects of heavy metals. Hg. Cd, Cu, Cr, Ni, Pb and detergent Linear Alkyl Benzene Sulphonate (LAS) on different parameters *viz.* (A) General Behavioural patterns (Shukla & Sharma 1990, Verma *et al.* 2005, Sharma & Shukla 2006, Lodhi *et al.* 2006, Tewari *et al.* 2007, Triapthi *et al.* 2007, Sharma *et al.* 2008) (B) Scaphognathite Oscillations (Tewari *et al.* 2008, Sen *et al.* 2008 a&b, Verma *et al.* 2007) and (E) Histopathology of Gill, Green gland, Hepatopancreas, Heart, Haemocyte morphology, sperm morphology & Neurosecretory cells (Shukla 1993, Shukla & Sharma 1993, Shukla & Sharma 2002, Lodhi *et al.* 2002, 2007 a&b, 2008, Sen *et al.* 2008, Lodhi 2009).

Above mentioned studies as well as studies of other workers from India (Nagabhushanam *et al.* 1979, Nagabhushanam & Kulkarni 1981, Narshimha *et al.* 1981, Sarojini *et al.* 1982, 1988, Sarojini & Gyananath 1983, Kulkarni *et al.* 1991) and from other countries (Dobkin 1967, Parveen *et al.* 2002, Bhuiyan *et al.* 2007, Saifullah 2005, Sharma & Subba 2005, LeBlanc 2007) are indicative of their suitability as laboratory model. Like other crustaceans, smaller fresh water prawns can also serve as better

"Bio-indicators" of worsening aquatic bodies (Phillips 1980, Vijayraman *et al.* 1999, Farkas *et al.* 2003, Verslyeke *et al.* 2004). After proper standardization various parameters *viz.* Behavioural patterns, heart beat rate, scaphognathite oscillations, THC, DHC, Haemocyte morphology & Sperm morphology can also be utilized in health assessment & monitoring programs of these animals as well as in environmental monitoring of freshwater aquatic bodies, a depleting resource of the country. *M. lamarrei, M. dayanum* and other small freshwater prawn are valuable resource of the country & their aquaculture potential has not been exploited yet. Researches in field of their biology, development of new culture techniques, generation of awareness, and government support can promote the culture of these economically important prawns in eco-friendly patterns. This will also be useful in improving nutritional status and economy of poor people of northern and central provinces of India like U.P., M.P., Bihar, Punjab, Harayana etc.

ACKNOWLEDGEMENTS

Authors are thankful to Head, Department of Zoology, University of Lucknow, Lucknow and Sri B. Singh, Head, Department of Zoology, BSNVPG College, Lucknow (U.P.), India for providing necessary laboratory facilities and encouragements throughout the work.

REFERENCES

Barrackpoor (1973, 1974) Technical Reports cited In. Jhingaran V.G.: (1991) Fish & fisheries in India- Hindustan Publishing Co. New Delhi.

Bhattacharjee P.C. and Das Gupta M. (1989) fecundity of the freshwater prawn, *Macrobrachium dayanum* (Henderson) from ponds & swamps of Kamrup, Assam, India. Environment & Ecology **7(3)**: 724-726.

Bhuiyan A.A., Arzu G. and Bhuiyan S.S. (2007) The correlation between fecundity with length & weight of *Macrobrachium dayanum* (HALL) from the river Padma, Rajshahi, Bangladesh. J. Bio-Sci **15**: 173-174.

Biswas R. & Qureshi T.A. (1994) Annual reproductive cycle of the freshwater prawn, *Macrobrachium dayanum* (Henderson 1893) Naturalia (Naturalia SAO Paulo) **19**: 131-47.

Brown J.H. (1991) Freshwater Prawns: In: Production of Aquatic animals, world animal science, C_4 (Ed. By C. Nash): 31-43. Elsevier science publishers. Amesterdam, Netherlands.

Chandrasekharan V.S., Singh V.K. & Singh R. (1997) Pond culture of the freshwater prawn, in the tarai region of Uttar Pradesh: in (Eds) Daliwal G.S., Arora R., Randhawa N.S. Dhawan A.K.- Ecological agriculture & sustainable development Vol.-I Procedings of international conference on ecological agriculture: Towards sustainable development, Chandigarh India 15-17, November, 1997.

Charlse P.M. (1999) Seasonality on the egg production and energetics of egg development of *M. lamarrei* In: (S.R. Misra eds.) Limnological Research in India, PP-84-102, Daya Publishing House New Delhi, India.

Dobkin. S. (1967) Abbreviated Laval development in Ceridean Shrimps and its significance in artificial culture of these animals. Proceeding of the world scientific conference on Biology & culture of shrimps & prawns. FAO Fisheries report **M. 57** Vol. 3.

FAO (2002), Fishery Statistics-Aquaculture production 2000: food & agriculture organization, Rome.

Frakas A, Saolanki & Istvan Varanka (2003)- Crustaceans as biological indicators of heavy metal pollution in lake Balaton (Hungary) Hydrobiologia **10**: 1-6 (2003).

Holthuis L.B. (1980) Shrimps and prawns of the world: An annotated catalogue of species of Interest to fisheries, FAO Fisheries Synopsis **125(1):** 1-272.

Jalihal D. R. and K.N. Sankoli (1975) on the abbreviated metamorphosis of the freshwater prawn, *Macrobrachium dayanum* in the laboratory. Karnatak Univ. J. Science **20**: 283-291.

Jaychandran K.V. (2001) Palaemonid Prawns: Biodiversity, Taxonomy, Biology & Management. Oxford & IBH Publishing Co. Pvt. Ltd. New Delhi.

Jaychandran K.V. and N.I. Joseph (1992) A key for field identification of commercially important Macrobrachium Spp. of India with a review of their bionomics in: Silas E.G. (Eds) Freshwater prawns. Kerala Agriculture University 72-77.

Jhingaran V.G. (1991) Fish and Fisheries of India. Hindustan Publishing Company New Delhi.

Kanoujia D.R. (2003) Indian river prawn, *Macrobrachium macolmsonii* and minor species of commercial importance In: souvenir. International symposium on freshwater prawns. 2003 (Aug 20-23, 2003) College of fisheries, Kerala Agriculture university, Cochin pp. 51-58.

Kanoujia D.R. (2006) Freshwater prawn breeding & culture In: Handbook of fisheries and Aquaculture (S. Ayyappan Ed.) PP-293-306: Directorate of Information & Publication in Agriculture (ICAR), New Delhi.

Kartre S. (1977) The relation between bodysize & number of eggs in the freshwater prawn, *Macrobrachium lamarrei* (H. Milne Edwards) (Decapoda Caridea) Crustaceana **33**: 17-21.

Katre S. And Pandian T.G. (1972) on the hatching mechanism of freshwater prawn, *Macrobrachium dayanum*. Hydrobiologia **40(1)**: 1-17.

Khan A.U., Sharma V.P. and Azami H.K. (1994) The Medicinal use of entomological drugs used by tribal communities of Eastern Uttar Pradesh, Uttar Pradesh J. Zool. **14(1)**: 55-59.

Kulkarni G.K., Glade L. Fingerman M. (1991) Oogenesis and affects of neuroendocrine tissues on in-vitro synthesis of protein by the ovary of red swamp crayfish, *Procambarus clarkii* (Girard) J. Crust. Biol. **11**: 513-522.

Kurian C.V. and Sebastian V.O. (1993) Prawn & Prawn fisheries of India. Hindustan Publishing Corporation, New Delhi.

Kutty M.N. (2003) Sustainable Aquaculture: a comparative view of freshwater prawn and marine shrimp : In sovenior - International Symporium on freshwater prawns 2003 (August 20-23, 2003) College of fisheries, Kerala Agricultural University Cochin pp. 14-20.

Le Blanc G.A. (2007) Crustaceans Endocrine toxicology: A review Ecotoxicology **16**: 61-81

Lodhi H.S. (2009) Effect of Copper sulphate on the Kidney or Green gland of freshwater prawns (Crustacea - Decapoda). Poll. Res. **28(2)**: 187-192.

Lodhi H.S., Ali. H., Verma D.R., Tewari K.J., Shukla S. and sharma U.D. (2008) Physiological & Histopathological alterations in heart of freshwater prawn, *Macrobrachium lamarrei* (Crustacea - Decapoda) exposes to Copper Sulphate. 28th session of Academy of Environmental biology Deptt. of Botany, Jamia Hamdard University New Delhi (Nov 20-20) pp. 49-50.

Lodhi H.S., Khan M.A. Verma R.S. and Sharma U.D. (2006) Acute toxicity of Copper Sulphate to freshwater prawns. J. Environ. Biol **27(3)**: 585-588.

Lodhi H.S., Khan M.A., Verma R.S. and Sharma U.D. (2004) seasonal fluctuations in the total Haemocyte counts (THC's) of freshwater prawn, Him. J. Env. Zool **18(1)**: 1-5.

Lodhi H.S., Shukla S. and Sharma U.D. (2002) Copper induced behavioural and histopathological alterations in freshwater prawn (Crustacea - Decapoda). Proc. Acad. Environ. Biol. 9:38.

Lodhi H.S., Shukla S. and Sharma U.D. (2009) Studies on structure of heart of two freshwater prawns under light microscope. Uttar Pradesh J. Zool. **29(1)**: 1-10.

Lodhi H.S., Shukla S., & Sharma U.D. (2008) Light microscopic & histochemical studies on haemocytes of freshwater prawn, *Macrobrachium lamarrei* (Crustacea - Decapoda). Aquacult **9(2)**: 121-134.

Lodhi H.S., Tewari K.J., Shukla S. and Sharma U.D. (2008) Alternations in total Haemocyte Counts (THC's) of freshwater prawn, *Macrobrachium dayanum* (Crustacea - Decapoda) after Copper Sulphate exposure. J. Adv. Zool. **29(2)**: 92-98.

Lodhi H.S., Tewari K.J., Shukla S. and Sharma U.D. (2008) Copper sulphate induced fluctuations in total Haemoyte counts (THC) of freshwater prawn, *Macrobrachium lamarrei* (Crustacea - Decapoda) J. Env. Bio-Sci. **22(2)**: 135-142.

Manna A.K. and S.K. Raut (1991), Clutch size in two shrimp species. *Macrobrachium lamarrei* and *M.dayanum*. Environment & Ecology **9(2)**: 349-351.

Mariappan P., Balamurugan P., and Balasundaram C. (2003) Freshwater prawns, *Macrobrachium nobilii*, a promising candidate for rural nutrition Aquaculture Asia **8(4)**: 13-14.

Martin E.A. (1966) "Nutrition in Action" II Edition. Holt Rinchort and Winston, New York.

Mathavan S. and Murugadass S. (1988) In improved design for in-vitro hatching of *Macrobrachium* eggs. Asian Fisheries Sciences **1**: 197-201.

Nagabhushnam R. & Kulkarni G.K. (1981) Freshwater Palaemonid prawn, *Macrobrachium kistenensis* (Tiwari) effect of heavy metal pollutions. Proc. India Nat. Sci. Acad. **B47**: 380-386.

Nagabhushnam R., Hanumante M.M., Mirajkar M. & G.K. Kulkarni (1979) Effect thermal stress on the neurosecretory cells of the freshwater prawn, *Macrobrachium kistenensis* (Tiwari). Hydrobiol. **62(3):** 269-273.

Nakata K. and Goshima S. (2004) Fecundity of the Japanese crayfish, *Cambaroides japonicus*: Ovary formation, egg. number & egg size. Aquaculture **242**: 335-343.

Nakata K. Matsubara H. Goshima S. (2004) Artificial incubation of Japanese crayfish (*Cambaroides japonicus*), egg by using a simple, easy method with a microplate. Aquaculture **230**; 273-279.

Narsimha R., Katre S. & Reddy R.S. (1981) - Moult reproduction relationship in freshwater prawn, *Macrobrachium lanchesteri* (De Man) Proc. Indian Acad. Sci. **90**: 39-52.

Nazari E.M., Costa-simoes M.A., Muller Y.M.R., Ammuar D. & Dias M. (2003) Comparison of fecundity Egg size & Egg mass volume of the freshwater prawns, *Macrobrachium potiuna* and *Macrobrachium Olfersi* (Decapoda - Palaemonidae) J. Crust. Biol. **23(4)**: 862-868.

New M.B. (2003) The role of fresh water prawn, in sustainable aquaculture: In souvenir - International Symposium on fresh water prawns-2003, College of fisheries, Kerala Agricultural University, Cochin. PP-10-13.

New M.B. and W.C. Valenti (2000) Freshwater prawn culture, Blackwell Science oxford.

Phillips D.J.H. (1980) Quantiative Aquatic Biological Indicators Applied science publishers - London- 488-493.

Prasad S. and Kanoujia D.R. (2006) Availability & breeding behaviour of Ganga river prawn, *Macrobrachium gangeticum* (Bate) and *Macrobrachium malcolmsonii* (H.M. Edwards). Asian Fisheries Science. **19**: 377-388.

Praveen S., Ahsan M.K., Ali A.A., Rahman M.S. and Hossian M.A. (2003) Distribution & production potential of small prawn species in Rajshahi area Bangladesh PP-97-101 In:wahab M. A., Thilsteds H. & Hoq. M.E. (eds.) Small indigenous species of fish in Bangladesh Tech Proc. of BAU-Enreca / Dandia workshop on potential of small indigenous species of Fish (SIS) in Aquaculture & rice field stocking for improved food & nutrition security in Bangladesh. 30-31 October 2002. BAU, Mymensingh, Bangladesh, Bangladesh Agricultural University, Mymensingh 2202, Bangladesh P-166.

Proudfit F.T. and Robinson C.H. (1955) Nutrition & Diet therapy "11th Edition, McMillan Company, New York.

Rajagopal A. (1995) Intensive shrimp culture and its impact on local economy and environment in India. Development education & exchange papers (DEEP) FAO. Rome.

Rajyalakshmi T. (1961), Larval development of *Palaemon lamarrei* (H.M. Edwards) and leander flumnicola Kemp. J. Zool. Soc. India **13(2)**: 220-237.

Rao K.J. and S.D. Tripathi (1993)- A Manual of Gaint freshwater prawn hatchery. Manual ser.2. Central Institute of Freshwater Aquaculture, Bhubaneswar **38** P.

Saifullah ASM., Rahman M.S., Jabber S.M.A., Khan Y.S.A., and Nazimuddin (2005) Study on some aspects of biology of prawns from north east & northwest regions of Bangladesh. Pakistan J. Biological sciences **8(3)**: 425-428.

Sarojini R. & Gyananath G. (1983) Behavioural dysfunctions of the fresh water prawn, *Macrobrachium lamarrei* exposed to pesticides. Uttar Pradesh J. Zool. **3**: 145-148.

Sarojini R. & Gyananath G. (1984) Gametogenesis in the freshwater prawn, *Macrobrachium lamarrei*. Acta. Physiological. Academic Scientiarum Hungaricae **63(1)**: 63-76.

Sarojini R., Mirajkar M.S., Nagabhushnam R. (1982) Annual reproductive cycle and effect of temperature on the reproduction of the freshwater prawn, *Macrobrachium kistenensis.* Biology III (3):1-5.

Sen P., Tewari K.J., Shukla S., Shukla R. and Sharma U.D. (2008) Effect of Cadmium on ventilation and oxygen consumption of freshwater prawn, *Macrobrachium dayanum* (Crustacea - Decapoda). Aquacult **9(1)**: 95-100.

Shamra U.D., Khan M.A. Lodhi H.S., Tewari K.J. & Shukla S. (2008) Acute toxicity and behavioural anomalies in freshwater prawn, *Macrobrachium dayanum* (Crustacea - Decapoda) exposed to Chromium. Aquacult **9(1)**:1-6.

Sharma A. and Subba B.R. (2005) General Biology of freshwater prawn, *Macrobrachium lamarrei* (H-Milne-Edwards) of Biratnagar, Nepal. Our nature **3**: 31-41.

Sharma U.D. & Shukla S. (1990) Behavioural dysfunctions of freshwater prawn, *M. lamarrei* (Crustacea - Decapoda) following exposure to a synthetic detergent Linear Alkyl benzene sulphnoate. Biol. Memoirs **16(1&2)**: 58-61.

Sharma U.D. and Shukla S, (1990) SEM studies of the Architecture of the thoracic chalets appendages of freshwater prawn, *Macrobrachium lamarrei* (Decapoda - Palaemonidae). Proc-XII ICEM International Congress for Electron Microscopy, Seattle USA **(3)**: 508-509.

Sharma U.D. and Shukla S. (2006) Acute toxicity of Heavy Metals & detergent to freshwater prawn, *Macrobrachium lamarrei* (Crustacea - Decapoda). Him. J. Env. Zool **20(1)**: 1-6.

Sharma U.D., Shukla S. and Lodhi H.S. (1997) A Report on *Macrobrachium dayanum* (Henderson) (Decapoda - Palaemonidae) form river Gomti, Lucknow (U.P.) India. Him. J. Envi. Zool. **11**: 21-24.

Shukla and Sharma U.D. (1992) Aprocrine secretion in hepatopancreas of Indian fresh water prawn, *Macrobrachium lamarrei* (Decapoda - Palaemonidae) J. Adv. Zool. **13(1&2)**: 60-62.

Shukla S. & Sharma (U.D.) (2000) Histopathological alterations in gills of freshwater prawn, *Macrobrachium lamarrei* (Crustacea - Decapoda) exposed to heavy metal. Proc. Acad. Environmental Biology PP-30.

Shukla S. & Sharma U.D. (1990) Scanning Electron Microscopic Structure & function of foregut of freshwater prawn, *Macrobrachium lamarrei* (Decapoda - Palaemonidae) Proc-XII International Congress for Electron Microscopy, Seattle USA. **(3)**: 516-517.

Shukla S. (1988) Studies on histopathology & behaviour of freshwater prawns, *Macrobrachium lamarrei* (Crustacea-Decopoda) infected with a helminth parasite *Phyllodistomum lucknowensis*, (Pandey 1970). Biol. Memories **14(1)**: 53-56.

Shukla S. (1993) Histopathological effect of Linear Alkyl benzene sulphonate (LAS) on Gill, Green gland and Hepatopancreas of fresh water prawn, *Macrobrachium lamarrei* (Crustacea - Decapoda). Abstracts of National Conference on Environmental & Applied Biology. pp. 43.

Shukla S. and Sharma U.D. (1993) Mercuric Chloride induced nephropathy in freshwater prawn, *Macrobrachium lamarrei* (Crustacea - Decapoda). Abstracts of National Conference on Environmental & Applied Biology pp. 44.

Shukla S. Lodhi H.S., Shukla R. and Sharma U.D.(2009) Fresh water prawns (Crustacea: Decapoda) Good laboratory models for fundamental and toxicological research. National Symposium on Animal models in Biomedical Research: Ethics, Welfare & Issues, CDRI, Lucknow, Feb. 25-26, 2009 PP-120.

Singh D. (1977) In. Unani Dravyaguna Par-III, Ayurvedic & Tibbi Academy, Lucknow, India.

Singh H. and Quereshi T. (1997) Production of fresh water prawn, *Macrobrachium lamarrei* at different stocking densities. Pakistan J. Zoology **29**: 101-105.

Tewari K.J. (2008). Heavy m*etal* toxicity studies on freshwater prawn, *Macrobrachium dayanum* (Crustacea - Decapoda). Ph.D. Thesis, University of Lucknow, Lucknow- India.

Tewari K.J. (2009) Lead induced alternations in scaphognathite oscillations of freshwater prawn, *Macrobrachium dayanum* (Crustacea - Decapoda). Environment & Ecology **27(1A)**: 330-333.

Tewari K.J., Lodhi H.S., Shukla Sandeep, Shukla S. & Sharma U.D. (2008) Effect of Lead nitrate on oxygen consumption of freshwater prawn, *Macrobrachium dayanum* (Crustacea - Decapoda). 28th session of Academy of Environmental Biology held at Deptt. of Botany, Jamia Hamdard University, New Delhi-India (Nov. 20-22, 2008) pp. 49.

Tewari K.J., Lodhi H.S., Tripathi R., Shukla S. and Sharma U.D. (2008) Effect of Lead on Heart beat Rate of freshwater prawn, *Macrobrachium dayanum* (Crustacea - Decapoda). Environment & Ecology **26(2A)**: 807-810.

Tewari K.J., Verma R.S., Kasherwani D., Lodhi H.S., Shukla S. & Sharma U.D. (2007) Lead Induced behavioural anomalies in freshwater prawn, *Macrobrachium dayanum* (Crustacea - Decapoda). Environment & Ecology **25S(3A)**: 802-804.

Tewari K.K. (1955) Distribution of the Indo-Burmese freshwater prawns of the genus *Palaemon*. Fabr. and its bearing on satpura hypothesis. Bull. National Inst. Sci. India **7**:230-239.

Tewari N.P., Rao A.P. and Singh R. (1997) Production of fresh water prawn, *Macrobrachium lamarrei* with fishes in fish culture pond. Third Agriculture Science Congress, PAU Ludhiana, March 12-15 1997, pp. 358.

Tewari R. and Qureshi T.A. (1998) Comparative studies on the fecundity of Macrobrachium dayanum (Henderson 1893)s Collected from two ecologically different aquatic system. In: (Thomas P.C. eds) current & emerging trend in the aquaculture. PP-225-234. Daya Publishing House, New Delhi-India.

Tripathi R. (2007), Effects of Cadmium on reproduction of freshwater prawn, *Macrobrachium dayanum* (Crustacea - Decapoda). Ph.D. Thesis University of Lucknow, Lucknow-India.

Tripathi R., Tewari K.J. Shukla S. and Sharma U.D. (2007) Sex relates acute toxicity and behavioural alterations in fresh water prawn, *Macrobrachium dayanum* (Crustacea - Decapoda) exposed to Cadmium chloride. Aquacult **8(2)**: 251-257.

Tripathi R., Tewari K.J., Lodhi, H.S., Shukla S. and Sharma U.D. (2007) Cardiac responses of fresh water prawn, *Macrobrachium dayanum* (Crustacea - Decapoda) exposed to cadmium chloride. 27th Annual session of Academy of Environmental Biology, Deptt. of Zoology, CCS, University Meerut, India (Oct., 26-28) pp.-50.

Upadhyay A.S., Kulkarni B.G. & Pandey A.K. (2006) Recent advances in giant freshwater prawn culture in India In: Proceeding of Recent Advances in Applied Zoology (Singh H.S., Choubey A.K. & Bharadwaj S.K. Eds). Deptt. of Zoology Ch. Charan Singh University, Meerut. PP 116-132.

Verma R.S., Khan M.A., Tripathi R., Shukla S. and Sharma U.D. (2005) Heavy m*etal* toxicity to fresh water prawn, *Macrobrachium dayanum* (Crustacea - Decapoda). Aquacult **6(1)**: 57-62.

Verma R.S., Lodhi H.S., Shukla S., Sharma U.D. (2007) Effect of Mercuric chloride on total haemocyte counts (THC's) and haemocyte morphology of freshwater prawn, *Macrobrachium lamarrei* (Crustacea - Decapoda).18th All India Congress of Zoology (SCIAZE) University of Lucknow, Lucknow (Dec., 7-9) pp. 55.

Verslycke T.A., Fockedey N., Mc Kennay C.L., Roast S. Jr., Jones M.B., Mees J., Janssen C.R. (2004) Mysid crustaceans as potential test organisms for the evaluation of endocrine disruption: A review Environmental Toxicology & Chemistry **23(5)**: 1219-1234.

Vijayraman K., Jhon George, P. Siva Kumar and R. Rafi Mohammad (1999) Uptake & loss of heavy metals by freshwater prawn, *Macrobrachium malcolmsonii.* J. Envrion. Biol. **20(3)**: 217-222.

Yaqoob M. (1980) Rearing of an economical freshwater prawn, *Macrobrachium lamarrei* (H.M. Edwards under lab conditions Pak. J. Agri. Res. **1(2)**: 139-141.

Bioresources for Food Security and Rural Livelihood, Pages 205–214
Edited by : G.K. Kulkarni and B.N. Pandey

BIORESOURCES AS A TOOL FOR FOOD SECURITY AND RURAL LIVELYHOOD WITH AN EXAMPLE OF *SCYLLA SERRATO* FISHING, MAJOR THREATS AND NEED OF SCIENCE COMMUNICATION FOR MASS EDUCATION

Prabhat K. Bajpai and Ashok K. Saxena

ABSTRACT

All food is obtained from one or the other bioresource and so naturally the security of food is one of the securities of life and since most of such food comes from rural areas the security of rural livelihood is an important concern. There are two major threats - 1. The rapid population growth and 2. Indiscriminate, unfettered exploitation of the resources, that poses a serious danger to the said securities in their future. The imprudent fishing of Scylla *serrata* presents a crucial example, where each year around *50%* of the total catch is the females, and even among the total catch a 2/3rd part is composed of juveniles and sub-adults. The only way to cope up with these problems is the mass education by way of communication. The major problems of science communication in regional languages are considered here, say as the uncertain number of dialects and regional languages, want of correct translation to express correct *concept* etc. Then two schemes to communicate and educate people at large are suggested in which the scientific facts are proposed to be percolated from science workers and experts to teachers, to media persons, to students and to common people and thus to largely expand in the society to curb the fast growing threats.

Keywords : Bioresource, food security, *Scylla serrata,* population, threats to resources, public education.

INTRODUCTION

All food in one way or the other is obtained from some bio-resources, as is popularly said that all flesh is grass. So is told to us through the stories of food chain and energy flow.

The food items are selected both from flora and fauna. In this regard the data on fishing of *Scylla serrata,* the mud crab is presented here to show as to how this example species is being indiscriminately fished. It is followed by two possible reasons that may threat the food security as a whole. The data on *Scylla serrata* has been provided to us by Dr. K.K. Pappu (Pappu, 2009).

FISHING OF *SCYLLA SERRATO*

Among sea foods the demands for crustaceans especially shrimps, lobster, and mud crab (*Scylla* sp.) are strongly rising. The mud crabs have succeeded to occupy an important place among crustaceans for their great nutritive value, besides the delicious taste and good proportion of meat and so are of high economic value in the national and international markets of present day. Mud crabs inhabit the estuaries, backwaters, burrows and mangroves, and according to an estimate about 80 per cent of the total annual landings of mangrove crab consist of this species S *.serrala.*

The mud crab fishing is practiced along almost entire east coast of India, bordered by Bay of Bengal i.e. West Bengal, Orissa Andhra. Pradesh, Tamil Nadu and Kerala. The data from landing station of Andhra Pradesh are given in Tables. 1-7.

Table 1 : Average daily catches from 25 landing stations of Andhra Pradesh during the year 2006.

S.No	Name of the station	Year 2006	Males %	Females %
1	Baruva	13-16 (444)	58	42
2	Isakalapalem	12-15 (390)	50	50
3	Vajrapukotturu	18-20 (575)	36	64
4	Naupada	8- 10 (244)	70	30
5	Bhayanapadu	22-28 (696)	41	59
6	Guppadapeta	5-8 (200)	40	60
7	Ampalam	11-14 (374)	45	55
8	Kalingapatnam	23-27 (755)	38	62
9	Bandaruvanipeta	20-25 (705)	75	25
10	Komaravanipeta	5-7 (199)	60	40
11	Mogadharapadu	18-22 (608)	42	58
12	Srikurrnam	5-7 (180)	75	25
13	Kallepalli	8-10 (285)	65	35
14	Kuppili	5-9 (194)	75	25
15	Konada	11-15 (397)	65	35
16	Chintapalli	13-17 (447)	50	50
17	Bhimili	15-20 (528)	62	38
18	Visakhapatnam harbor	4-9 (186)	25	75
19	Vadachepurupalli	3-5(132)	70	30
20	Bangarammapalem	19-23 (628)	65	35
21	Kakinada	320-450 (I 0395)	60	40
22	Amalapuram	350-450 (11494)	55	45
23	Machilipatnam	16-20 (551)	85	15
24	Ongole	11-13 (372)	70	30
25	Krishnapatnam	14-19 (461)	76	24

Values in parentheses indicate monthly catches.

Table 2 : Average daily catches from 25 landing stations during the year 2007.

S.No	Name of the station	Year 2007	Males %	Females %
1	Baruva	13-15 (407)	65	35
2	Isakalapalem	10-12 (332)	59	41
3	Vajrapukotturu	16-20 (529)	26	74
4	N a\lpada	8- 9 (2]7)	70	30
5	Bhavanapadu	20-24 (649)	27	73
6	Guppadapeta	5-7 (176)	35	65
7	Am pal am	7-9 (246)	38	62
8	Kalingapatnam	21-25 (704)	40	60
9	Bandaruvanipeta	21-23 (671)	38	62
10	Komaravanipeta	5-8 (182)	65	35
11	Mogadharapadu	18-20 (568)	45	55
12	Srikurmam	4-6 (154)	70	30
13	Kallepalli	7-9 (240)	70	30
14	Kuppili	3-7 (168)	68	32
15	Konada	10-13(335)	55	45
16	Chintapalli	12-14 (385)	58	42
17	Bhimili	15-17 (466)	67	33
18	Visakhapatnam harbor	4-6 (162)	36	64
19	Vadachepurupalli	3-5 (107)	61	39
20	Bangarammapalem	17-21 (573)	65	35
21	Kakinada	250-350 (9154)	73	27
22	Amalapuram	325-425 *(IQI55)*	70	30
23	Machilipatnam	15-)9 (503)	65	35
24	Ongole	10-12 (327)	55	45
25	Krishnapatnam	13-15 (414)	67	33

Values in parentheses indicate monthly catches.

Table 3 : Averagedaily catches from 25 landing stations during the year 2008.

S.No	Name of the station	Year 2008	Males %	Females %
1.	Baruva	11-13 (277)	63	37
2.	Isakalapalem	6-8 (211)	80	20
3.	Vajrapukotturu	13-15 (396)	39	61
4.	Naupada	4-6(150)	58	42
5.	Bhavanapadu	15-17(554)	36	64
6.	Guppadapeta	3-5 (113)	42	58
7.	Ampalam	6-8 (165)	48	52
8.	Kalingapatnam	5-9 (520)	31	69
9.	Bandaruvanipeta	14-19 (493)	44	56
10.	Komaravanipeta	5-7 (144)	72	28
11.	Mogadharapadu	8-10 (492)	34	66
12.	Srikurmam	3-5 (891)	70	30

[*Contd. Table*

[*Table Contd.*]

S.No	Name of the station	Year 2008	Males %	Females %
13.	Kallepalli	3-5 (187)	56	44
14.	Kuppili	5-7 (128)	77	23
15.	Komida	12-14 (256)	58	42
16.	Chintapalli	5-9 (284)	85	15
17.	Bhimili	11-13 (359)	77	
18.	Visakhapatnam harbor	4-6 (103)	45	55
19.	Vadachepurupalli	2-4(70)	56	44
20.	Bangarammapalem	18-22 (493)	57	43
21.	Kakinada	300-400 (8381)	64	36
22.	Amalapuram	325-430 (9334)	83	17
23.	Machilipatnam	19-21 (362)	70	30
24.	Ongole	11-13 (246)	66	34
25.	Krishnapatnam	29-31 (349)	80	20

Values in parentheses indicate monthly catches.

Table 4 : Average landings of different size groups in percentage Year - 2006

S.No	Name of the station	Juveniles	Sub adults	Adults	Extra large
1.	Baruva	41	33	17	9
2.	Isakaiapalem	39	34	14	13
3.	Vajrapukotturu	46	23	23	8
4.	Naupada	55	20	15	10
5.	Bhavanapadu	51	30	13	6
6.	Guppadapeta	40	40	15	5
7.	Ampalam	37	27	24	12
8.	Kalingapatnam	34	30	22	14
9.	Bandaruvanipeta	39	32	17	12
10.	Komaravanipeta	49	27	15	9
11.	Mogadharapadu	41	38	11	10
12	Srikurmam	49	34	14	3
13	KallepaUi	39	38	16	7
14	Kuppili	42	37	13	8
15	Konada	44	35	14	7
16	Chintapalli	39	38	11	12
17	Bhimili	56	22	13	9
18	Visakhapatnam harbor	54	22	17	7
19	Vadachepurupalli	39	45	13	3
20	Bangarammapalem	39	40	15	6
21	Kakinada	47	15	24	14
22	Amalapuram	42	29	]7	.12
23	Machilipatnam	49	19	25	7
24	Ongole	35	44	13	8
25	Krishnapatnam	41	35	12	12

Table 5 : Average landings of different size groups in percentage Year - 2007

S.No	Name of the station	Juveniles	Sub adults	Adults	Extra large
1.	Baruva	42	30	19	9
2.	Isakalapalem	37	35	12	16
3.	Vairapukotturu	39	39	14	8
4.	Naupada	52	27	16	5
5.	Bhavanapadu	49	30	12	9
6.	Guppadapeta	34	29	28	9
7.	Ampalam	42	20	24	14
8.	Kalingapatnam	40	23	22	15
9.	Bandaruvanipeta	45	22	16	17
10.	Komaravanipeta	51	19	19	11
11.	Mogadharapadu	42	29	20	9
12.	Srikurmam	53	11	21	15
13.	Kallepalli	40	34	16	10
14.	Kuppili	43	29	19	9
15.	Konada	41	38	16	5
16.	Chintapalli	40	27	17	16
17.	Bhimili	49	21	19	11
18.	Visakhapatnam harbor	46	28	20	6
19.	Vadachepurupalli	42	36	19	3
20.	Bangarammapalem	34	38	19	9
21.	Kakinada	32	35	16	17
22.	Amalapuram	37	24	24	15
23.	Machilipatnam	49	16	27	8
24.	Ongole	38	32	21	9
25.	Krishnapatnam	39	18	25	18

Table 6 : Average landings of different size groups in percentage Year-2008

S.No	Name of the station	Juveniles	Sub adults	Adults	Extra large
1.	Baruva	41	22	24	13
2.	Isakalapalem	43	22	18	17
3.	Vajrapukotturu	40	24	27	9
4.	Naupada	43	25	24	8
5.	Bhavanapadu	46	27	13	10
6.	Guppadapeta	39	23	29	9
7.	Ampalam	37	21	23	19
8.	Kalingapatnam	34	23	27	16
9.	Bandaruvanipeta	36	28	22	14
10.	Komaravanipeta	40	27	24	9
11.	Mogadharapadu	.34	29	26	11
12.	Srikurrnam	49	23	14	14
13.	Kallepalli	36	30	25	9

[Contd. Table

[*Table Contd.*]

S.No	Name of the station	Juveniles	Sub adults	Adults	Extra large
14.	Kuppili	38	26	28	8
15.	Konada	40	24	26	10
16.	Chintapalli	41	15	27	17
17.	Bhimili	42	23	26	9
18.	Visakhapatnam harbor	50	25	18	7
19.	Vadachepurupalli	43	28	21	8
20.	Bangarammapalem	48	26	17	9
21.	Kakinada	45	11	25	19
22.	Amalapuram	48	7	29	16
23.	Machi 1 ipatnam	49	19	24	8
24.	Ongole	35	29	29	7
25.	Krishnapatnam	39	26	17	18

Table 7 : Dead or Injured crabs from different landing stations in Kgs

S.No	Name of the station	2006	2007	2008
1.	Baruva	16	18	23
2.	Isakalapalem	8	12	14
3.	Vajrapukotturu	5	8	21
4.	Naupada	11	16	13
5.	Bhavanapadu	22	20	26
6.	Guppadapeta	14	14	6
7.	Ampalam	6	18	0
8.	Kalingapatnam	23	18	16
9.	Bandaruvanipeta	15	17	14
10.	Komaravanipeta	11	15	20
11.	Mogadharapadu	10	8	8
12.	Srikunnam	19	20	15
13.	Kallepalli	26	24	20
14.	Kuppili	13	6	21
15.	Konada	7	9	11
16.	Chintapalli	.9	14	20
17.	Bhimili	25	18	22
18.	Visakhapatnam harbor	7	10	9
19.	Vadachepurupalli	5	7	4
20.	Ban,garammapalem	12	16	14
21.	Kakinada	23	27	32
22.	Amalapuram	19	25	29
23.	Machilipatnam	7	4	10
24.	Ongole	11	5	12
25.	Krishnapatnam	14	7	11

GRADING

Extra large	-	1-1.5 Kg and above (20% of the catch)
Large	-	500g - 1 Kg (30-35% of the daily catch)
Medium	-	300-500 gms (20% of the daily catch)
Smsal	-	200-300 gms (30% and above of the daily catch)

The threats

1. The table 1-3 show that the fished crabs comprise of both males and females whereas it is of utmost importance that females should be spared for the secured continuity of further generations whole the tables 4-6 show that the catch contains a small number of extra large crabs and also the adult crabs and a considerably larger part of the catch are the juvenile and sub adult crabs. Such unfettered and indiscriminate fishing is obviously a danger for the possibility of the sustained livelihood of fishermen, at some locations may be today and at other locations may be tomorrow.
2. There is a 70,000,000 net world population increase each year. Addressing environmental issues without intercalating the overpopulation factor os like trying to sweep sand against the tide (yahoo answers, 2009) Besides this there is good large number of deleterious consequences of population increase and that certainly escalates pressure on the bioresources In geometric ratios (answers.yahoo, 2009).

Suggested Remedy to Minimise the Threats

There is only one feasible solution in a democratic set up that the general people are persuaded to understand the consequences of these threats, in case they are not checked and are allowed to continue with the prevalent degrees. Higher is the amount of threat sooner the consequences shall precipitate. The simple equation is that as long as the 'development' is going on, the natural balance of food resources will keep on shrinking and this will be further escalated by population growth i.e. larger the population, more mouths are to be fed, higher is the demand for the civil amenities, job opportunity, standard of life etc that ultimately ask for more development and thus adding more pollution, more exploitation of resources and consequently a gradual dwindling of the food, and related economic affairs.

To save future from such adverse situations a massive programme for mass education in people's language is an inevitable prerequisite.

The Problems of Science Communication

Linguistic Diversity in India

1. Sir George Grierson (1903-1923) in his *Linguistic Survey of India* (12 *vols.*) recorded 179 languages and 544 dialects.

2. Census - 192 I recorded 188 languages and forty nine dialects.
3. Census - 1961 recorded 184 mother tongues with less than 10.000 speakers each.
4. Census - 1981 reports 112 mother tongues with more than 10,000 speakers each.
5. Eighth Schedule of Constitution of India (1992 amendment) enlists 18 languages.
6. Times of India (Education Times, Dee 17, 2007) reported 22 official languages and over 2000 dialects.

Besides such wide variations in linguistic diversity and uncertainty of number of languages there are certain challenges that have to be resolved for successful result oriented efforts.

The Challenges

Patairiya (2002) recognizes following challenges :

1. Mass Media interest is not being attracted by Science.
2. Science communication and scientific temper largely suffers at the hands of illiteracy; besides efforts made in 18 official languages, those in local languages are yet wanting.
3. Training Science Communicators.
4. Systematic planned effort yet requires more attention to be paid.

Marzia Mozzonetto (2005) has reviewed Science Communication in India, its current situation, history and future with similar impressions.

Besides the above mentioned challenges there are certain other problems and perspectives that deserve to be enumerated here as well (Bajpai, 2008).

- Mass inquisitiveness is highly wanting.
- Most of students study with examination oriented psyche rather than curiosity with intentions to disseminate the knowledge that they gather.
- The science communication is still largely limited to traditional methods like print media, sometimes radio and TV but on very limited and usually on topics not concerned with the issues high lighted above and so fails to inspire response.
- Government funded schemes like NSS and NCC and local chapters of ISCA (as there are 24 of them in the country) could be used that have not been taken into consideration so far.
- The human resource management too is to be considered for the purpose.

Suggested Ways to Address the Problems of Science Communication

The primary requirement is the correct translation so that the message is properly transmitted in a form as it should be. As far as possible the translation has to be so precise that if some text is translated from a regional language to another one, if not from original source, the source matter should be perfect enough to provide nearest possible correct concept. For this the principles from *Upnishads* give a very good guide line in following two aphorisms :

1. *Gavam anek varnanam kshirasyasti ekvarnata* - cows are of many colours but milk's colour is one. So let there be words from different languages adapted to express a thought but let they all mean the same.
2. *Ekam Sat viprah vahuda vadanti* - The truth is one but the learned ones speak the same in diverse ways. So if there is a term with one truth or meaning or sense it has to be spoken in different languages i.e. in different ways, but with same sense. This has to be followed by correct way of presentation so that the communication may not end in to a fiasco of non receptive attitude of the people, as inquisitiveness is highly wanting.

There are other ways to convince the people about population growth and its dangers. The materials available in print or electronic media, including websites are to be brought to common people. For example the e-articles like 'Top 20 thought provoking effects of over population' (Paggu.Com.) that tells almost every thing about the threats imposed by overpopulation and such materials are to be widely publicised so as to let people realize the hard facts. This can better be achieved, particularly for rural areas, by audiovisual means of communication, because the communicator or anyone 'else concerned with public' education must keep in mind the old dictum - seeing is to believe. In the same way the people may be taught about the consequences of the imprudent, unfettered and indiscriminate exploitation of the resources without keeping in mind the futuristic consequences i.e. today's effect on tomorrow.

For the regional purposes it is highly important that the regional personnel recruited or selected for communication must be thoroughly well versed to talk to the rural masses in an acceptable manner that must be convincing to the people in the perspectives of the prolorged assurance of livelihood and a better opportunity to continue with the profession of food generation, not only for current generation but also to leave legacy for the cocningones.

Two related Schemes are being Mentioned here below

Science Communication: Scheme I - Scientist and science workers to teachers trained to communicate in regional linguistic medium with press notes to social circles and to Students (A class from various different socioeconomic groups) who should be trained to communicate it to the people, peasants and others from related neighbourhood.

Science Communication: Scheme II. Science communicators, Journalists, and teachers (specially from junior levels of teaching) should simultaneously be exposed to the views of resource persons *for* the intended information and then the three recipients should go to disseminate this through there own channels to common people either directly or through secondary means. Thus Joth of these schemes in nut shell are directed towards science communication by way of percolation from top resources to bottom recipients and then its expansion from person to person on grass root level.

CONCLUSIONS

It is obvious that the over population and indiscriminate exploitation of nature resources are posing a potential threat to the food security when considered as a long term affair. Where ever both of these are combining to affect the people, as is the case with our country, the future situations are further threatened to become grimmer than is being expected today. Therefore, the mass communication and public awareness remain to be only means to rectify the mistakes of past and present, because it is the people who are solely responsible for the population growth and also for the indiscriminate exploitation.

How governments can help the situation? The answer to the question is - by educating its people, regardless of their age and socioeconomic *status.* Thus a great deal of responsibility lies with the educational system including those who formulate education policies, frame syllabi, those who are responsible for formulating policies for information and broadcasting, with those who are there to protect and maintain the environment, with those who are" supposed to protect, preserve and assure a continuum of wild life and above all with those who claim to be teachers as they are said to be architects of future, by inspiring their students to think wider beyond curricula and examinations.

REFERENCES

Bajpai, Prabhat K. (2008). Challenges in science Communication in regional languages: an educationist's view. . First *Rashtriya Vigyan Sancharak Sammelan* (Science Communication Meet). Parallel session to 95th Ind. *St.* Congo Held at Andhra University, Vishakhapatnam, p.96.

Grierson George Abraham (1923). *Linguistic Survey of India,* 11 Vols. in 19 Parts Delhi, Low Price Publ. (2005)

Mazzonetto, Mama (2005) : Science communication in India, current situation, history and future developments. J.Science Communication. (jcom.sissa.itlcontentsof?Authors=Marzia%20Mazzonetto)

Open Question (2009). answers.y ahoo. com/question/index?qid = 20090828\\5655AAh 7\nw

Patairiya, Manoj (2001).Science Communication in India:Perspectives and ChalleJ?ges (2001). British Association for the Advancement of Science Festival, Glasgow, September, 2001.

Pappu, K.K. (2009). Fishery, Conservation, Feeding and Fecundity of mud crab *Scylla serrata* (FORSKAL, 1775) from coastal Andhra Pradesh with special reference to Srikakulam District. Ph.D. Thesis, Andhra University, Vishakhapatnam, A.P.

Bioresources for Food Security and Rural Livelihood, Pages 215–243
Edited by : G.K. Kulkarni and B.N. Pandey

INTEGRATED FISH FARMING IN SCARIFYING THE RURAL POVERTY WITH SPECIAL REFERENCE TO EASTERN U.P.

J.P. Shukla

ABSTRACT

Integrated Fish Farming (IFF) is a system that focuses diversified agricultural production with emphasis on fishes. Our survey report revels that roughly each village of eastern (U.P.) exhibits at least one pond. IFF possesses the potential of bridging wide gap between fish demand and supply. The fish integrated culture is discussed emphasizing its importance and relevance to poverty alleviation / hunger eradication. Its potential has to be fully exploited particularly in eastern U.P. in concrete terms of immense fish production from agriculture. The theme undertaken reviewed the various IFF practices that exists roughly allover the world and groups them under distinct systems that may well flourish in the climatic conditions of North- India in general and Eastern U.P. in particular.

Keywords : IFF; Aquaculture Integration; Fish cum Rice farming; Fish cum Horticulture; Fish Cum cattle farming; Fish cum duck farming ; Fish cum poultry farming.

INTRODUCTION

Integrated fish farming (IFF), also called as agropisci culture or integrated agriculture aquaculture, has a long history in Asia dating back to more than 1500 years in India (Coche 1967) and more than 2400 years in China (Willman *et al.,* 1998).In China , IFF started as a complex system combining poultry, livestock and corp farming with fish polyculture. It also involved the application of grass, crops and manure as feed and fertilizer , all in a complex, mutual beneficial system. Such a long practice of IFF in china partly accounts for the fact that China produces more than 60% of world's total aquaculture production (Rana *et al.*, 1998). In view of Asala (1994), the idea behind integrated fish farming is to create a mutually beneficial system, which will lead to maximization of fish productivity through optimum resource use.

Integrated fish farming (IFF) incorporates two or more than two production technologies running simultaneously on parallel footing to boost each other's farming. This results to increase the total productivity, additional employment, reduced investment on fish farming., enhanced efficiency in resource utilization and reduced risk factor through diversification of crops. In the words of K.Ven Kataramanujam (2002), the major capital investments in a fish farm. These costs and other costs are increasing regularly throughout the world. Therefore, it is imperative that fish production per unit area has to be increased to minimize the effective cost and thereby mitigating the world's "protein-hunger" especially that of developing countries. One such means has been adopted in which a combination of some form of animal husbandry with fish farming has been successfully employed. Out of all these forms of cultures, fish-cum-duck culture appears to be the most useful, however, the practice of other combinations like poultry or pigs with fish culture may also serve the desired purpose.

IFF and agriculture is quite compatible to each other by balancing the economy of natural resources, and thus to reduce the cost of production. Pond embankments could be used for growing napier grasses for grass carp. Vegetables such as tomatoes grow well on bundhs which are fertilized with pond silt which is rich in plant nutrients. Pond silt from fertile ponds, in many ways, is equal to good quality compost which can also be used as a fertilizer in fields and gardens. The pond embankments could also be planted with fruit trees which provide shades , thus providing regular additional income. In short, IFF may be defined as a mixed farming system in which along with the fish production. It also provides cereal, fruit, milk, eggs, meat etc. Thus IFF has initiated a new way towards the improvement of the financial condition of the farmers. A systematic representation about the central theme of IFF is shown in Fig. 1.

Indeed Asia is the world's foremost Continent in terms of IFF. Vast areas of land (particularly rice fields) in Chin, India, Japan, Indonesia, Vietnam, Thailand, Bangladesh and Philippines are used for integrated rice-fish farming.

IFF in Africa has been reported in nations like Benin, Nigeria, Madagascar, Malawi, Camerron and Zambia but chiefly at subsistence level. Brown (1983) reported the use of vast areas of land in Hungary.

FAO (1979) and WFC (2003) has listed the benefits of IFF in view that jet millions of people who depend on fishes to live are every day faced by the fear of food shortage. It shall not be out of place to mention that Indian women may play a key role in a achieving food security as in Africa's agricultural sector (Moehl 2003).

PRINCIPLE OF INTEGRATED FARMING

The basic principle involved in integrated farming is the utilization of the synergetic effects of inter-related farm activities and the conservation, including the full utilization of farm wastes. It is based on the concept that "there is no waste" and "waste is only a misplaced resource which can become a valuable material for another product" (FAO, 1979).

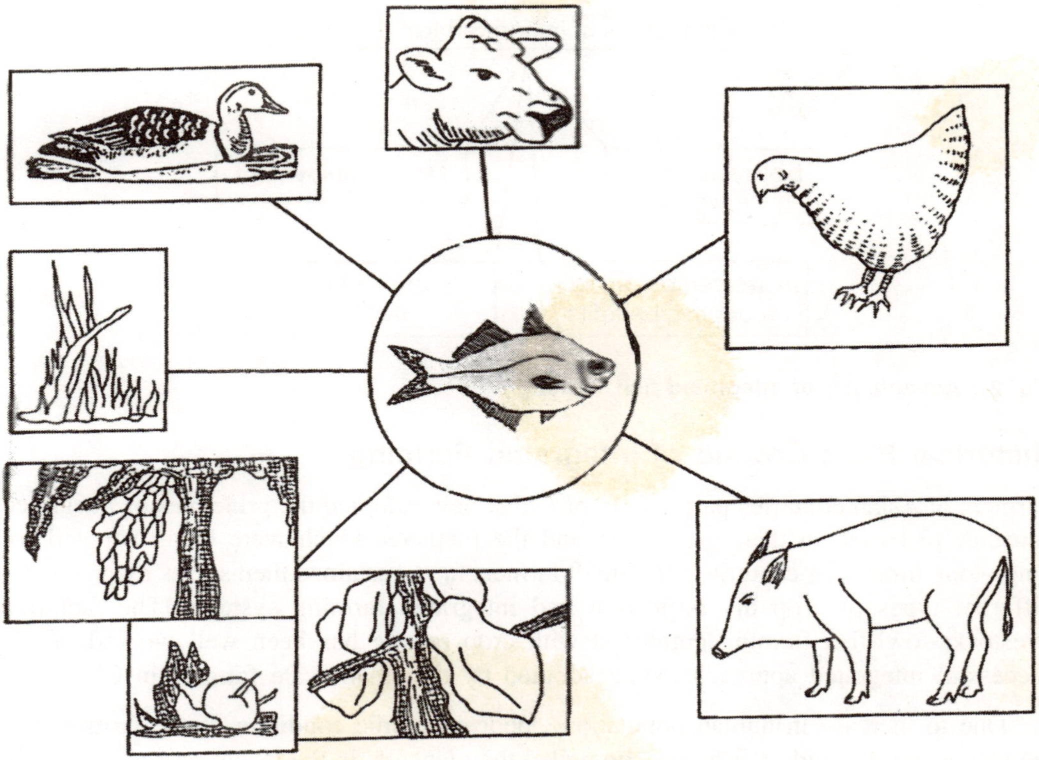

Fig. 1 : Integrated fish farming includes fish, crop, livestock, horticulture, sericulture, poultry etc. The wise integration of these farming systems promotes full utilization of land area and recycling of wastes and by-products, minimiZes the operational expenses infeed and fertilizer and maintains a balanced ecosystem.

SCOPE OF INTEGRATED FISH FARMING (IFF)

Today's growing emphasis is on maximization of production through systematic utilization of all available resources. Multicommodity integrated farming systems involving crop farming, livestock raising and pisciculture, maximize production through recycling of wastes, proper utilization of resources which yields higher return to the farmers. Advantages of IFF in precise form is shown in figure 2.

The concept of integrated fish production with other activities is in contrast to the relatively recent development of intensive pisciculture. In which, emphasis is laid on a more complex technology and high inputs of resources and energy.

Practically, the extensive aquaculture is the most common form of fish production involving the traditional techniques, which yields low production of fish pr unit area by putting maximum inputs. While the integration with agriculture or other activities may improve the fish production . This system is more viable and better option , thus giving relatively high yields with fairly low input cost.

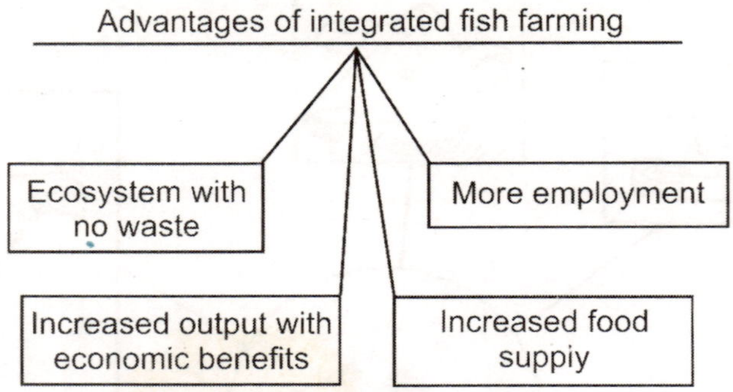

Fig. 2 : Advantages of integrated fish farming

Historical Back Ground of Integrated Farming

Farmers of Asian countries particularly of China have adopted the principles of integrated aquaculture based on their experience and the methods which were being adopted for long –long times. For centuries , the small farmers have sustained themselves by practicing different types of crop diversification and integrated farming systems. The fact that livestocks-fowl-fish farming combined with crop raising has been well worked out as successful integrated approach, being adopted by the small scale farmers in China.

Due to increase in human population , poor economic condition of the farmers and the increasing demand of fish, has compelled the planners, in the recent past to adopt IFF system in South East Asian countries including India.

Practice of catching wild fish, prawns and shrimps for food from paddy fields is seemingly old, this is an example of the combination with the paddy cultivation. Now, it has been combined and scientific methods are being adopted in South East Asian countries. Paddy cum fish farming in China is being carried out science ninth century.

Besides China, countries viz. Taiwan, Vietnam, Thailand , Philippines, India, Bangladesh etc. have adopted IFF as an aquaculture activity. IFF system is not being followed in the countries like USA , Germany, Israel and Hungary. The IFF system has also contributed in providing employment opportunities to the people. Besides, many developing Asian countries, the countries in Africa and South America have also introduced this system on a large scale.

SALIENT FEATURES OF INTEGRATED FARMING SYSTEM

1. In Integrated farming, pisciculture is practiced as a basic activity which is supplemented by other crops. In some cases it is an additional or secondary activity to aquaculture animal husbandry, depending upon the prevailed conditions.
2. South east Asian countries are rich in animals resources . For example, in India alone, about 321 million tons/year of agricultural residue; animal wastes in the form of urine and excreta and agriculture based raw materials are available for recycling in IFF system.

3. Since IFF system involves lower cost and is environment friendly, it is economically sustainable and is being freely adopted.
4. This system is most suitable for small and petty farmers, possessing small land base but large man power.
5. It promotes the full use of the available natural resources.

   Potential of IFF system in India with special reference to North- East

   Fish plays avital role and is a staple diet for the large population in India and particularly for the people of North- East region. Farming is the principal major economic activity of this region. In North- East region there are vast opportunities for IFF, this will be clear from the table 1 (N.K. Datta And S.M. Kumar, 2002).

Table 1. Showing North-East Region's Aquatic Resources for I.F.F.

States	Rivers (Km)	Reservoir (ha)	Beels, Lakes etc (ha)	Ponds/tanks (ha)	Paddy cum Fish (ha)
Arunanchal Pradesh	2000	160	2500	150	575
Assam	5050	10,730	1,00,000	22,800	—
Manipur	2000	100	19,150	5000	—
Meghalaya	5600	8430	375	500	85
Mizoram	1700	32	—	1,752	120
Nagaland	1600	—	275	500	2000
Tripura	1200	4500	500	10,264	—
Total	19,150	23,792	1,43,740	40,809	2,780

IMPORTANT INTEGRATED FISH FARMING SYSTEMS

Pisciculture in India has undergone a progressive evolution during the past few decades and there has recently been made some vigorous efforts to integrate livestock and crop farming with it. Barash *et al* ., (1982) divided integrated animal- fish farming into two:

1. Semi- integration i.e. raising livestock on land and transporting the manure to the fish pond . Complete or vertical integration i.e. raising livestock directly above the fish pond so that animal manure can drop directly into the fish pond. However, V. Sundarraj and B. Santhakumar (2002) have grouped important IFF systems into two principal types;
   1. Aquaculture integration – agriculture
   2. Aquaculture – livestock integration.

   In aquaculture integration , following farmings are practiced.

   a. Rice- fish integrated farming

   b. Rice – azolla- fish integration

c. Fodder- cattle- fish integration.

d. Horticulture- crop- fish integration, and

e. Milberry- silkworm- fish integration.

In aquaculture livestock integration, following farmings are practiced.

a. Fish- duck, integrated farming

b. Fish- cattle , integrated farming

c. Fish- poultry, integrated farming

d. Fish- pig, integrated farming , and

e. Fish- goat/ sheep, integrated farming.

FISH PRODUCTION IN IMPORTANT INTEGRATED FARMING

In intensive fish farming, the indigenous and exotic are cultured. Among the major carp, *Hypophthalmichthys molitrix* (silver carp) records best growth, followed by *Ctenopharyngodon idella* (grass carp), *Labeo rohta* (rohu), *Cirrhinus mrigala* (mrigal) and *Cyprinus carpio* (common carp). *Catla* is also cultivated in integrated fish farming revealing good growth.

M.K. Datta and S.M. Kumar (2002) has recorded following fish yield in important integrated farming systems.

Fish cum rice farming. Both Rabi and Kharif- Extensive fish farming is adopted . Fish production approximately 250-600 kg/ha/yr.

Fish cum horticulture. Fish farming extensive and semi- intensive . Fish production approximately 2-3 mt/ha/yr. (1mt=1000kg).

Fish cum cattle farming. Fish and cattle both farming semi-intensive . Fish production approximately 2-3 mt/ha/yr.

Fish cum duck farming. Fish and duck both farming semi-intensive . Fish production approximately2-3mt/ha/yr.

Fish cum pig farming. Fish farming semi- intensive whereas pig farming intensive, Fish production approximately 3-6 mt /ha/ yr.

Fish cum poultry farming. Fish farming semi-intensive whereas [poultry farming intensive. Fish production approximately 3-5 mt /ha/yr.

PROSPECTS OF IMPORTANT IFF IN INDIAN CONTEXT

S. Sundarraj and Santha Kumar (2002) has focused silent ideas over prospects of important IFF in Indian context.

Paddy –cum- Fish culture in India as well as other S.E. Asian countries is an age – old practice. This farming is being adopted in areas where paddy fields retain water for three to eight months a year and the culture of fish in fields which remain flooded even after the paddy is harvested, this cold also serve as an off season occupation for farmers. In the recent years the paddy cultivation involving the use of pesticides has greatly impeded fish culture in the paddy fields. Despite these constraints, there is still scope of paddy-cum-fish culture, especially in the low-lying inland areas where the paddy fields get inundated and the water depth exceeds 75 cm.

A traditional paddy cum prawn culture system is popular in the paddy fields of adjoining tidal back-waters and estuaries of rivers. During the monsoon months (June to September) when the water in the fields is almost fresh, paddy is cultivated and for the rest of the year , the fields are utilized for prawn culture.

There is about 26,000 hectares area of total brackish water, of which only 5000 hectares are currently utilized for growing paddy during the south- west monsoon season and prawn for the rest of the year. Thus from 5000 hectares of paddy fields in central Kerala, nearly 6000 tons of prawn catch is estimated in addition to the paddy crop. Moreover, in brackish water areas of West Bengal and Kerala where paddy fields are traditionally used for prawn culture and a production of 50 kg /ha of *penaeus monodon*, 250 kg/ha of mullets and 3000 kg of *Tilapia* have been achieved in addition to 2.4/h of paddy (Sundararaj and Santha Kumar, 2002). Some important IFF systems are discussed below.

1. Paddy-cum-Fish Culture in Freshwater

Cheryl (2002) reported about 2.3mha of low-lying paddy-growing tracts found in the Eastern and Northeast provinces of India covering Assam, Manipur, Tripura, West Bengal, Orissa, Northern Bihar and Eastern U.P., where one crop (tall deep water variety) of paddy is only cultivated.

For a duly prepared paddy field of 1.02 ha, a perimeter canal of 0.27 ha is excavated inside the field. During **kharif** season a deep water paddy variety viz. **Jaladhi-2** is sown directly in the paddy field (0.75 ha) which grows fastly in deep waters. Fingerlings of Indian major carps rohu, catla and mrigal are stocked at the rate of 6000 nos. / ha in the ratio of 3:3:4 in the same field. These are fed with rice bran and GNOC at 1:1 ratio at the rate of 2-3% of body weight. After harvesting the deep water paddy during early December, the main paddy field dries up gradually and the fishes move instinctively to the perimeter canal. In the **Rabi** season, high yielding paddy varieties namely **Jaya** and **Ratna** are cultivated in the same field. Insecticides like BHC and Demecron may be applied to destroy paddy insect pests. Before application, care is taken to prevent drainage of insecticides washing into the perimeter drainage of insecticides washing into the periphery of the paddy field.

Procedures of raising fishes in paddy fields

The management of pisciculture in paddy field is almost similar to that of pond fish farm, but since it is closely related with agronomical base of paddy culture, certain deviations are obvious as described below.

Layout : The paddy plots of 25 ×10 m are regarded ideal for paddy cum pisciculture. They may be cut on either side of a central line which is dug deep to act as a main supply or drain canal. Each plot is bordered on all the sides by bundhs (dike) of about 0.3 m height and 0.3 m in width. Towards the side of drainage canal, the dike is provided with inlet and outlet devices made suitably of hollowed bamboo pipe. These are placed in the middle of the length of the dike and are fitted with screens to avoid entry of predators and escape of the cultured fishes.

Each paddy plot is dug along its middle to form a long central canal of 0.5 and 0.75 m width and depth respectively. This canal connects with the main supply through the dike. It acts as a refuge canal for the fishes not foraging among the plants and as capture canal when the water level in the paddy field goes down.

Alternatively, marginal canals can be dug along side the bundh bordering the paddy plot. These canals may be connected with the several oblique canals running across the paddy field. At one side of the field, the marginal canal connects with water inlet pipe, from main supply channel and at the other side with the outlet pipe which drains into main water collection channel. The space between outlet channel of marginal canal and the collection channel may be dug deep to trap fishes during release of water from the paddy field. Both, the inlet and outlet pipes are fitted suitably with screens. Such arrangement ensures a constant flow of water through the field, which suits the fishes inhabiting the slow running water in the nature.

Production. The kharif paddy produced a crop of 1200 kg/ha while Rabi crop yielded 4.300 kg/ha making a total of 1200+4300=5500 kg/ha/yr. Within the 10 months (6 month along with kharif +4 months in the canal during rabi season) a production of 700 kg of fish/ha could also be achieved.

Advantage of paddy-cum-fish culture

The principal advantages of this integrating farming system are:

1. In low lying areas where irrigation facilities are not available and rabi crop is not possible, farmers may adopt this system for a second crop of paddy along with sizable crop of fish.
2. Operational economics of this integrated farming system shows a higher return than only paddy cultivation.

3. Fish feces serve as fertilizer for paddy and paddy yield is increased by 10-15%.

   [**Note.** Culture of prawn (*Macrobrachium rosenbergii*) in a saucer shaped deep water paddy field along with deep water rice variety Jaladhi-2 with silver carp 150/ha introduced in culture, to control excess algal bloom yield 1700 kg/ha for deep water paddy, roughly 500 kg/ha/8 months for prawns and 63 kg/ha/8 months for fishes.]

Yield : Considering the minimum production that can be obtained from paddy (5,500 kg/ha) and fish (700 kg/ha) plus other crops, the net profit will be around Rs. 3500/ha).

It is important to mention that the practice of shrimp production (*Penaeus indicus, Macrobrachium rude*) in the rice fields on the west coast of India is carried out in those areas where water is salty and only one crop (July-September) of a salt resistant variety of rice can only be grown.

2. Fish-cum-*Azolla* Integrated Farming

Azolla is a free floating aquatic heterosporous fern belonging to the family Azollaceae. In India, the genus is represented by *Azolla pinnata* which is widely distributed and occurs free floating on the surface of water in ponds, pools, reservoirs and lakes. It fixes atmospheric nitrogen through the cyanobacterium *Anabaena azollae* present in its dorsal leaves. It is annual and found in large clusters farming mats and reproduces both vegetatively and sexually. It may quickly cover the open surface of water bodies. Its thick mat often eliminates mosquitoes and hence it is also called mosquito fern. Under favorable conditions Azolla biomass doubles every 4-10 days. The high nitrogen fixing capacity, rapid growth and also decomposition rate of this fern results in quick nutrient release that makes it an ideal nutrient in farming systems.

There exists a symbiotic relationship between *Azolla* and *Anabaena azollae* i.e. *Azolla* provides shelter, nutrients and energy to the alga while *Anabaena* in turn fixes atmospheric N_2. Though *Azolla* is capable of absorbing N_2 from the atmosphere, *Anabaena* meets the entire N_2 requirements of *Azolla-anabaena* association. The average N_2 fixing rate of a developed *Azolla* mat is in the range of 1.0-2.6 kg/ha.

Chemical composition of Azolla : Freshly harvested *Azolla* contains:

Nitrogen	:	3.4%
Phosphorus	:	1.0%
Potassium	:	2.3%
Fat	:	6%
Protein	:	13.23%
Carbohydrate	:	60%

This composition makes *Azolla* useful either as a source of nitrogen or feed.

Commercial application of *Azolla* : *Azolla* has been utilized for centuries as livestock feed in Indo-China. Being high in Potassium, its ash may be mixed with fat to synthesize soap. It can be used in water purification containing N, P and heavy metal contamination.

Application of *Azolla* in fish farming : *Azolla* in aquaculture, is primarily useful as a nitrogenous bio-fertilizer. Due to its high decomposition rate, it becomes a suitable substrate to enrich the detritus food-chain or for microbial processing such as composting prior to application in ponds. *Azolla* biofertilization fulfills the nutrient requirements of composite carp culture which could be met through its application @ 40t /ha/yr which supplies over 100 kg of nitrogen, 90 kg of potassium and 25 kg of phosphorus in addition to about 1500 kg of organic matter.

The care must be taken during the integrated use of *Azolla* in fish pond, *Azolla* should cover only 20% of the surface area of the pond so that light penetration is normal.

Advantages of Azolla cum fish integration

1. Azolla is a good food for grass carp.
2. Azolla acts as a bio-fertilizer.
3. During hot days, Azolla provides shade to the fishes.
4. Dried Azolla may be used as food for cattle.
5. Decayed Azolla also becomes feed for common carp and mrigal.

3. Duck-cum-Fish Integrated Farming

Based on ICAR experimental findings of 1985, for one ha pond area 200 numbers of ducks were reported to be sufficient for integration. In these experiments 3.7 tons of fishes, 16,000 numbers of duck eggs and 500 kg of duck meat obtained from one ha area in one year. Gross and net income obtained through such experiments during 1985 were Rs. 54,400.00 and 18,175.00 for duck and fishes respectively.

The first scientific experiment on duck-cum-fish farming was made by Probst in Germany in 1934, but due to World War II the results remained unutilized. After the war, when there was serious shortage of protein in European countries, large scale experiments were initiated in Hungary (1952), Czechoslovakia (1953) and East Germany (1955) to determine optical husbandry methods for raising ducks in fish ponds in the climatic condition of central Europe.

Economical justification of duck-cum-fish culture

Combined culture of ducks and fishes can be justified economically as they are mutually beneficial and the profitability of the integrated system enhances manifolds. The prime advantaged is to gain larger quantity of animal protein from limited area. Construction cost of fish pond is recovered quickly and the ducks are benefited from the fish pond and fish growth is accelerated by the duck manure, which maintains a continuous supply of organic matter containing C, N and P. These elements boost the biological production in the water column and increase the natural food supply for the stocked fishes.

Food of ducks in fish pond. Ducks utilize natural food from the pond and consume wide variety of organisms e.g. aquatic weeds, insects and their larvae, snails, tadpoles and frogs. After feeding roughly 50-60% of their excreta falls into the pond and manures it.

Composition of duck manure. According to **T. Francis** (2002), each duck produces roughly 7.0 kg fresh manure over 36 days period. If 500 ducks are present in a fish pond then total manure produced in 36 days would be approximately 3.0-3.5t. The manure contains about 57% water and 26% organic matter. One hundred kg of duck manure would contain about 10 kg carbon, 1.4 kg P_2O_5, 1.0 kg Potash (K_2O), 1.08 kg calcium 2.8 kg of other materials and rest water etc.

Stocking density of ducks in fish pond. There is no fixed number of ducks to be cultured in fish pond, however, in India this number is generally 100-150 ducks/ha.

The composition of fish population and the stocking density determines the fish yield. Phyto and zooplankton production increases significantly in the pond due to the duck manure. The fish chosen should be able to crop this food; e.g., a phytoplankton feeder like the silver carp and a zooplankton strainer like the big head carp (*Aristichyhys nobilis*). The natural food of a common carp (a bottom feeder) mostly increases indirectly. The high plankton population increase depends upon the organic matter content of the pond bottom which in turn boosts the benthic population.

Management of ducks in fish pond

For stocking day old or 10-14 days old ducklings are ideal. The breeders must be in age of 1 to2 years with sex ratio 1 male to 4 to 6 females. Egg laying is up to 4 years but with decreasing return. Sexual maturity is achieved at 5th month and egg laying commences at 6th to 7th months. A suitable protein rich feed (20% protein), drinking water and proper illumination are essential. The daily feed ration is about 9-10% of the body weight. The egg laying 120-140 eggs/female. The weight of normal egg is about 80-90 gm and 8 cm long and 5cm wide. The incubation period is 28 days. The survival rate from incubation day old ducklings is 75%. Losses during the nursing periods (10-14 days) and rearing periods are 8% and 2-3% respectively. Day old ducklings required special care and a controlled environment with respect to temperature, feed, drinking water, light, space etc., up to an age of 10-14 days, after which they can be stocked in the ponds. During the first week, about 50-55 ducklings/sq meter are kept in a heated room on a screen floor (0.2cm thick wire: 1.5 cm mesh size) to allow manure to fall through. The pelleted starter feed is offered in simple demand feeders and clean tap water is provided in troughs, which permit access only to the beak to prevent duckling from becoming wet. The air temperature must be in the range of 30-32°C as suggested by **T.Francis,** (2002).

Transferring the ducks in fish ponds. In earlier commercial duck-cum-fish integrated farming practice, the ducks were allowed to room freely on large ponds, dispersing their excreta uniformly over the entire pond area while foraging for natural food. Wire enclosures are provided for the ducks in the fish pond (called " water runs") with proper resting and feeding places on the shore or over the water. The enclosures keep the ducks in , but allow the fishes to entre the " Water runs". Wave action carries the well manured water out the " water runs". Approximately 0.5-1.0 m^2 of enclosed dry run 2-4 m^2 of 'water run' is allowed per duck.

Duck Feed : The general composition of feeds for the different stages of duck rearing is shown in Table 2 (T. Francis, 2002).

Ingredients applied for ducks feeds are shown in Table 3.

Apart from the above feed, concentrate feed may be given at the rate of 100gm/duck day.

Table 2 : General composition of important feed for different stages of duck rearing (From: Duck-cum-fish culture by T. Francis)

Components (Value in %)	Type of feed			
	Starter	Rearing	Fattening	Layers
Dry Matter	86.0	86.0	86.0	86.0
Carbohydrates	68.0	68.0	69.0	67.0
Digestible Protein	18.0	15.0	11.5	16.8
Crude protein	20.5	18.0	14.0	19.2
C_aCO_3	3.0	3.0	3.0	4.5
$C_{a3}(PO_4)_2^+$	0.8	0.7	0.7	1.6
Mineral premix	0.5	0.5	0.5	0.5
Vitamin premix	0.5	0.5	0.5	0.5

Table 3 : Ingredients applied for duck feed

Ingredient	Starter feed (%)	Rearing feed (%)	Layers feed (%)
Wheat	57.9	53.9	40.9
Maize	20.0	20.0	20.2
Fish meal (64% crude protein)	10.0	2.0	2.5
Soya grit(47% crude protein)	8.0	9.8	12.3
Meat meal(50% crude protein)	0.9	4.0	5.7
Wheat bran	2.5	8.7	5.0
Wheat germ	—	—	1.5
Milled sunflower seed cake	—	—	3.1
Food lime	—	0.3	5.3
Methionine premix	—	0.5	0.5
Salt	0.2	0.3	0.2
Mineral premix	—	—	2.5
Vitamin premix	0.5	0.5	0.5

(**Source:** Techniques and Management of IFF).

The ratio composition for laying ducks is given below.

Broken rice	:	50 Parts
Sunflower oil cake	:	15 parts
Wheat bran	:	15 Parts
Fishmeal without salt	:	10 Parts
Gingelly oil cake	:	5 Parts
Mineral mixture for poultry	:	3 Parts
Shell grit	:	2 Parts

Feeding time. 8:00 AM and 4:30 PM

Duck health care. Duck must be given plague vaccine. They must be deformed using piperazine adopted 20% solution at the rate of 60 ml/100 ducks through drinking water. It must be repeated once in six months.

Fish and duck yield. Experiments conducted in German Democratic Republic show an average increase in carp production of 100kg/ha with 300 duck/ha kept in the pond for 10 months. In Hungary, 300-500 ducks/ha give fish yield of 500-800 kg/ha in 150 days. The number of ducks may range from 2500 to 3500/ha/yr to yield 5-6 t/ha of duck meat and 2750-5640 kg/ha of fish.

In Vietnam, raising 1000-2000 ducks/ha on pond increased the average fish yield to 5.0 t/ha/yr compared to 1.0 t/ha/yr without ducks.

In India experimental trials have yielded 4323 kg/ha/yr fish with 1000-150 ducks/ha.

Ducks must be left on ponds until they reach to a suitable size. It is desirable to have 4-6 duck production cycles/yr. The expected yield/ha from duck fish farming is approximately 4 ton of fish and 2.0-2.5 ton of ducks.

The number of ducks raised per ha of pond is about 1500 in tropics and subtropics. The highest fish yield is achieved by using suitable fish species stocked at optimum numbers and by continuous harvest of marketable fish and then replacement by young stock.

Economics of fish-cum-duck integrated farming

The economics of fish-cum-duck integrated farming produced by R. Suresh and P. Selvary (2002) considering the assumptions mentioned ahead in 11 points is as follows:

A. Capital investment

(i) Cost of construction of Duck pen 200×1 sq. ft/duck @ 100/sq. ft	:	Rs.	20,000
(ii) Cost of pumping set	:	Rs.	20,000
(iii) Cost of inlet-outlet structures	:	Rs.	10,000
Total Capital investment	:	Rs.	50,000

B. Costs

(a) Fixed costs

(i) Depreciation on capital items @ 10%	:	Rs. 5,000
(ii) Interest on capital @ 12%	:	Rs. 6,000
(iii) Repairs and upkeep	:	Rs. 2,000
(iv) Insurance premium @ Rs. 1.5 /bird	:	Rs. 300
(v) Rental Value of land/ha	:	Rs. 10,000
(vi) Wages to permanent labor @ Rs. 1000 /month	:	Rs. 12,000
Total fixed costs	:	Rs. 35,300

(b) Variable costs

(i) Pond preparation cost/ha	:	Rs. 2,000
(ii) Cost of fish seeds 10,000 nos. @ Rs. 500/100	:	Rs. 5,000
(iii) Cost of 200 ducks @ Rs. 80/duck	:	Rs. 16,000
(iv) Cost of 200 duck feed × 30 kg/duck @ Rs. 7 /kg	:	Rs. 42,000
(v) Veterinary and electricity expenses for 200 ducks @ Rs. 2 /duck	:	Rs. 400
(vi) wages for fish harvesting for 1 ha	:	Rs. 5,000
Total Variable costs	:	Rs. 70,000
Total Costs (a+b)=35,300+70,400	:	Rs. 1,05,700

(c) Returns

(i) By sale of fish (500 kg @ Rs. 20 / kg)	:	Rs. 1,00,000
(ii) By sale of duck eggs (200×125 eggs/duck × Rs. 2 / egg)	:	Rs. 50,000
(iii) By sale of culled ducks (200×1.5 kg / duck @ Rs. 25 / kg)	:	Rs. 7,500
(iv) Total Return	:	Rs. 1,57,000
Net Return= (Total return total costs)	:	Rs. 51,800

Assumption : In Fish-cum-Duck Farming

1. Fish farming is taken up for 10 months in the existing 1 ha water spread area to which duck rearing is integrated.
2. Interest on capital @ 12%.
3. Insurance premium @ Rs. 1.5 / duck.
4. Stocking density of fish seeds @ 10,000 / ha.
5. Cost of 18 weeks old ducks @ Rs. 80 / duck.
6. Duck feed required @ 30 kg / duck @ Rs. 7 /kg.

7. Veterinary and electrical expenses @ Rs. 2 / duck.
8. Sale price of fish is @ Rs. 20 / kg.
9. Sale price of duck egg is @ Rs. 2 / egg.
10. Sale price of culled duck is @ Rs. 25 / kg.
11. Monthly wages for 1 permanent labor @ Rs. 1000 / months.

Ecological significance of duck- fish culture

1. Raising ducks over fish ponds fits very well with fish poly culture system.
2. Duck dropping act as fertilizer and reduce about 60% of total input cost producing planktonic food for fish.
3. Ducks check aquatic weeds, earthworms, leeches, insect and larvae.
4. 10-20% of feed wasted by ducks in utilized in ponds as fish feed and fertilizer.

4. Fish-cum-pig Integrated Farming

Integrated fish -pig farming in the country can be selectively due to personal reasons.

This system can improve socioeconomic status of the weaker section of society and the people living in rural areas. The tribal's who traditionally raise pigs can take up fish-pig farming easily. FAO (1979) report reveals that Chinese Consider pigs as free fertilizer factories on hooves. Pig manure is a complete manure, its phosphorus content is much higher than cow dung. Pig manure contains major inorganic nutrients like N, P, K in addition to trace elements such as Ca, Cu, Zn, Fe and Mg. The major portion of pig urine is nitrogen in the form of urea and decomposes easily. It has been reported that a pig excretes about 1000 kg of faeces and 1200 kg of urine in a culture period of 8 months i.e. from piglet to an adult swine (Anon, 1991). Normally a pig excretes about 7 to 20% of its body weight.

From 1979-1981, fish-pig integrated farming in Eastern India yielded fish ranging from 6,644.5 to 6,792.4 kg/ha/yr, without supplementary feeding and fertilizers. This data (4250-4, 448 kg /ha / yr) is much encouraging than achieved after using intensive feeding and fertilizer application. In addition pork production ranging from 563.28 to 563.32 kg / yr has been achieved. The cost of fish production was calculated to be Rs. 2.4 / kg as compared to Rs. 5.00 kg in conventional fish culture.

Application of pig manure

In fish ponds, pig manure is applied as fertilizer in raw or fermented form, the fermentation of pig manure takes place by compositing.

The application of fresh and untreated pig manure to fish ponds are possible where pig sites are located over the ponds. The use of fresh and untreated manure to fish ponds has given high fish yields, but excessive amounts may cause fish mortality because of O_2 depletion in the pond water. Daily application of pig manure has been found to raise the efficiency of conversion of manure to the fish body protein thus leading to higher fish yields. Hence daily fresh pig manure (not in excess) application to fish pond gives maximum yields than compost form. The chemical composition of pig excretory products is given in Table 4.

Table 4 : Chemical composition of pig manure and urine in fresh matter Source : Utilization of Piggery in Fish culture by S. Athithan.

Components		Faces (%)	Urine (%)
1	Moisture	70-77	90-95
2	Dry matter	23-30	5-10
3	C/N ratio	14:1	- - -
4	Organic matter	15	2.5
5	Inorganic matter		
	(i) Total carbon	2,72-4.00	- - -
	(ii) Total nitrogen	0.60	0.40
	(iii) NH_3 nitrogen	0.24-0.27	- - -
	(iv) Phosphorus	0.50	0.10
	(v) Potassium	0.40	0.70
6	pH (not in %)	6.6-6.8	7.6

Varieties of pigs used for integration with fish culture. The following varieties of pigs may be used for integration with pisciculture.

1. Large white Yorkshire
2. Middle white Yorkshire
3. Berkshire
4. Hampshire
5. Tam worth
6. Saddle back, and
7. Landrace

From the above mentioned varieties, large white York shire and landrace are being widely used in India; while others are being used in some selected areas.

Large white shire : This breed is actually native of U.K. and is being used as an extensively exotic breed in India. The body color is white with occasional black pigmentedspots. Ears are erected. The snout is of medium length and has a dished face. It is prolific breeder.

Landrace : This breed is native of Denmark. It is white with black skin spot freclies. The body is long and large with long snout.

Pig housing : The pig housing is collectively called as 'Pigsty'. The housing system may be indoor or open air system (if large area). Basic considerations in providing proper structures are necessary for housing pigs, particularly for exotic variety and should provide proper hygienic conditions to maintain the health of pigs.

Pigsty has one or more pens. In each pen, one or more pigs may be kept. In indoor system, the pigsty is usually constructed in one corner of the dyke of the fish pond. The pigsty generally consists of the following components, namely (i) Pig accommodation room, (ii) Feeding compartment, (iii) Side, and (iv) main door. In pigsty over heating should be controlled.

Feed of pigs : Balanced quality feeding plays an important role in successful pig production. Pigs are the most rapidly growing livestock but are more susceptible to nutritional deficiencies than the ruminants. Pigs are also fed with chopped grasses once in a week to prevent mineral deficiency. A complete diet for pig consists of protein (not less than 20 %), carbohydrates, lipids, minerals, good water and vitamins. Table 5 depicts various requirements in the formation of creep, grower and finisher ratio for pig.

Table 5 : Nutritional requirements for the creep, grower and finisher.

	Components (up to weaning) (20-40 kg)	Creep feed	Grower ratio	Finisher ratio
1.	Protein Supplements (%) (i) Oil cakes (ii) Animal Protein	16-188-10	14-164	13-142
2	Grains, maize, millets, sorghum (%)	60-65	50-55	40-50
3	Rice/wheat bran (%)	5	10	20
4	Mineral mixture (%)	0.5	0.5	0.5
5	Antibiotic supplement (mg)	40	20	10

Source : Utilization of piggery waste in fish culture by S. Athithan (2002)

Further, the amount of feed according to the weight to be given to pig / day is shown in Table 6.

Growth of pigs. The time taken for pig to reach 70 kg body weight is 8-9 months in winter and 9-11 months in summer. At weaning time, the piglets weigh about 15-16 kg. Under proper management practices, pigs may reach 70-80 kg body weight in 6 months time.

Table 6 : Amount of feed according to the weight to be given to pig / day

Weight of pig in kg		Daily consumption of feed (kg/Pig)
1	7.5-8.0	0.7
2	10-12	1.2
3	20-25	1.3
4	30-35	1.5
5	40-50	1.6
6	70-80	1.7
7	90-100	1.8
8	120-130	2.0

Source : Utilization of Piggery waste in fish culture by S. Athithan. Weight of pig in kg Daily consumption offeed (kg/Pig)

Yield of integration of pisciculture with pig farming

The integration of pig with pisciculture in India, was attempted for the first time in 1979 at Krishinagar, Nodia district of West Bengal. The yield of 7.3 tons / ha / yr from pig-fish culture experiment without the application of supplemented feed and fertilizers is high when compared to the average yield of 3.232 t/ha/yr. The above yield is also significantly higher than the yields obtained from pig-fish rearing in other countries (3.625 and 4.14 t/ ha/ yr in USA; 3.667 t /ha/yr in Penang and 5.5 t/ha/yr in France.

Experiments and field trials in India regarding pig-fish farming have given encouraging results. During 1985, ICAR team has conducted fish culture experiments by integrating two batches of piglets (each) which resulted in 4.2 tons of pig production in 6 months time, in addition to 6 tons of fishes/ha/yr with no expenditure towards feed cost.

Culture Practices: The ponds measuring about 1000 m^2 should be located near the farmer's house so that he may take care of the fishes and pigs and can discourage poaching . Dikes should be constructed properly. The pond should be deep enough so as to retain more than 1m water depth during the dry period.

Pond Preparation: Farmer should drain and dry the pond to remove all the weeds and fish fauna remaining in the pond. If it is not possible to drain the pond, all the fishes may be killed by applying 15 kg of both bleaching powder and urea for a 1000m^2 pond. Alternatively, 250 kg Mahua oil cake in a pond of 1000 m^2 can be applied, this also kills all the fishes and acts as organic pond fertilizer. Pigs are brought to the pond before stocking the fish, so basal application of manure is hardly required.

Cultivable fish species stocking density with integration

The most favored species combination in composite fish culture with pig integration 1000 m^2 pond, includes.

1. Catla catla (Surface feeder) – 160
2. Labeo rohita (Column feeder) – 160
3. Mirgal (Bottom feeder) – 120
4. Common Carp (Bottom feeder) – 120
5. Silver Carp (Surface feeder) – 160
6. Grass Carp (Feeding on macro vegetation and terrestrial grasses) – 90

For4 3 species the stocking rate is 320 catla; 240 rohu and 240 mrigal. The stocking densities vary from 5000-10000 nos./ha/yr and species rates of 40% surface feeders, 20% column feeder, 30% of bottom feeders and 10% macro vegetation feeders is preferred for high fishy production. Farmers should stack the pond with fingerlings 7 days after the treatment with bleaching powder.

Note: About 35-40 pigs would be sufficient to fertilize one hectare of fish pond water.

Economics of fish-cum-pig integrated farming: The economics of fish-cum-pig integrated farming as produced by R. Suresh and P. Selvaraj (2002) considering the assumptions mentioned ahead in 13 points is as follows:

(A) Capital investment

1.	Cost of construction of Pigst 35×16 sq.ft/pig@100/sq.ft	:	Rs.	56,000
2.	Cost of Equipments @ Rs. 100/pig	:	Rs.	5,500
3.	Cost of pumping set	:	Rs.	20,000
4.	Cost of inlet-outlet structures	:	Rs.	10,000
	Total Capital Investment	:	Rs.	91,500

(B) Costs

(a) Fixed costs:

1.	Depreciation on capital items @ 10%	:	Rs.	8,950
2.	Interest on capital @ 12%	:	Rs.	10,740
3.	Repairs and upkeep	:	Rs.	2,000
4.	Insurance premium @ 6% on the value of pigs	:	Rs.	945
5.	Rent or Rental Value of land/ha	:	Rs.	10,000
6.	Wages to permanent labor	:	Rs.	12,000
	@ Rs. 1000/month×12	:		
	Total fixed costs	:	Rs.	44,635

(b) Variable costs:

1.	Pond preparation cost / ha	:	Rs.	2,000
2.	Cost of fish seeds 10,000 nos.@ Rs. 500/1000	:	Rs.	5,000
3.	Cost of piglets of 15 kg size 35nos/cycle×2 cycle @ 450/piglets	:	Rs.	31,500
4.	Cost of pig mash @ 300kg/animals/cycle@ Rs. 5/kg	:	Rs.	1,05,000

5.	Cost of fodder @ 400kg/animal/cycle for 35 animals/cycle for cycle @ Rs. 50 p/kg.	: Rs.	14,000
6.	Veterinary Expenses @ Rs. 100/pig×70	: Rs.	7,000
7.	Electricity expenses @ Rs. 200/months ×12	: Rs.	2,400
8.	Wages for fish harvesting for 1 ha	: Rs.	5,000
	Total Variable cost (b)	: Rs.	1,71,900
	Total cost (a+b)= 44,635+1,71,900=2,16,535	: Rs.	2,16,535
(c)	***Returns:***		
1.	By sale of fish (500 kg @ Rs. 20/kg)	: Rs.	10,000
2.	By sale of pig (5600 @ Rs. 30/kg)	: Rs.	1,68,000
	Total return (TR)	: Rs.	2,68,000
	Net return (TR-TC)	: Rs.	51,465

Assumptions in fish-pig farming

1. Fish farming is taken up for 10 months in the existing 1 ha water spread area to which piggery integrated.
2. Number of pigs required to manure 1 ha water spread area in 2 cycles of six months each.
3. Pigsty @ 16 sq. ft/pig for which cost of construction is Rs. 100/sq feet/
4. Depreciation @ 10% on capital items.
5. Interest on capital @ 12%.
6. Insurance premium @ 6% of the value of the pig.
7. One permanent labor @ Rs. 1000/month.
8. Stocking density of fish seeds @ 10,000 /ha
9. Cost of 2 months old piglets weighing 15 kg each is Rs.450.
10. Pig mash required @ 300 kg/ pig/cycle in addition to 400 kg of fodder/pig/cycle.
11. Veterinary expenses @ Rs. 200/month.
12. Sale price of fish is @ Rs. 20/kg and sale price of pig is @ Rs. 30/kg.

Advantage of fish-cum-pig farming: Fish-cum-pig farming system has obvious advantages:

1. The pig dung acts as an excellent pond fertilizer and raises the biological productivity of the pond and consequently increases fish production.
2. Some of the fishes feed directly on the pig excreta which contains 70% digestible food for the fishes.
3. No supplementary feed is required for the fish culture , which normally accounts for 60% of the total input cost in conventional fish culture.
4. The pond dikes provide space for the erection of animal housing units.
5. Pond water is used for cleaning the piglets and for bathing he pigs.

5. Fish-cum-Cattle (Dairy) Farming

For certain groups of rural folk in India, cattle raising is a way of life and the application of cattle manure for fertilizing fish ponds is an ancient practice. Consequently, the fish dairy integrated farming is easily adopted by such farmers. For example, raw cattle dung could be used for fertilizing fish ponds and humus from the ponds, in turn, would make a good fertilizer for growing cattle fodder on dykes along the fish ponds. Recent attempts to estimate the optimum number of animals required to maximize productivity per unit of water when raising small livestock and fish have shown that five cows can produce 9500 kg of milk along with 3,500 kg of fish, resulting in a net profit of Rs. 63,350 /ha/ yr. The cattle when grown in the same farm in close proximity, make it easy and inexpensive to utilize their waste materials for fertilizing the fish ponds. For example , the cattle sheds are located adjacent to the fish ponds and wastes and washing from the sheds are sent through pipes directly into the pond water. Part of the fodder required for the cattle may be grown on the pond banks.

In fact, cattle and fish integration is a simple and most common mode of integration adopted since long. The cattle urine , dung and washings can be disposed easily which saves energy, labor and money, but can also supply fish and milk both to the market.

Significance of cattle used in fish cattle farming

1. Cattles provide organic manure to fish ponds.
2. Cattle dung is nutritionally rich.
3. Cattle dung is very fine due to repeated digestion in the cattle.
4. Cattle dung can suspend longer in fish pond.
5. The suspend ability of dung not only enables fish to get more feed, but also reduces oxygen consumption caused by manures, this also avoids the formation of harmful gases.
6. The BOD of cattle dung is relatively less than other livestock manures, because the cattle forage decomposes already by microbes in the cattle stomach itself. This causes no risk for the fish ponds using cattle manure.

Table 7 : Nutritional components of cattle dung and urine in fresh matter source: Fish-cattle agricultural integration by N. Ramanathan

Components	Quantity in %
Organic matter	(i) In cattle dung - 14.0
	(ii) In cattle urine - 2.3
Inorganic matter	(i) In cattle dungNPK = 0.3 : 0.3 : 0.1
	(ii) In cattle urineNPK = 1.0 : 0.1 : 0.4

The cattle wastes like dung, urine and wasting when stagnate in the vicinity of cattle farm, it creates a breeding ground for mosquitoes which are the well known of several human disease. But if these wastes are utilized for fertilizing the pond, this will boost the fish culture on the hand and also at the same time the growth of disease causing organism would by prevented.

Stocking density of cattle: For one hectare water spread area of fish pond, six members of cattle (buffalo/cow) are found to be ideal. The cattle shed constructed on the fish pond bundh should a concrete flooring so as to drain the washing along with cattle dung into the fish pond. The fooder raised on the pond bundhs may be used for feeding the cattle. In this type of integrated fish farming the inputs on fertilizers and supplementary feeds are not required which leads to less cost of fish production.

Production of fishes in cattle-fish integration: Experiments made on this type of integration reveal a production up to 7420 kg of fishes/ ha/ yr.

Importance of cattle-fish integration

1. Minimizing wastes leading to improved environment quality of the area.
2. Reducing need for fertilizer but with increased fertility.
3. Improves soil nutrients and increased its fertility.
4. Increases fish production leading to improved nutrition and income of rural population.
5. Reduces dependence upon outside inputs leading to increase stability.
6. Increasing employment opportunities.
7. Cattle are raised for milk and manure on pond banks and their washing are drained directly into the pond . Under this system cost of fish culture is reduced by 50%.

Economics of fish-cattle integrated farming

Considering the assumption mentioned ahead in 13 points by Suresh & Selvaraj, the economics of fish-cattle integration is as follows:

A. Capital investment

1.	Cost of cross-bred cows 6 no. @ Rs. 12,000/cow	:	Rs.	72,000
2.	Cost of construction of cow-shed 6×50 sq. ft /cow ×Rs.150/sq. ft	:	Rs.	45,000
3.	Cost of equipments 6×Rs. 500/cow	:	Rs.	3,000
4.	Cost of pumping set	:	Rs.	20,000
5.	Cost of inlet-outlet structures	:	Rs.	10,000
	Total Capital Investment	:	Rs.	1,50,000

B. Costs

(a) Fixed costs:

1. Depreciation on capital items @ 10% except for coes	: Rs.	7,800
2. Interest on capital @ 12%	: Rs.	18,000
3. Repairs and upkeep	: Rs.	2,000
4. Insurance premium @ 6% on the value of cow	: Rs.	4,320
5. 5. Rent or Rental Value of land/ha	: Rs.	10,000
6. Wages to permanent labour2×12 month@ Rs. 1000/ labor/months	: Rs.	12,000
Total fixed costs	: Rs.	54,120

(b) Variable costs:

1. Pond preparation cost / ha	: Rs.	2,000
2. Cost of fish seeds 10,000 nos.@ Rs. 500/1000	: Rs.	5,000
3. Cost of cattle feed 4kg/cow/day for 300days and 2kg/cow/ day for 65 days 6×1330 kg @ 1/kg	: Rs.	47,880
4. Cost of fodder grass 10kg/cow/day 6×3650 kg @ 1 kg	: Rs.	21,900
5. Veterinary Expenses @ Rs. 100/cow/yr × 6	: Rs.	1,200
6. Electricity expenses @ Rs. 100/cow/yr × 6	: Rs.	1,200
7. Wages for fish harvesting for 1 ha	: Rs.	5,000
Total Variable cost	: Rs.	84,180
Total cost (a+b)	: Rs.	1,38,300

(c) Returns:

1. By sale of fish (5000 kg @ Rs. 20/kg)	: Rs.	1,00,000
2. By sale of milk (10 1/cow/day×300 day)	: Rs.	1,53,000
Total return	: Rs.	2,53,000
Net return	: Rs.	1,14,700

Assumptions

1. Number of cross-bred cows required to manure 1 ha water spread area is 6.
2. Cost of cross-bred cow yielding 101 of milk/day is Rs. 12,000.
3. Cattle shed @ 50 sq. ft/cow for which cost of construction of Rs. 150/ sq. feet.
4. Cost of equipments @ Rs. 500/cow.
5. Depreciation @ 10% on capital items.
6. Interest on capital @ 12%.
7. Insurance premium @ 6% of the value of the cows.

8. Two laborers required for maintance for whom monthly wages are to paid @ Rs. 1000.
9. Stocking density of fish seeds @ 10,000/ha.
10. Cattle feed @ 4kg/day/cow for 300 days & 2kg/day/cow for a day period of 65 days in addition to 10kg/day/cow of fodder grass.
11. Veterinary expenses @ Rs. 200/cow/yr and Electrical expenses @ Rs. 100/months.
12. Milk production 10@ 1/cow/day/300 days.
13. Sale price of fish is @ Rs. 20/kg and milk @ Rs. 8.50.

6. Fish-cum-Poultry Integrated Farming

A sample and economically viable system for integrating pisciculture and poultry farming without the use of any supplementary feed or fertilizer have been developed by the zoologists and agricultural scientists.

This type of farming system has been found to produce annually 4,500-5,000 kg of fish, more than 70,000 eggs and about 1,250 kg of poultry using 500-600 birds per ha of pond area. An annual net income of roughly Rs. 50,000 /ha/yr is possible at the current rate.

Integrated fish-poultry farming is practiced in many countries of the world more particularly in Asia. Such integrated farming in not only an efficient way of recycling of farm wastes but also yields high economic returns. The integration of aquaculture with poultry results in a more efficient use of resources . The cost associated with pisciculture operation are reduced by approximately 70% when integrated with chicken. It is because fish farming recycles chicken wastes and spilled chicken feed as food and fertilizer for the fishes. Consequently, supplemented food and fertilizer is hardly required and thus it is not provided. Birds are reared to produce eggs and meat. Birds raised for meat and called broilers and that for eggs are called layers. Scientific poultry management aims at maximizing returns with minimum expenditure and management inputs.

Selection of birds : The birds used in fish-poultry integrated farming may be classified on the basis of utility, financial gain or for fancy purpose. The kind of bird to be raised along with fish may be chosen with care from the meat type (broilers) or egg type (layers) birds. The birds of Babcock BV-300 of Leghorn breed are suitable for this purpose.

Stocking density. About 500-600 birds (layers) produce litter, which is adequate enough to fertilize one hectare of fish pond water area. About eight week old chicks, after vaccination against viral disease along with other necessary prophylactics, are kept in poultry house near the pond.

Each pen of laying birds is provided with nest boxes over laying eggs. Egg production commences at the age of 22 weeks and then gradually declines. The birds are usually kept as layers up to the age of 18 months. The egg laying is 210-250 eggs/ birds/ year.

According to the reported research findings of ICAR 1985 , one can get 5.2t of fish, 70,000 eggs,and 1250 kg poultry meat integrating 550 layer birds in one hectare water spread.

Poultry farming and management : Chicken production has undergone great technological advancement and expansion during the last 3 decades. Broadly , it can be divided into two general classifications.

1. Layers for commercial egg production
2. Broiler production for meat.

Poultry housing and waste collection facilities : Poultry regardless of species, is generally housed in similar types of shelter with primary differences occurring in feeder or eater design and stocking density.

Two distinct housing structures , based on management , are being used presently. These are cage system and litter or floor system.

Cage system is used predominantly for egg laying operations but there is a growing trend towards caged broiler production. Cage contain 1-25 birds depending upon the size.

Litter systems are used predominantly for chicken broiler and layer feeding and also for breeding . With floor operations, 4 to 6 inch of an absorbent litter material like leaves, sand, sawdust, rice bran, wheat straw etc., can be used for litter with choice. Litter life can be extended with greater ventilation to remove moisture.

Feeding : Feeding is the most important aspects and it accounts for 60-70% of production cost. Feed should be filled in the feeder three or four times in a day . There should not be any left over in the night to avoid rats and wastage . The requirement of protein and energy for different categories of chick are furnished in Table 8.

Balanced poultry feeds under different trade names are available in the market. The poultry bird under caged system or deep litter system should be fed regularly with balanced poultry feed according to their age.

Table 8. Protein and Energy Requirement in Chicken.

Category of chicken	Age in weeks	Protein %	Energy in k cal/kg of
Layer chick	0-8	22	2600
Grower	9-20	18	2500
Layer	20-72	17	2600
Broiler			
Starter	0-4	23	2800
Finisher	5-8	20	3000

Source: Livestock management by S Balachandran.

An ample supply of water is made available at all times otherwise egg production is adversely affected.

Nutritional status of poultry manure : Depending upon conveniences, the poultry shed can be constructed over the pond in such a way that birds droppings may reach the pond water directly or on the pond bundhs. Each bird produces about 40 g of manure/ day. Poultry manure is thought to be a complete fertilizer having both the organic and inorganic contents. The conversion coefficient for chicken manure is 6.8

Health care : A strict adherence should be followed in the schedule of vaccination against some of the contagious of vaccination disease, as there is a prime need to prevent the loss of the birds.

Advantages of integration of fishes with poultry farming : Raising chicken over ponds has various advantages:

1. The chicken houses constructed on ponds maximize use of land needed for other purposes.
2. Hygienic conditions are better in the chicken houses constructed over ponds, as the faeces fall directly into the pond's water.
3. Chicken excreta fertilizers the pond and boosts planktonic bloom.

Fish stocking : Carp fingerlings such as Catla, Rohu, Mrigal Grass carp, Silver carp and Common carp are stocked at a density of 5000 numbers/ ha in a ratio mentioned earlier in this chapter (Fish-pig integration).

Yield : According to 1985 research finding of ICAR , one can get 5.2 tons of fish, 70,000 eggs and 125 kg poultry meat by integrating 550 layer birds in one ha water spread.

Economics of fish-poultry integrated farming : Considering the assumptions as mentioned ahead by Suresh & Selvaraj (2002), economics of fish-poultry integrated farming is as follows:

[A] Capital investment

1.	Cost of construction of poultry shed 500×1 sq. ft/.birds@ Rs.100/sq/ft	:	Rs.	50,000
2.	Cost of 500 cages @ Rs. 45/cage.	:	Rs.	22,500
3.	Electrical installations and overhead water provision 500×Rs.10/bird	:	Rs.	5,000
4.	Cost of pumping set	:	Rs.	20,000
5.	Cost of inlet-outlet structures	:	Rs.	10,000
	Total Capital Investment	:	Rs.	1,07,000

[B] Costs

(a) Fixed costs:

1. Depreciation on capital items @ 10%	:	Rs.	10,750
2. Interest on capital @ 12%	:	Rs.	12,900
3. Repairs and upkeep	:	Rs.	2,000
4. Insurance premium 500×Rs. 1.50/bird	:	Rs.	750
5. Rent or Rental Value of land/ha	:	Rs.	10,000
6. Wages to permanent labor @ Rs. 1000/ months×12 month	:	Rs.	12,000
Total fixed costs	:	Rs.	48,400

(b) Variable costs

1. Pond preparation cost / ha	:	Rs.	2,000
2. Cost of fish seeds 10,000 nos.@ Rs. 500/1000	:	Rs.	5,000
3. Cost of layer birds 500× Rs. 90/ bird	:	Rs.	45,000
4. Cost of poultry feed 500×40 kg/bird @ 6.25/kg	:	Rs.	1,25,000
5. Veterinary and electricity Expenses 500× Rs. 3/bird	:	Rs.	1,500
6. Wages for fish harvesting for 1 ha	:	Rs.	5,000
Total Variable cost	:	Rs.	1,83,500
Total cost (a+b)	:	Rs.	2,31,500

(c) Returns

1. By sale of fish (5000 kg @ Rs. 20/kg)	:	Rs.	1,00,000
2. By sale of eggs (500×280 eggs/bird @ 1/kg)	:	Rs.	1,40,000
3. By sale of called birds (500×1.5kg/bird @ Rs.300/kg)	:	Rs.	22,5000
Total Return	:	Rs.	2,62,500
Net Return (TR-TC)	:	Rs.	30,600

Assumptions

1. Fish farming is taken up for 10 months in the existing 1 ha water spread area to which poultry farming is integrated.
2. Number of birds required to manure 1 ha water spread area is 500 stocked at egg laying (pullet) stage of 20 weeks reared for 52 weeks in cages.
3. Poultry shed constructed at 500 sq. ft @ Rs. 100/ sq. ft provided with 500 cages @ Rs. 45/ cage and electrical and water provisions @ Rs. 10/ brid.
4. Depreciation @ 10% on capital items.
5. Interest on capital @ 12%.
6. Insurance premium @ Rs. 1.50 / bird

7. One permancent labor is required for farm maintenance for whom monthly wages are to be paid @ Rs. 1000.
8. Stocking density of fish seeds @ 10,000/ ha.
9. Cost of 20 weeks old layer birds (pullets) Rs. 90/ bird.
10. Poultry feed required @ 40 kg/ bird @ Rs. 6.25/ kg.
11. Veterinary and Electrical expenses @ Rs. 3 birds.
12. Egg production @ 280 egg/ bird and weight of culled birds 1.5/kg/bird.
13. Sale price of fish is @ Rs. 20/kg sale price of egg is @ Rs. 1/egg and sale price of culled bird is @ Rs. 30/kg.

Besides the above described integrated fish farming systems, integration of fruit and vegetable farming on the fishpond embankment has also been experimented on India. It has also several advantages:

1. The farmer gets additional income from growing fruits and vegetables on the pond embankment.
2. The nutrient-rich pond mud is used as fertilizer for growing crops, eliminating the cost of organic manures.
3. Manured pond water is used for irrigation of plants.
4. Fruit and vegetable residues are used as feed for the fishes.
5. The plants on the embankment strengthen the dikes.

INFERENCE

IFF mat thus be concluded as a system which the fish culture being the target commodity combined with the livestock or crop cultivation . Thus it is a sort of dual farming system which has been attempted as a new kind of non traditional approach to meet and supplement the growing economy of several Asian countries. Since this technology is being practiced only in limited areas, it needs to be promoted and should be intensified in much larger water bodies in larger areas of our country. This would help to boost the economy of petty farmers and also simultaneously improves the socio-economic condition of fishermen.

REFRENCES

Anon. R (1991). Quoted from Fish & Fisheries by K. Pandey & J.P. Shukla. Vol 1(2006) Rastogi Publication, Meerut.

Asala, G.N (1994). Principles of integrated aquaculture. In: A.A Olatunde;

Ayeni J.S.O. and I.M Ogunsuyi (eds). Proceedings of the Nat. Fisheries workshop on aquaculture dev. Fish seed production and post harvest tech. NIFFR-FACU 1994. Pp. 206-220.

Athithan. S. (2002). Quoted from Fish & Fisheries by K. Pandey & J.P. Shukla. Vol 1(2006) Rastogi Publication, Meerut.

Balachandran S. (2002). Quoted from Fish & Fisheries by K. Pandey & J.P. Shukla. Vol 1(2006) Rastogi Publication, Meerut.

Barash, H; I. Pbavnick And R. Moav (1982). Integration of duck and fish farming;

Experimental results. Journal of aquaculture. 27(2). Pp. 129-140.

Brown E.E (1983). World fish farming : cultivation and economics. Avi publishing

Co. (USA). Pp 263-426.

Coche, A.G (1967). Fish culture in rice fields- A world-wide synthesis.

Cheryl. S. (2002) Quoted from Fish & Fisheries by K. Pandey & J.P. Shukla. Vol 1(2006) Rastogi Publication, Meerut.

Datta. N.K. and S. Kumar (2002) Quoted from Fish & Fisheries by K. Pandey & J.P. Shukla. Vol 1(2006) Rastogi Publication, Meerut.

FAO, Fresh water aquaculture development in China. FAO-UNDP Study tour report.

FAO Rome. 1979 pp. 29-32.

Fransis. T. (2002) Quoted from Fish & Fisheries by K. Pandey & J.P. Shukla. Vol 1(2006) Rastogi Publication, Meerut.

Moehi John (2003): Gender and aquaculture development in the African Region , FAO Aquaculture News Letter, No. 29 pgs 35-36.

Rana, K. Perotti, M; Montanuro,S and A. Inmink (1998). Aquaculture productin in China . FAO aquaculture news letter. Dec. 1998. No. 20 pp. 9-10.

Sundraraj. S and S. Kumar (2002) Quoted from Fish & Fisheries by K. Pandey & J.P. Shukla. Vol 1(2006) Rastogi Publication, Meerut.

Suresh. R. and P. Selvary (2002) Quoted from Fish & Fisheries by K. Pandey & J.P. Shukla. Vol 1(2006) Rastogi Publication, Meerut.

World Food Center (2003) Fish: An issue for every one A concept paper for fish for all 10 pp.

Willman, R. Halwatt, M and U. Barg. Integrating fisheries and agriculture to enhance fish production and food security: FAO Aquaculture News Letter. Dec. 1998. No. 20. pp. 3-9.

Bioresources for Food Security and Rural Livelihood, Pages 245–250
Edited by : G.K. Kulkarni and B.N. Pandey

INNOVATIVE UTILIZATION OF WHEY

J.N. Khedkar and A.A. Phadatare

INTRODUCTION

Whey is the serum or watery part of the milk that remains after separation of curd that results from the acid or proteolytic enzyme mediated co-agulation of milk. It is major by-product of dairy industries manufacturing products like channa, paneer, cheese and chakka (Sharma and Bhatia, 1999). In India near about one million tonne of whey annually drived and which is mostly thrown away as a waste which create environmental pollution. But, in that whey there are about 70,000 tonnes of nutritious solids (Puranik, 1999) are lossed.

Whey is a source of carbohydrates, minerals, protein and water soluble vitamins. Whey is also rich in calcium, phosphorus, essential amino acids (Parekh, 1997). It contains most nutritious whey protein viz., a lactolabumin, β-lactolabumin, serum albumin and immunoglobulin, lactose, milk salts (Zadow, 1986). Whey received from about 80-92 per cent of the volume of milk used is for conversion in to different product. It retains about 45 to 55 per cent of the whey derived milk nutrients (Horton, 1995). Most of the whey solids are usually drained off causing great loss of valuable nutritious solids and creating problem of environmental pollution. It has been observed that Biological Oxygen Demand (BOD) of whey is very high (35000-45000 mg/L). Because of the presence of high organic matter content in whey, it's disposal creates a serious problem. Therefore, many dairy organizations treat the whey before disposal, which is rather expensive (Gandhi, 1989). To avoid pollution as well as utilization of nutritious solids needs to converts whey in to various beverages.

NUTRITIONAL AND THERAPEUTIC IMPORTANCE OF WHEY

a. Nutritional Importance

Prendergast (1985) reported following important benefits of whey :

1. Whey drinks are light and refreshing
2. Less acidic than fruit juice
3. They provide health and nutritional benefits
4. Good source of carbohydrates, minerals, protein

Types of whey and its chemical composition

Sr. No.	Types of whey (%)	Chemical constituents Fat (%)	Protein (%)	Acidity (%)	Ash (%)	TS (%)	Lactose	pH	References
1.	Chakka whey	0.01-0.39	0.35-1.03	0.55-1.13	0.33-0.65	4.98-7.03	3.97-5.26	3.94-5.06	Kulkarni (1984)
		0.1-0.22	0.53-1.13	1.38-1.99	- -	4.19-6.92	- -	- -	Shaha (1986)
2.	Channa whey	0.54	0.37	0.46	0.53	6.9	5.08	-	Singh and Mathur (1973)
3.	Paneer whey	0.25	0.45	0.20	0.63	6.75	5.42	-	Kapil (1991)
4.	Cheese whey	0.1-1.63	0.57-0.92	- -	0.46-0.80	- -	3.50-5.57	4.6-6.6 -	Racotta *et al* (1979)

b. Therapeutic Importance

Whey has been an important food. The father of modem medicine, Hippocrates, insisted on a daily dose of two litre of whey. By the year 1600, whey was used to treat jaundice infected lesions of skin, arthritis, anemia, liver complaints and epilepsy (Prendergast, 1985).

Various products from whey

Sr. No.	Types of whey	Products	References
1.	Channa whey	Mango R TS Beverage	Sikdar and Ghatok (2003)
		RTS Hot soup	Singh *et al.* (1994)
2.	Paneer whey	Kokum juice	Rupnar (2006)
		Cold coffee drink	Nanaware (2007)
		Pineapple juice	Kesarkar (2002)
		Guava whey beverage	Singh and Kapoor (1999)
3.	Cheese whey	Kinnow juice Beverage	Khamrui and Rajorhia (1998)
		Chocolate drink	Singh (1991)
4.	Chakka whey	Fruit jelly	Joshi (1985)
		Mango whey beverage	Shende (2006)

Storageability of whey beverages

Sr. No.	Products	Addition! Treatment	Storageability	References
1.	Kokum juice	UHT and aseptic packaging	6-8 week	Prendergast (1985)
2.	Kinnow RTS beverage	Thermal treatment	Negligible changes in pH during 24 week store	Rantoe et.al., (1992)
3.	Guava fruit juice	Pasteurization 74°C/10 Sec.	Flavour stable for 14 week at	Branger *et al.* (1999)
4.	Pineapple flavour juice	Carbonation	3°C Acceptable upto 30 days	Roberts and Torry (1988)
5.	Whey based kinnow juice (KMS) (350 ppm)	Potassium metabisulphide	4 months	Khamrui and Pal (2004)
6.	Mango based RTS	0.05 % Ashwagandha and 8.22 mg Ascorbic acid/100ml	12 WK	Chakkar and Kadam (2006)

PRACTICAL USE

Manufacture of such beverages would be the income sources to the SHGs, NGOs and private canteen. This would be the save food nutrients and reduce the environmental pollution.

CONCLUSIONS

In conclusion, the whey derived from milk is economically be converted into various beverages like kokam juice, pineapple flavour juice, kinnow RTS beverage, mango based RTS, chocolate drink, cold coffee drink, by addition *of* fruits, nuts, flavour etc. Using the whey we can save the environmental pollution and nutritious solids which losses in whey as waste. The whey with fruit juice combinations seem to hold good promise in the manufacture *of* value added nutritious beverages. Such beverages have been found to be highly acceptable as refreshing drinks.

Utilization of whey : A technique Whey

Koham juice
Paneer whey → Cooling at room temperature → Defatting → Deprotenization → Cooling at room temperature → Addition of sugar (15%) → Addition of jerra powder (0.1 %) → Addition of Koham juice → Addition of Colour → Mixing and bottling → Refrigeration and storage (5-7°C)

RTS soup
Channa whey → Vegetable pulp (Tomato, Carrot, Potato) → Mix pulp in whey → Addition of maida (1%) → Adjust pH 4.8 (Citric acid) → Adddition of onion (1%), Ginger (0.2%), garlic (2%)-Frying → Salting of material with salt (1%) → Heat (2 min.) → Storage

Mango whey beverage
Chakka whey → Neutralization (0.75% acidity) → Addition of sugar (14%) → Pasteurization at 60°C for 30 min → Cooling to room temperature → Addition of mango pulp → Mixing → Filtration → Bottling → Storage

Chocolate drink
Cheese whey → Heat treatment → Removal of clear supernatant of whey → Blend of whey and milk (equal volume) → Heating → Homogenization → Mixing sugar (60 g/lit) stabilizer (0.5 g/lit) → Filtration → Addition of coca powder (5 g) → Packaging → In bottle pasteurization → Cold storage

REFERENCES

Brange, E.B., Slims, C.A. and Schmidt, R.M. 1999. Sensory characteristics of cottage cheese whey and grape fruit juice changes during processing. J. Food Sci. **64(1)** : 180.

Chakkar, S.M. and Kadam, S.S. 2006. Development of ready to serve beverage containing Ashwagandha extract. Beverage Food World. **43(2)** : 49-50.

FAO. 1995.0 Production Year Book. Food and Agriculture Organization of the United Nations, Rome, Italy.

Gandhi, D.N. 1989. Whey utilization for Beverage Production Indian Dairyman. **41(1)** : 35.

Horton, B.S. 1995. Whey processing and utilization. Bull. IDF. 308 : 2.

Joshi, P.S. and Waghmare, P.S. 1985. Utilization of curd whey for the preparation of fruit jelly. Indian Food Packer. **39(2)** : 38-42.

Kapil, V. 1991. Manufacture of lactose: Effect of processing parameter on yield and purity. J. Fd. Sci. **23(3):** 167.

Kesarkar, T.V. 2002. Studies on preparation pineapple flavoured paneer whey beverage. M.Sc. (Agri.) Thesis, M.P.K.V., Rahuri.

Khamrui and Pal. 2004. Effect of potassium metabisulphite and storage condition on physico-chemical and microbial changes in Whey based Kinnow Juice. J. Food Sci. Tech. **41(4)** : 41-44.

Khamrui and Rajorhia, G.S. 1998. Formulation of ready to serve whey based kinnow juice beverage. Indian J. Dairy Sci. **51(6):** 413-419.

Kulkarni, M.B. 1984. Studies on chemical composition and utilization of chakka whey. Ph.D. Thesis submitted to M.P.K.V., Rahuri.

Nanaware, H.M. 2007. Studies on preparation of cold coffee drink. M. Sc. (Agri.) Thesis submitted to M.P.K.V., Rahuri.

Parekh, J.V. 1997. Milk processing and downstream products science and technological needs. Beverage and Food World, **53(1)** : 62-64.

Prendergast, K. 1985. Whey Drinks - Technology of Processing and Marketing. J. Society Dairy Techn. **38(3)** : 103.

Puranik, D.B. 1999. Whey and whey protein in Geriatic Nutrition. Indian Dairyman. **51(5)** : 5-10.

Racotta, V., Bovrges, H. and Navareete, A. 1979. Use of whey proteins for-supplementing tortilla. J. Agril. Food Chern. **27(4)** : 668-672.

Rantoe, P.S., Bawa, A.A. and Saini, S.P.S. 1992. Thermal process for Kinnow RTS. Ind. Fd. Packer. **46(2)** : 16.

Roberts, R.F. and Torrey, G.S. 1988. Inhibition of psychrotrophic Bacterial Growth in Refrigerated Milk by addition of CO2. J. Dairy Sci. **71(2)** : 52-60.

Rupnar, P.S. 2006. Studies on preparation of paneer whey beverage using Kokamjuice. M.Sc. (Agri.) Thesis, M.P.K.V., Rahuri.

Shaha, P.S. 1986. Studies on chakka whey. Alfa lactobumin status. M.Sc. (Agri.) Thesis submitted to M.P.K.V., Rahuri

Shanna, A. and Bhatia, K.L. 1999. Functional properties of whey protein concentrates - A Review. J. Dairying, Food and Home Sci. **18(3 & 4) :** 141.

Shende, D.H. 2006. Studies on preservation of Mango whey beverages. M.Sc. (Agri.) Thesis submitted to M.P.K.V., Rahuri.

Sikdar, B. and Ghatak, P.K. 2003. Fonnulation of mango whey beverage. Beverage and Food World. Dec. 2003. pp. 11-12.

Singh, G.P. and Mathur, M.L. 1973. Channa whey its composition. Indian J. Dairy Sci. **41(3) :** 322-325.

Singh, S. 1991. Utilization of cheese whey for the preparation of a chocolate Drink: Beverage and Food World. Jan. 1991. pp. 25.

Singh, S., Ladkani, B.G. and Abhay Kumar. 1994. Whey utilization for the manufacture of ready to serve soups. Indian J.Dairy Sci. **47(6) :** 501-504.

Singh, W. and Kapoor, C.M. 1999. Standardization of Technology for the manufacture of Guava Whey Beverage. Indian J. Dairy Sci. **52(5) :** 1999.

Zadow, J.G. 1986. Utilization of milk components: Whey in Modem Dairy Technology. Vol. R.K. Robinson (ed.) Elsevier Applied Sci. Publ. London, U.K. pp. 273.

Bioresources for Food Security and Rural Livelihood, Pages 251–258
Edited by : G.K. Kulkarni and B.N. Pandey

FISH PRODUCTION ENHANCEMENT AND SUSTAINABLE DEVELOPMENT USING FISH CUM LIVESTOCK FARMING IN SEWAGE FED BHERIES IN EAST KOLKATA WETLAND, W.B.:

AN INTEGRATED FARMING APPROACH

Mousumi Saha, M.N. Saha, B. Bandhophadhyay and Agniswar Sarkar, Mitalli Dhimani and S.S. Mishra

ABSTRACT

Wastwater aquaculture, as practiced in the East Kolkata Wetlands, has attracted much international attention as a model system for the reuse of urban wastewater and resource recovery. At present the multifunctional wetland ecosystem covers approximately 12,500 ha, and is comprised of mainly of 254 fisheries managed for wastewater aquaculture, agricultural land, horticultural plots and residential areas. It constitutes a unique system of resource recovery, in which nutrients are extracted from the city's wastewater through fish farming and agriculture. However, in recent years, there has been gradual decline in fish production with occasional fish kills, which has become the major area of concern for the fishers as they fully dependant on these sewage-fed *bheries* for their livelihood. It has been reported that maximum profit can be achieved by integrating fish farming with livestock raising and horticulture. Attempts have been made to assess the possible impact of bioremediation in reducing stress and enhancing fish production potential of the culture system, besides integration of animal husbandry and horticulture to have synergistic effect in maximizing fish production and economic upliftment of the fisher groups involved in East Kolkata Wetland. Integration of agriculture with livestock has been shown to reduce 50% expenditure for inorganic fertilizer (Faroda *et al.,* 1978). Various studies on integrated farming system revealed that integration of two or more components ensured sustainability in production, economic return, recycling of wastes and livelihood improvement to the farming community over any single component (Singh *et al.,* 1993, 1997). Integration offishery with crop and livestock leading to higher economic benefit have bee reported by several workers (Jayanthi *et al.,* 2003; Singh *et al.,* 2004). In experimental trials in aquaculture sewage treatment plant using various duckweed and fish ponds, for utilization and treatment of domestic sewage at Cuttack, Orissa, Ayyappan (2000) reported reduction of BOD levels of the effiuents by 60-80%, total coliforms by 70-100% during different periods. The concept of sustainable integrated fish farming in sewage-fed bheries with livestock and duckery and possible impact on economic gains and profitability have been elaborated in the present paper

Keywords : East Kolkata wetland, Sewage Fed Aquaculture, Integrated - fishery, livestock and horticulture, Bioremediation.

INTRODUCTION

The East Kolkata wetlands (22024/-22036/ N to 88023/-88032/ 8) are a series of wetlands & swamps in the eastern fringe of the city Kolkata. East Kolkata Wetland was declared as a Ramsar site on November 2002 as per Ramsar Guidelines. It was a rare example of combination of environmental protection and development where the local farmers have adopted a complex ecological process by mastering resource recovery activity. It is the largest sewage-fed aquaculture in the world. (Gautam gupta, 2000). Wastewater flows through fishponds covering about 4,000 ha, within the area known as East Kolkata Wetlands, with these ponds facilitating a wide range of physical, biological and chemical processes which help improve the quality of the water. Consequently this wetland system is popularly known as the kidney of the city and has been described as one of the rare examples of environmental protection and development management in which a complex ecological process has been adopted by the local farmers for mastering resource recovery activities. The wetland also supports the livelihoods of around 60,000 residents through the fisheries and other socio-economic activities.

Fish cultivation in Kolkata sewage-fed fisheries is a unique feature. There are more than 154 big fisheries or bheries, as they are known locally, although fish culture is also practiced innumerous small ponds or jhils spread throughout the region. (Bunting, S., Kundu, N., Punch, S. and Little, D, 2001). Kolkata wetlands, which have been operational since about 1930, are the world's largest wastewater fisheries, comprising an area of approximately 4000 ha. and providing employment for 17000 families and producing about 38.68 ton of fish per day (Nandeesha, 2002; Strauss, 2003). Out of these, in Gomukpata bhery with area of 133.33 ha (effective water area 117.5 ha), a total number of 355 families are dependant for their livelihood. More than 1350 m gallons of domestic waste water per day is discharged from the city through open canals. The high fertility value of these effluents have been effectively utilized by fish farmers through sewage-fed fisheries. Various types of carps and *Oreochromis mossambicus* (Tilapia) are grown in these ponds. Sustainable production of 5-7 tonnes of fish has been achieved in some sewage-fed experimental farms by utilizing only raw sewage, without treatment. (Jhingran and Ghosh, 1988, Ghosh, 2002, Ghosh *et al., 1989)*. However, in recent years, there has been gradual decline in fish production with occasional fish kills, which has become the major area of concern for the fishers as they fully dependant on these sewage-fed *bheries* for their livelihood. It has been reported that maximum profit can be achieved by integrating fish farming with livestock raising and horticulture. The droppings and fecal matter from farm animals not only be utilized to produce fish-food and vegetables, but can also be used to enhance productivity of the system, minimizing fish kills. The basic principle involves proper utilization of the synergistic effect of interrelated farm activities and full utilization of farm wastes to enhance the productivity and all-round economic development. This is essential because the integration would not only make the farm economically viable but can also serve as model for recycling wastes water with sustainable fish production.

The scientific opinion does not favour the use of untreated sewage. Possible health hazard from wastewater aquaculture has been a subject of serious debate. Wastewater contains correspondingly high concentrations of excreted pathogens like bacteria, viruses, protozoa and helminthes and responsible for occurrence of disease outbreaks (Buras, 1988). WHO guidelines on the safe use of wastewater and excreta in agriculture and aquaculture aimed to protect the health of the workers involved and the public at large". According to WHO and World Bank, hygiene standards applied to wastewater reuse is based on the presence and survival of pathogens (WHO, 1989, 2006). Hence it is desirable to study the public health aspects of sewage recycling and take up appropriate measures to screen such problems through treatment and engineering (Ghosh *et al.,* 1988; Jhingran and Ghosh, 1988). Thus assessment of microbiological quality of produce and possible health hazard impact on consumers and fishers associated, assumes importance.

Another important factor in sewage-fed system is stress to cultured animals. The environment in the wastewater-fed ponds is stressful. It is therefore necessary to treat the sewage prier to its application in ponds to reduce the putrescible organic matter of sewage responsible for oxygen depletion and accumulation of ammonia, toxic to fish, causing fish kill (Jhingran and Ghosh, 1988). In this regard, several models of aquaculture for treatment of domestic sewage has been proposed employing biotic components, that include bacteria, algae, duckweed, microphytes, fish and shellfish. The principle of all the models has primarily been dilution, oxidation, reduction of BOD, COD and suspended solids along with nutrient recovery in terms of biomass (Ayyappan, 2000). The main principle of application is to allow the fish pond to remain aerobic all through the day and night during the period of culture 'with less to no pathogen load. It is where the role of probiotics and bioremediation becomes important in sewage treatment. Addition of these selected harmless or beneficial bacteria to competitively exclude potential pathogens from larval culture media has been suggested as early as 1991 and now popularly being used in fish and shrimp culture (Verschuere, *et al.,* 2000) and found useful for enhancement of water quality (Browdy, 1998). Several bacteria like *V. alginolyticus, Bacillus* sp., *Lactobacillus* spp., *Altermonas* sp., *Pseudomonas* spp., etc. have been used in aquaculture system (Moriarty, 1998; Verschuere, *et al.,* 2000). But the same has not been exploited in the existing sewage-fed bheries in East Kolkata wetlands. This may be the main reason of under-production of fish in such culture systems, thus warrants urgent attention.

In India, several Institutes have carried out research on various aspects of wastewater aquaculture and various information have been published (Bhowmik *et al.,* 1993, 1994, 1996; Pandey *et al,* 1994, 1995; Sarkar *et al.,* 2000). Integration of agriculture with livestock has been shown to reduce 50% expenditure for inorganic fertilizer (Faroda *et al.,* 1978). Various studies on integrated farming system revealed that integration of two or more components ensured sustainability in production, economic return, recycling of wastes and livelihood improvement to-the farming community over any single component (Singh *et al.,* 1993, 1997). Integration of fishery with crop and livestock leading to higher economic benefit have bee reported by several workers (Jayanthi *et al.,* 2003; Singh *et al.,* 2004). In experimental trials in aquaculture sewage treatment plant using various duckweed and fish ponds, for utilization and treatment of domestic sewage at Cuttack, Orissa, Ayyappan (2000) reported reduction of BOD levels of the effluents by 60-80%, total coliform by 70-100% during different periods. However, Bhowmik *et al.,* (2000) in their studies in four sewage-fed fish farms in Calcutta isolated 24 bacterial species belonging to 5 families.

Traditional practices of recycling sewage through agriculture, horticulture and aquaculture have been in vogue in several countries (Edwards, 1996). Contamination of fish and fish products with pathogenic microorganisms like *Vibrio cholerae, Salmonlla* spp. etc. and their possible transmission to human being has been studied (Daalsgard, 1995). Application of bioremediation has been found useful in fish and shrimp culture. Browdy (1998) observed that these probiotics may also enhance larval nutrition by supplying essential nutrients, improving digestion through supply of essential enzymes, mediating direct uptake of dissolved organic material and producing substances which may inhibit the growth of opportmristic pathogens. Accordingly several bacteria like *V. alginolyticus, V.harveyi, V.pelagius, Bacillus* sp., *Lactobacillus* spp., *Altermonas* sp., *Aeromonas* spp., *Pseudomonas* spp., etc. have been used as bioremediation in fish culture.

Methods of integrated approach: Integrated farming is a popular concept of utilizing wastes for food production and efficient use of farm space for sustainable development. Integration of livestock farming with fisheries and horticulture is an excellent method of recycling organic wastes for the production of highly digestible proteins at low cost and also serve as model of water recycling. This would help to add on farm wastes and dung in the ponds increasing the nutrient level in the ponds there by optimizing the components of food-web, leading to enhancement of fish production. Again chances for fish kills becomes less with such integration. Cow dung and urine are known to have some beneficial effects on the filter feeding and omnivorous fishes like silver carps and *Catla catla.* Certain fish species like Indian major carps, Common carp and Tilapia can directly feed on the undigested components in domestic animal waste. The level of phytoplankton and zooplankton is increased in the system which are predominant food for Silver carp and grass carp. Application of suitable microbial bioremediation mix can not only reduce the toxic level of ammonia and hydrogen peroxide produced in the system. they also reduce the level of pathogenic microbial load, thus increasing the shelf-life and consumer acceptability. Besides testing commercial products, microbial mix can easily be prepared in the laboratory or at farm level using indigenous but beneficial microbes, which would be economically feasible to apply in the system.

Integration of duck farming gives add on benefits as they feed on insects and parasites harmful to the fishes, duck droppings provide essential nutrients and importantly, they increase the DO level in the pond with their frequent movement thus reduce stress to the animals. In mixed fish culture system, stocking of Rohu, Catla, Mrigal, grass carp and common carp in the ratio of 15:20.:15: 30: 20 at a density of 25,000 fish/ha would be ideal to produce fish up to 25,000 -70000 kg/ha/yr. Besides these some tilapia species *(Oreochromis mosambicus)* can also be stocked in the *bheries.* Multiple stocking with multiple harvesting model will be adopted in the proposed *bhery* as is normally practiced in the area Again integration of vegetable farming with above system would be highly beneficial for meeting the daily demands of fishers. The silt of the pond bottom on the dykes/ bunds would be the ideal place for growing vegetables. Not only this would make the fishers engaged during their off-time, but the fruits and vegetables can be the source for their domestic demands. The plant wastes can be recycles in the pond/soil and can be the source of food for grass carp and tiapia species. Besides these the income generated through sale of milk. meat and vegetables would give additional economic benefits for the resource-poor fishers.

The proportion of different animals/birds/plants to be taken in the integrated farming in 3 ha area in the *bhery* (as per NABARD 2007 guidelines) is given below:

Sl. No.	Component	Varieties/Species	Numbers
1.	Fish	IMC (Rohu, Catla, Mrigal), Grass carp, common carp and Tilapia	35: 15: 20: 30 at a density of 25,000 fish/ha
2.	Buffalo	Murrah buffalo	6-8
3.	Duck	Khaki Campbell/Indian runner	150-200
4.	Vegetables	Cabbage/cauliflower, brinjal, tomatoes etc	As required
5.	Plants	Fruits/coconut plants	80-100

MEASUREMENT PROPOSED

Following measurements have been proposed under the project to evaluate the utility of the project in enhancing economic condition, profitability and improved livelihood of the fishers group involved in the project. The measurement monitoring indicator would broadly include

(i) Reduction of stress in bhery in terms of reduction in level of Ammonia and hydrogen peroxide and level of microbial pathogen load.

(ii) Optimal physicochemical parameters of sediment and water

(iii) Enhancement of fish productivity

(iv) Reduction in fish disease occurrence

(v) Enhancement of profitability and improved livelihood

(vi) Increase in awareness level and their capacity

SOCIETAL COMPONENT OF THE PROPOSAL

The concept of integrated fishery with livestock-vegetable farming to be introduced in the locality for enhancing the benefits to the fisher groups involved. The age-old fish farming knowledge along with livestock rearing and cultivation of vegetables on the dykes/ bunds could be a good source of additional income and meeting the demands of household daily needs. The diary farming would not only enhance the fertility of the system, but would help to generate additional resources for these needy fishers. Diversification of fishery-livestock farming would serve as the mantle for poverty alleviation and food security in the region.

Bioremediation bacterial mixture thus prepared at farm level will be utilized in the bhery to reduce toxic level of ammonia, increase dissolved oxygen and reduce pathogen load in the culture system, which would help to reduce stress and enhance fish production.. In this regard following interventions in sewage-fed bheries have been proposed:

(i) Integration of livestock (buffalos, duck) and vegetable farming with fishery in sewage-fed systems with mixed fish species culture

(ii) Application of laboratory/ farm made bioremediation mixture in the bhery to reduce toxic level of ammonia/ nitrite and microbial pathogen load.

The outcome of this project would serve as a model of fish cum livestock-horticulture farming in sewage-fed *bheries,* which will have a direct impact on the community of adjoining areas, where the present system would be replicated for gaining maximum benefits. The personnel trained would motivate and also serve as resource persons or guide for other interested *bhery* groups. This would lead to overall economic development of fisher groups in the region

The bioremediation mixture thus developed will be applied in the *bheries* to reduce toxic level of ammonia, thereby reducing stress to animals and enhancing production and economy of the persons involved.

The microbiological analysis would give quality status of fish and farm produce so as to increase the shelf life, better consumer acceptability and higher market price, thus more economic gain.

REFERENCES

Ayyappan, S. & Jena, J.K. 2000. Coastal fisheries in Orissa, an important economic activity. *In* VK Sharma, ed. *Environmental problems of coastal areas in India.* New Delhi, Bookwell.

Bhowmik ML, Chakrabarti PP, 2000 Microflora present in sewage fed system and the possibility of their transmission. Wastewater recycling in developing world, Kalyani, University of Kalyani, International Ecological Society, pp 71-77.

Bunting, S., Kundu, N., Punch, S. and Little, D. (2001) East Kolkata Wetlands and Livelihoods Workshop Proceedings. Stirling, UK: University of Stirling.

Bunting, S., Kundu, N. and Mukherjee, M. (2002) Situation Analysis of Production Systems and Natural Resources in Peri-Urban Kolkata. Stirling, UK: University of Stirling.

Booting, S.W., Koodu, N. and Mukherjee, M. (2001) Renewable Natural Resources use in livelihoods at the Calcutta peri-urban interface: li terature review. Stirling, UK: University of Stirling.

Buras N. Microbial safety of produce from wastewater-fed aquaculture. In: Pullin RVC, Rosenthal H, MacLean JL, eds. Environment and aquaculture in developing countries. Proceedings of the 31st I CLARM Conference. Manila, International Centre for Living and Aquatic Resources, 1993: 285-295.

Edwards, P. and RS.V. Pullin. 1996. Wastewater-fed aquaculture. Proceedings of the international Seminar on Wastewater Reclamation and Reuse for Aquaculture, Calcutta, India, 6-9 December, 1988, Environmental Sanitation information Center, Asian Institute of Technology, Bangkok.

Faroda, AS. and Singh, S. 1978. Desertification - Causes, impacts and future trends in Indian arid zone. Pages 1-18in Desertification Control in the Arid Ecosystem of India for Sustainable Development (Surendra Singh, S. and Kar, A., eds.). Bikaner: Agro Botanical Publishers (India).

Gautam Gupta Department of Economics Jadavpur University Kolkata 700 032 Urban Wastewater: Livelihoods, Health and Environmental impacts in India: The Case of the East Calcutta Wetlands, 2000

Jayanthi, C., Baluswamy, M., Chinnusamy, C and Mythily, S., 2003, Integrated nutrient supply system of linked components in lowland integrated farming system. Indian J. Agron., **48 :** 241-46

Jhingran A G and Ghosh A Productive Utilization of Sewage Effluent Through Aquaculture-Case Studies in India. Wastewater -fed Aquaculture edited by Edwards and Pullin R S V, Proceedings of the international Seminar on Wastewater reclamation and reuse for Aquaculture, Calcutta, India 6-9 December 1988.

Nandeesha M.C., Gangadhara B., Manissery J.K., 2002. Further studies on the use of mixed feeding schedules with plant-and animal-based diets for common carp, *Cyprinus carpio* (Linnaeus). Aquaculture Research **33 :** 1157-1162.

Sarkar S 1999 Wilderness Preservation and Biodiversity Conservation-Keeping Divergent Goals Distinct; *Bio-Science* **49 :** 405-412

Singh, H.P. 1997. Role of farm research and farming systems perspective in sustainable development of Thar desert In Desertification and control in the arid eco-system of India for sustainable development (Singh Surendra and Kar Amal, eds;), Agro-Botanical Publishers, New Delhi, pp. 380-386

Strauss, M. 1996. Health (pathogen) considerations regarding the use of human waste in aquaculture. In: J. Staudenmann, A. Schonbom and C. Etnier (Eds.), Recycling the Resource, Proceedings of the 2 international conference on ecological engineering for wastewater treatment. School of Engineering Wadenswil-Ziirich, September 18-22, 1995. Transtec Publications, Zurich. pp. 83-98.

Verschuere, L., H. Heang, G. Criel, S. Dafuis, P. Sorgeloos, and W. Verstraete. 2000. Protection of *Artemia* against the pathogenic effects of *Vibrio proteolyticus* CW8T2 by selected bacterial strains. Appl. Environ. Microbiol. **66 :** 1139-1146

Bioresources for Food Security and Rural Livelihood, Pages 259–269
Edited by : G.K. Kulkarni and B.N. Pandey

DAIRY FARMING IN PUNJAB: AN ALTERNATE TO WHEAT RICE CROPPING SYSTEM

A.K. Jain and M.L. Mehra

ABSTRACT

The green revolution in Punjab especially Wheat- Rice rotation has lead to ecological imbalance in the state. The declining soil fertility, depleting water table and static/declining cereal yield per hectare has lead the planners and scientists to think of the diversification to other rural professions in the state. The thrust of the Government and increased demand of milk and consequently rumernative price offered by the milk plants, the progressive farmers of Punjab are taking up dairying on a large scale. Punjab with only 1.5% of the geographical area of the country and with 3% of the national bovine population produces about 10% of the total milk produced in the country. The per capita availability of milk in Punjab is 961 gm per day in comparison to national average of 246gm. There is pragmatic shift from livestock farming as a way of life to economic dairy farming in Punjab. Many progressive dairy farmers of Punjab after getting superior germplasm, training in A.I., P.D. and herd management from the university have established herds of crossbred cows and buffaloes with more than 4000 kg and 2500 kg milk per lactation, respectively. These milk production levels are comparable to milk productivity in dairy developed countries. The strategies for the improvement of Cows and Buffaloes have been discussed in the paper. The facilities which need to be provided on the part of Govt. agencies/ Breeders /NGO's for the improvement of livestock; like development of bull mother farms, training of livestock farmers/ professionals, coordination among different agencies working for the improvement of livestock, establishment of breeder's associations, herd registration and organization of livestock shows have also been discussed which will result in fast improvement of livestock productivity.

Keywords : Milk production, Cows, Buffaloes progency testing.

INTRODUCTION

In the Indian subcontinent, agriculture is the main occupation of 70% of the rural population. About 60% of the workers engaged in agriculture, possess an average land holding of 0.39 hectare and for them animal husbandry/allied profession is the main source of sustenance. The Indian agricultural production system especially of Punjab, which was sustained by emphasis on wheat-rice crop production, has started to experience the so called fatigue of green revolution and has created ecological disorders. Soil fertility deterioration and alarming depletion of water table has necessitated the shift from Wheat –Rice rotation to diversification to other rural based professions like Dairy, Poultry, Fishery, Apiculture, and Horticulture. Rising income and living standards particularly in urban and semi urban areas has promoted a paradigm shift in demand for livestock products like milk, meat and eggs, thus promoting their increased production. The Poultry farming has already taken the shape of industry and is mostly controlled by large entrepreneurs. Dairy farming remains an important source of food and social security of rural families and rural employment and have symbiotic relationship in Man- Land-Livestock ecosystem. Till the end of twentieth century most of the increase in livestock production in India/Punjab was achieved due to horizontal growth of buffalo and crossbred cows population, rather than by substantial increase in per unit production. This has been so because the livestock rearing and production is largely in the hands of resource poor farmers in rural areas who can not afford to adopt the capital intensive technology of livestock rearing nor has the training to manage such capital/skill oriented enterprise.

The contribution of these landless and small land holding families to the country's milk production is still quite substantial. These categories of resource poor livestock owners hold about 70% of the bovine population and contribute to about 60% of the total milk production mostly depend upon a Cow or a Buffaloe for their day today livelihood.

These landless and marginal farmers try to maintain their animals at zero cost input. The housing is in the open under the trees, or in the house of the farmer and grazing on village panchayat land is a common practice. Since these dairy animal owners can not afford to optimally feed their animals, obviously the milk yield of these animals is low. Communication with this type of farming community clearly brings out differential requirement of type and size of animals and the suitable genetic improvement model to be arrived at in the light of these considerations.

The small and medium types of farmers (holding 2 to 4 hectares of land) having 3 to 10 animals do construct a temporary grass thatched cattle shed and feed some concentrate to their milking or pregnant animals. The expectation of these farmers is to have a moderate milk producing animals to meet home needs of milk and milk products. Surplus milk production if any is sold in the market but there is no aggressive commercialization in these herds.

The large farmers are the ones, who rear more than 20 milking cows or buffaloes and are dairy entrepreneurs or potential dairy entrepreneurs. Some of these farmers own more than hundred of high yielding milch animals and are actively involved in the milk/ animals sale business. The number of these farmers is increasing rapidly and in most villages of Punjab and in peri-urban areas dairy farms with more than hundred milch animals is a very common sight. These farmers grow their own fodder, many of them milk the animal with machines, have installed the bulk milk cooling tank, make silage of the extra green fodder and make their concentrate mixtures at the farm itself.

Increasing the productivity of Dairy Cows and Buffaloes is a step towards:

- poverty alleviation of rural masses
- increasing the employment opportunities in villages
- empowering women with economic independence
- providing nutritional security to families of livestock owners
- conservation of scarce water resources of the state through diversification
- improving soil fertility by shift from Wheat-Rice rotation to fodders and addition of farm yard manure

Punjab has already made considerable progress in improving milk production in the state. *Punjab with only 1.5% of the geographical area of the country and with less than 3% of the National bovine population produces 10% of the total milk produced in the country.* The per-capita availability of milk in Punjab is 961 gm per day in comparison to National average of 246 gm. (Table 1).

Table 1 : Milk production and per capita availability in Punjab and India

Year	Annual Milk Production (million tonnes)		Per Capita Availability of Milk (gms/day)	
	Punjab	India	Punjab	India
1965-66	1.92	20.6	3 80	117
1970-71	2.20	22.5	439	112
1980-81	3.49	34.3	567	136
1990-91	5.14	53.9	682	174
2000-01	7.77	80.6	870	220
2006-07	9.16	100	961	246

The changing pattern of livestock population in Punjab state over the years indicates pragmatic shift from livestock farming as a way of life to economic dairy farming. The total milk production of the State, which was 1.92 million tonnes in the year 1965-66, has increased to 9.16 million tonnes in the year 2006-07 which comes out to be 477% increase during this period. But during this period the population of cattle has declined from 31.6 lac to 17.6 lac, a decrease of 44.3% (Table 2). However, buffalo population has increased from 29.8 lac to 50.0 lac, an increase of 67.8% during this period. The annual milk yield of cows in the state has increased from 750 kg in 1970 to 2700 kg at present, which is highest when compared to cattle milk yield average of other states of India but, which is much below the average 305-days lactation milk yield of 9455 kg. in Israel, 8235 kg in U.S.A. and 7930 kg. in Netherland. Some institutions and many progressive dairy farmers of the Punjab state have established herds of crossbred cows and buffaloes producing more than 4000 kg. and 2500 kg. milk yield per lactation, respectively.

The intervention with breeding nutritional, health cover, marketing and processing technologies at this stage can make Punjab a dairy developed state comparable to any dairy developed country in the world.

Table 2 : Livestock Population Trends in Punjab

(in Lacs)

Year	Cattle		Buffaloes	
	Total Cattle	Breedable cows	Total buffaloes	Breedable buffaloes
1966	31.61	8.25	29.83	14.86
1972	33.90	9.12	37.96	19.10
1977	31.20	9.28	41.10	21.17
1982	35.66	9.50	44.72	22.52
1990	28.32	11.27	55.78	32.70
1997	26.39	12.86	61.71	37.27
2003	20.39	9.17	59.95	32.79
2007	17.61	10.31	50.02	29.08

Breeding strategies for conservation and improving the productivity of a particular breed/genetic group of animals depends on agro-climatic conditions of the area of inhabitation, agriculture and livestock system in vogue, crop by- products available, literacy rate, herd size, marketing structure available, animal genetic resources, their production potential and utilization pattern, socio-economic level of farmer/livestock owner, available infrastructure and development facilities, human resources, political and administrative will of state for bringing about change.

The improvement strategies of cattle and buffaloes can be discussed under the following sub heads:

IMPROVEMENT OF COWS

(a) **Zebu breeds by grading-up with Sahiwal breed:** The low producing non -descript Zebu cattle which constitute 20% of the total cattle population of the Punjab state is the target population in this scheme. The Livestock owners who are resource poor viz; landless laborers, small and marginal land holding farmers, who cannot grow sufficient fodder and in regions of the state lacking in infrastructure and marketing facilities, their non-descript cows shall be improved by grading up by using semen of improved Sahiwal breed. Herds of well defined Sahiwal breed with very high genetic variation in production traits are available at state Govt. farms/gaushalas and with individual farmers. These herds shall be used for production of bulls and supply of semen for upgrading local non descript cows. The bulls to be used for grading up, should be produced from superior dams, which have lactation yield of more than 2500 kg. Within 5- 6 generations of grading up, the non-descript population can be transferred into well defined milch breed.

(b) **Improvement of Sahiwal cows by selective breeding:** Sahiwal breed needs to be improved genetically by selective breeding. The existing herds of the well defined breed needs to be strengthened further. By selective breeding it is expected that genetic improvement will be achieved ranging from 1 to 1.5% per annum. Such breed improvement programme must include formation of coordinating bodies fcr monitoring the germplsam production, performance recording, evaluation and selection of young bulls and testing there genetic merit both at organized herds including progressive gauashalas maintaining sahiwal breeds as well as farmers' herds under field conditions.

(c) **Improvement of Zebu Cows by Crossbreeding:** The other approach to improve non-descript Zebu cattle is through crossbreeding with *Bos taurus* dairy cattle breeds of temperate origin especially in areas where adequate green fodder can be grown, livestock owners are resourceful to purchase feed ingredients and can spend on health care, housing and demand for cow milk exists in that area. Crossbreeding with Holstein-Friesian and Jersey can be effective tool for increasing milk production rapidly in such areas. The optimum level of exotic germplasm needs to be maintained between 50 to 75%. The crossbreeding of non-descript Zebu cows with semen of progeny tested bulls of exotic breeds will result in enhancing milk production by 2-3 times to that of non-descript cows, reduce age at first calving by 12-15 months and shorten the calving interval by 2-3 months. This type of programme is being under taken on large scale in all the districts of the state.

It is now fairly well established that the crossbred cows with around 62.5% exotic germplasm are most suited under the conditions prevalent in most parts of Punjab. The cattle breeding policy of the state envisages 75% exotic inheritance in crossbred cattle. The results of some studies also show that a considerable deterioration in performance is likely to occur when the crossbreds are bred through inter se mating without conducting proper selection. Now that a substantial part (80%) of the cattle in Punjab are crossbred, it has become necessary to take immediate measures to select crossbred bulls of 75% exotic inheritance of high genetic potential for the breeding of crossbred cows. Selection on the basis of progeny-performance is recognized to be the most potent tool for selection in cattle. The major factor affecting the accuracy of progeny testing is the number of daughters on which the bull has been evaluated. The physical facilities available with the institutional farms do not permit the recording of large number of daughters. The progeny testing programme, therefore, have to be extended to the field conditions, where large number of daughters per bull can be recorded. However, for effective implementation of the progeny testing programme under field conditions the requisite infrastructure must be created.

BUFFALO IMPROVEMENT

Buffaloes are preferred over cows because of their adaptability to harsh argo-climatic conditions, high fat (6.5-8.0%) and SNF (9.0- 10.5%), resistance to tropical bovine diseases, and ability to utilize low grade roughages. The easy disposal of uneconomical and unproductive buffaloes and surplus male calves for meat has the added advantage because culling of such animals will bring about faster genetic improvement through selection and bring additional economic returns to the farmers from their sale for meat. Late sexual maturity, low milk production, poor reproduction efficiency, prolonged calving interval and high calf mortality are some of the problems of this species of livestock which requires immediate attention.

Since the best buffalo germplasm is available in India, improvement in the genetic potential of buffaloes for milk production can not be brought about by introducing gremplasm from outside the country. Selection among native buffaloes (both males and females) is therefore, the only breeding method available for bringing about genetic improvement in their productivity. Selection among the females on the basis of milk yield may not be possible because most of the female stock is required for herd replacement. As compared to this, only small numbers of males are required for breeding purposes through A.I. There is therefore, considerable scope for doing intense selection among the males. It is well known that the sire of the replacement stock and of the future sires account for about 70% of the genetic progress. The breeding strategy for bringing about genetic improvement in the milk yield of buffaloes should therefore, focus on accurate and intense selection of the breeding bulls and their extensive use.

PROGENY TESTING OF BUFFALO BULLS/CROSSBRED CATTLE BULLS

The progeny testing of buffalo and crossbred bulls and the extensive use of selected bulls is the only method available for fastest genetic improvement of buffaloes and crossbred cattle respectively. As the institutional farms can progeny test only a small number of bulls and requirement of the state is very large, the progeny testing of bulls needs to be initiated on large scale under field conditions. The methodology developed for progeny testing under field conditions by dairy developed countries is not as such applicable in Punjab/ India because buffaloes are kept either as house hold dairying or as a subsidiary occupation to agriculture farming. Only few large farms of buffaloes are existing in the state and most of them are near to cities for supply of fresh/ unprocessed milk to consumers and such dairy owners are not interested in any breed improvement programme and normally do not rear the calves. The system of progeny testing under field condition has been modified and is in use in about 80 villages around Ludhiana by the experts of department of Animal Breeding & Genetics and can be copied for replication in all other states. The programme of field progeny testing is to be spread over all the blocks and in selected villages where inseminators trained in data recording, ear tagging, milk recording needs to be employed. A central agency can be developed to progeny test buffalo and crossbred bulls. Data may be maintained by the agency and bulls selected and progeny tested and semen of such bulls preserved during testing. Even if the programme is started now it will take 7-8 years to test the first set of bulls.

For effective implementation of the field progeny testing programme of crossbred cattle and buffalo bulls, the following methodology can be followed:-

1. **Procurement of bull calves:** Male calves/bulls for progeny testing should be selected from the milk-recorded herds of institutional farms or from progressive animal breeders. Bull calves may also be procured from the field area by recording their dams milk yield at least for two days i.e. once in the second month and once in the third month both in the morning and then in the evening. The dam whose calf is to be purchased for putting under progeny testing and male calf to be procured should be ear-tagged. One day before the dam is to be recorded, the animals should be completely milked in the preceding evening in the presence of a team appointed for the procurement of bull calves. Next day both in the morning and evening the dam is to be milked in the presence of team and its milk yield will be recorded. On the basis of two times milk yield, the lactation milk yield will be estimated and male calves with dam's milk yield of more than 3500 kg. in first lactation or more than 4500 kg. in other lactations in crossbred cattle and 2500 kg in first lactation and 3000 kg in other lactations in buffaloes may be purchased.

   The bull calves may be purchased at three months of age at the rate to be decided by the committee and then reared at the semen station under optimum feeding and management conditions. The data on feed intake, growth rate and other health aspects will be recorded at the station. The number of bull calves to be procured should be double the requirement of breeding bulls keeping in view the losses due to mortality, poor growth rate and poor semen quality, freezing etc. The bull calves, if fed well and managed properly will start donating semen at the age of about 2.0 years.

2. **Number of test inseminations**: In order to record first lactation milk yield on 40 daughters a minimum of 1000 AIs per test bull are to be performed and approx. 2000 straws per bull will be required .

3. **Semen freezing and storage:** At least 2000 doses of semen should be frozen before putting any bull under progeny testing. A minimum of 8000 doses of semen of each test bull should be stored for future use after evaluation.

4. **A. I.'s of test bulls:** The semen of all the test bulls should be used simultaneously in the project area by rotating the semen supply in the A.I. centres. At a time semen of only one bull is to be supplied to a centre to avoid the entry of wrong parentage and all the previously supplied semen may be taken back and bull number of semen supplied to the centre may be entered in the A.I. register of the centre. Matings of each set of test bulls should be completed in 18 months. Relevant records pertaining to animal number, date of A.I. and pregnancy records shall be maintained at the A.I. centers.

5. **Ear tagging of animals:** All the animals which are inseminated with the semen of test bulls should be eartagged. All the female calves born out of test matings may also be ear-tagged and tattooed within 15 days of calving and calving record may be entered in the calving register.

6. **Follow up of female progeny:** All the female progeny born out of the test matings should be followed for their growth rate at one year of age and at calving. After 1.5 years of age animals should be followed after every three months for pregnancy status and according to the pregnancy status it should be followed for the calving.

7. **Milk-Recording:** Field level milk recording system is extremely important for overall progress and development of dairy cattle and buffalo breed improvement. Unless we know the milk production ability of the animals, it would be very difficult to select parents with high transmitting ability for future generation. In case, we optimize feeding, management and other inputs on the basis of production ability of our animals, we can develop them into more efficient producers of milk. Once the female has calved it should be put under milk recording within a month of its calving. Milk should be recorded once a month, both in the morning and in the evening by a spring balance for ten months. Missing records will be estimated as the average of the preceding and following records. At the time of milk recordings various other parameters like quantity of concentrate, green fodder, dry fodder, milk fed to calf etc. may also be recorded. Milk samples may be collected for fat and SNF testing in 2nd, 5th and 8th month after calving.

8. **System of milk recording:** System of milk recording through casual milk recorders on contract basis may be adopted. Each milk recorder may be given 3-4 animals to be milk recorded per day both in the morning and evening. Milk recorder should be selected from the villages where milk recording is to be done. Farmers may be encouraged to do own milk recording and they should be paid for milk recording. The field production-recording system shall be designed in a manner that farmers shall be helped in improving overall herd management/feeding/recording and marketing of the produce to achieve higher level of profits from dairy farming.

9. **Supervision of milk recording :** At least one milk recording (preferably second or third) on each animal should be done in the presence of responsible officer or by Milk Recording Supervisor and monthly supervision report should be checked by the officers for any extra ordinary records and shall be checked immediately.

EVALUATION OF TEST BULLS

While evaluating the performance of the bulls under test no progeny of the bull whose milk record is available is to be rejected (except the progeny which have died during the lactation before completing 150 days of lactation). The 305-days milk yield of the progeny may be estimated from the 20 milk recordings of ten months. The bulls may be evaluated by using least square analysis of the milk production records of the daughters.

A system of *inter se* mating among the crossbreds after producing the crossbreds with 50-75% of exotic germplasm and use of progeny tested crossbred bulls will lead to the development of a new strain/breed of cattle with potential for high milk yield, adaptability to harsh climatic conditions and resistance to various tropical diseases.

The facilities which need immediate action on the part of Government agencies/ Breeders agencies/NGO/Charitable institutions/Animal welfare groups for improvement of cattle and buffaloes are:

a) **Development of bull mother farms:** Institutions shall not only breed and rear females, but also purchase male calves of outstanding dams of local breed, Crossbred Cattle, Murrah and other local breed of buffaloes from the field area which have been milk recorded and bring them to institutions, test them for various diseases, rear them to maturity, collect and freeze semen and supply it to livestock farmers for improvement of cattle and buffaloes. Bull mother farms are needed for crossbred bulls with 50-75% exotic germplasm, for local cattle, Crossbred cattle and different breeds of buffaloes.

b) **Training of livestock farmers:** Before any further extensive livestock improvement programme is taken up there is urgent need to train the livestock farmers on different aspects of dairy farming. Training/education of livestock farmers at following two levels:

   i) **Training for farmers with less than 10 milch animals:** This training shall be based on feeding and management practices of animals. The job of AI, PD, breeding, record keeping shall be managed by agency staff appointed at village level by the Government. Farmers may be asked to pay for the services provided. Alternatively a educated person from village may be trained by institutions in A.I, P.D, record keeping, health cover for animals and shall provide these services to farmers in the villages. This trained staff needs to be associated with some agency for supply of materials and guidance.

   ii) **Training for farmers with more than 10 milch animals:** Such livestock owners shall be given training in all aspects of livestock farming including AI, PD, herd management, milk marketing, vaccination, and treatment of animals. These farmers shall pay for training and all the supplies like semen, quality feed, mineral mixture, health cover, vaccination etc to the agency/institutions.

c) **Training of livestock professionals:** Most of the staff employed in the field are trained/interested in treatment of livestock. They have very less knowledge in breed improvement programme, importance of herd registration, feeding, management of livestock and marketing of milk. These professionals in the field need training to upgrade their knowledge. Alternately there is a need to develop a new department of Animal Production with in each state to look after the work of livestock improvement.

d) **Coordination among different agencies:** The scattered efforts made by different dairy development agencies operating in the same area may not produce the desired impact on the improvement of cattle and buffalo productivity. Collaboration among various agencies like State Animal Husbandry Department, Milk Cooperatives, Private agencies and NGOs etc. in the selected operation areas will be required for obtaining the desired results. Only one agency should have the control over the supply of semen of all the test bulls in that area.

e) **Establishment of Dairy Cattle and Buffalo Breeders' Association:** In the dairy developed countries, the work on the genetic improvement, performance recording, progeny testing, artificial insemination service are undertaken by the breed associations for the respective breeds. It is, therefore necessary that Dairy Cattle and Buffalo Breeders' Associations are also established in the state to undertake cattle improvement programme. One such organization i.e. Punjab Dairy Farmers Association with headquarters at GADVASU, Ludhiana is doing a very useful work in this area since 1972.

f) **Herd Registration:** Registration of cattle and buffaloes in a Herd Book is another important steps towards their improvement. The herd books are essentially data banks providing useful information about animals' performance for enabling proper selection and designing of breeding strategies.

The state should prepare a programme for herd registration and performance recording of cows and buffaloes both for progeny testing of bulls and for identifying genetically superior animals to be used as bull mothers. To begin with herd registration should be introduced in selected dairy-developed blocks of the state. The dairy farmers participating in this programme should be given special incentives in the form of semen from bulls of high genetic merit, artificial inseminations at the door steps of the farmers, seed of good quality fodders, feed quality evaluation facility and animal health cover.

g) **Livestock competition at district and State level**: Breed-wise livestock competition of different categories of the cows, buffaloes should be held at district and state levels annually.

h) **Incentive to farmers establishing mechanized farms of cows and buffaloes:** State must provide subsidy in mechanization of the large livestock farms, e.g. subsidies should be given for milking machines, construction of sheds, installation of Gobar Gas Plants, installation of milk processing machinery, purchase of fodder harvesters and other equipment.

CONCLUSION

The supply of high pedigreed male calves, semen of test bulls and progeny tested bulls from the university have changed the scenario of dairy farming in the area adopted under the field progeny testing programme in Ludhiana district of Punjab. The dairy animal rearing have changed the economic level of small, marginal and landless farmers in the villages. More and more farmers are adopting to the dairy farming and field progeny testing project have major contribution in changing the scenario of dairy farming by providing technical know how, germplasm and motivation of farmers. The facilities mentioned in the paper which need to be provided on the part of Govt. agencies/ Breeders /NGO's for the improvement of livestock will result in fast improvement of livestock productivity.

Bioresources for Food Security and Rural Livelihood, Pages 271–278
Edited by : G.K. Kulkarni and B.N. Pandey

HORMONAL CONTROL OF FINAL OOCYTE MATURATION IN *TENUALOSA ILISHA*

Pramanick Kousik

ABSTRACT

Hormonal control of final oocyte maturation (FOM) in *Tenualosa ilisha* was examined in vitro following stimulation of isolated ovarian follicles with shad pituitary extract (SPE), both obtained from vitellogenic (SPE-V) and post-vitellogenic (SPE-PV) stage fish. Result shows that germinal vesicle breakdown (GVBD) started after 6hr of incubation and shows maximum GVBD after 16 hr of incubation. SPE-PV is more potent than SPE-V to induce GVBD. Inhibitors of protein synthesis (actinomycin D and cycloheximide) completely blocked SPE-induced oocyte maturation. This result indicates the requirement of *de novo* transcription and translation of protein synthesis for FOM.

Keywords : *Tenualosa ilisha,* shad pituitary extract, oocyte maturation.

INTRODUCTION

The sequential physiological events of vitellogenesis, final oocyte maturation and ovulation during fish oogenesis are under the differential control of regulatory mechanisms involving principally the pituitary gonadotrophin and ovarian steroid regulators (Nagahama 1987, Yoshikuni and Nagahama *et al* 1991, Kagawa *et al* 1994). The vitellogenic phase is initiated and maintained by one of the gonadotrophins GTH-I or FSH (Kawauchi *et al.* 1989, Yoshikuni and Nagahama 1991, Swanson and Dittman 1997) and oocyte maturation and ovulatory events are initiated by the other gonadotrophin GTH-II or LH. Fully grown oocytes in most species get arrested at prophase of first meiosis. After exposure to maturation inducing hormone (MIH), oocytes undergoes a process called oocyte maturation. The first visible event of oocyte maturation in most of the fish species is the migration of germinal vesicle (GV) towards the animal pole and is often used as indicator of the onset of oocyte maturation. Following migration of GV, several processes occur in the nucleus and cytoplasm of the oocyte. These are germinal vesicle breakdown (GVBD) which indicates the end of prophase I, condensation of chromosomes, formation of a spindle body and extrusion of first polar body (Nagahama 1987, Paul *et al* 2009).

The Indian shad *Tenualosa ilisha* belongs to the subfamily alosinae of family clupidae and is largely an anadromous species. Two other ecotypes - a fluvial potamodromous type and a marine type have also been recognized. The *T. ilisha* occurs in the foreshore arens, estuaries, brakish water lakes and freshwater rivers of the western division of the Indo-Pacific faunistic region. Its marine distribution extends from Iran and Iraque, in the Persian Gulf to the west coast of India, in the Arabian Sea, and in the Bay of Bengal. This fish ascends almost all riverine system in the Bay of Bengal including the river Ganga and provides a rich fishery in the ganga and Hoogly estuaries. This fish is one of the most important food fish in this sub-continent. It ascends the river for breeding during the monsoon season and returns the sea after completion of spawning to marine habitats (Pillay and Rossa 1963). Unfortunately, the population of *T ilisha* declined drastically in the river Hoogly for last two decades due to over fishing and indiscriminate loss of habitat. To overcome the depletion of the gangetic hilsa in the Hoogly River, both restoration of the ecological factor and development of intensive production techniques are equally important. The success of this technological development requires a comprehensive understanding of hormonal regulation of gonadal development and maturation. Surprisingly, except some scanty information on the biology and fisheries of the *T ilisha* (Pillay 1958). Surprisingly, no report has yet been available on gonadotrophic regulation of oocyte maturation in vitro in any member of this clupidae family fish including *Tenualosa ilisha.* Therefore the aim of the present study was to investigate the effects of shad pituitary extract (SPE) on the ability of in vitro oocyte maturation in *Tenualosa ilisha.*

METHODOLOGY

Animals

T. ilisha were caught in the fisherman trap nets from the river ganga between Triveni and Naihati, twice in each month either in the evening or in the early morning depending on the tidal times. Fish might have spent up to two hours in trap nets at approximately 15 ft depth. During the collection of fish from trap nets only the live fish were selected for our experiment. Immediately after collections, ovaries from each fish were dissected out in aseptic conditions. The outer covering of the ovary was immediately cut from posterior to anterior region and peeled off to collect the follicles. Loose follicles from vitello genic and post-vitello genic stages fish were then immediately immerse in oxygenated ice cold Idler's medium containing streptomycin (100 Jlg/ml) and penecilline (100 IVlml) adjusting pH 7.4. All these were done on the fisherman boat. In case of the preparatory stage fish, the whole ovarian tissues were cut into small pieces and placed in Idler's medium. For microscopic observation, ovarian tissues and follicles were fixed in a clearing solution of acetic acid: ethanol: formalin mixture (1 :6:3 v/v) for 12 hrs. The maturity stages of female gonad of *T. ilisha* collected from river Ganga have been categorized into following different stages according to Chonder (1999).

Stage-I : Previtellogenic stage: Ovary becomes slightly broader, dull reddish yellow. Oocytes transparent with few yolk granules (diameter 0.05-0.2mm).

Stage-II : Vitellogenic stage: Reddish yellow in colour, diameter 0.2-0.5mm. Cytoplasm of the oocytes filled with densely packed yolk granules. The germinal vesicle (GV) was centrally located.

Stage-III : Postvitellogenic stage: Deep yellow in colour, diameter 0.5-0.65mm. Centrally located GV surrounded by cytoplasm full of with lipid droplets, periphery transparent. Preparation of homologous carp pituitary extract Pituitaries were collected from vitellogenic and postvitellogenic stage adult female shad. The average weight of the donor fish was 800-1000g. Immediately after collection, the pituitaries were kept in chilled acetone for 24 h. After 24 h, they were homogenized in phosphate buffer (pH 7.4) under ice and centrifuged in a cold centrifuge at 5000xg for 10 min. The supernatant was taken as shad pituitary extract (SPE). Shad pituitary extract prepared from pituitaries obtained from vitello genic stage fish has been designated as SPE-V and that from postvitellogenic stage fish as SPE-PV.

INCUBATION OF INTACT FOLLICLES

Immediately after dissection, ovaries were removed and placed in ice-cold Idler's medium containing streptomycin (100 ug*/ml)* and penicillin (100 IU*/ml)* adjusted to pH 7.4. Composition of Idler's medium is as follows: NaCl-1.080g, KCl-0.272g, Glucose- 0.108g, $CaCl_2$-0.32g, $MgCl.6H_2O$- 0.404g, Tris-0.96g, $NaH2PO_4$- 0.034g, $NaHCO_3$- 0.42g, in 1.0 liter triple distilled sterilized water. Idler's medium has a similar ionic composition and osmolality to the extra cellular fluid of teleosts. Individual ovarian follicles were separated from the remaining ovaries with forceps and either used for incubation of intact follicles or processed for isolation and separation of follicular cells. Follicles weighing approximately 100 mg were initially placed in individual wells of a 24-well culture plate (Tarson, India) for 2 h that contained 1.0 ml control medium. As determined earlier (Sen *et at.,* 2002), this 2 h pre-incubation time was required to waive the surgical shock. After 2 h, the medium was replaced with fresh medium containing the effectors. Cultures were placed in a metabolic shaker bath at 23 ±1° C under air. Viability of ovarian follicles was observed to be about 90% as detected using 0.1 % trypan blue dye exclusion. After termination of incubation intact follicles were fixed in acetic acid-formalin-ethanol (1:3:6, v/v) mixture for 12 hours and oocyte maturation were assayed by scoring GVBD using a stereomicroscope after addition of 100% glycerol.

Stastistical Analysis

All experiments were repeated in ovarian follicles collected from five different fishes. Data from each experiment were subjected to one-way analysis of variance (ANOV A), followed by Bonferroni multiple range tests. Differences were considered significant at $p<0.05$.

RESULTS

Oocyte Responsiveness to SPE

In *Tenualosa i/isha* ovary to determine the degree of maturation required to undergo final oocyte maturation (FOM) in response to fish pituitary extract prepared from both vitello genic (SPE-V) and postvitellogenic (SPE-PV) stage fish, a preliminary experiment was conducted. Table 1 show that previtellogenic and vitellogenic oocytes failed to mature when exposed to SPE- V but very few percent of postvitellogenic follicles underwent GVBD (15%). Previtellogenic, vitello genic and postvitellogenic follicles when exposed to SPE-PV, previtellogenic follicles failed to mature, vitellogenic follicles shows 40% GVBD, and most of the postvitellogenic follicles underwent GVBD.

Table 1 : Effects of SPE- V and SPE-PV on previtellogenic, vitellogenic and postvitellogenic stage ovarian follicles of *Tenualosa ilisha* after 24 hr of incubation.

Treatments	% GVBD		
	Previtellogenic oocytes	Vitellogenic oocytes	Postvitellogenic oocytes
Control	–	–	2.25
SPE-V (1 pit./ml)	–	–	15.5
SPE–PV (1 pit/. ml)	–	40.25	89.8

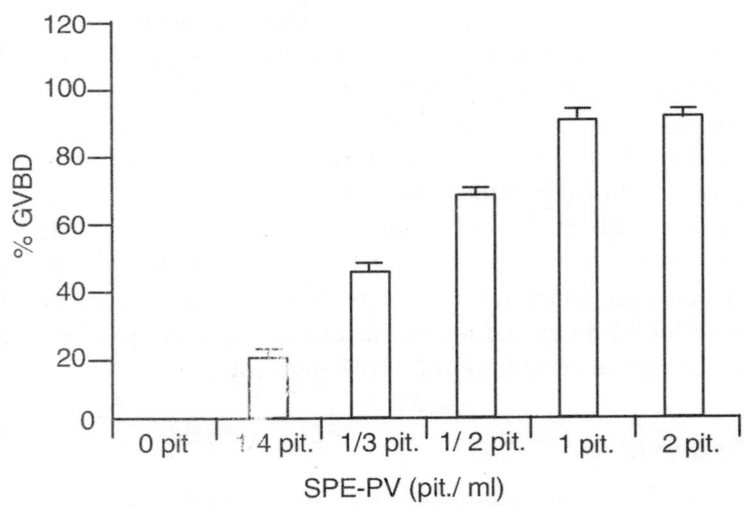

Fig. 1 : ***In vitro*** **effects of various concentrations of homologous shad pituitary extract (SPE) on the oocyte maturation in Tenualosa ilisha ovarian follicles after 24 her of incubation. Eachbar indicates themean (± SEM) values of five replicates from five donor fish.**

Effect of SPE-PV on the Induction of GVBD

As the postvitellogenic follicles only shows its responsiveness to SPE-PV, in the induction of GVBD, the effects of various concentration of homologous SPE-PV in the induction of GVBD of *T. ilisha* oocytes were examined *in vitro.* For this purpose postvitellogenic follicles with centrally located GV were incubated separately with increasing concentration of SPE-PV (1/4, 1/3, 1 and 2pit./ incubation) for 24 hr. Results show that (Fig. 1) a lower dose of SPE-PV (1/4pit./ ml) stimulated significantly more oocytes to complete GVBD compared to control incubates (without SPE-PV). Maximal GVBD (89%) were achieved with 1 pit.! ml SPE-PV. SPE-PV concentrations greater than 1 pit.! ml did not further enhance GVBD.

Time Course Effect

Time-course studies of *in vitro* oocyte maturation exposed to SPE-PV were conducted for different periods from 0-24 hr and results are depicted in Fig. 2. At each specific time, follicles were collected separately and GVBD were assayed. It appears from the figure that the process of oocyte maturation after SPE-PV stimulation was shown to commence after 4 hr and almost 88% oocytes underwent GVBD by 16 hr without further increase thereafter.

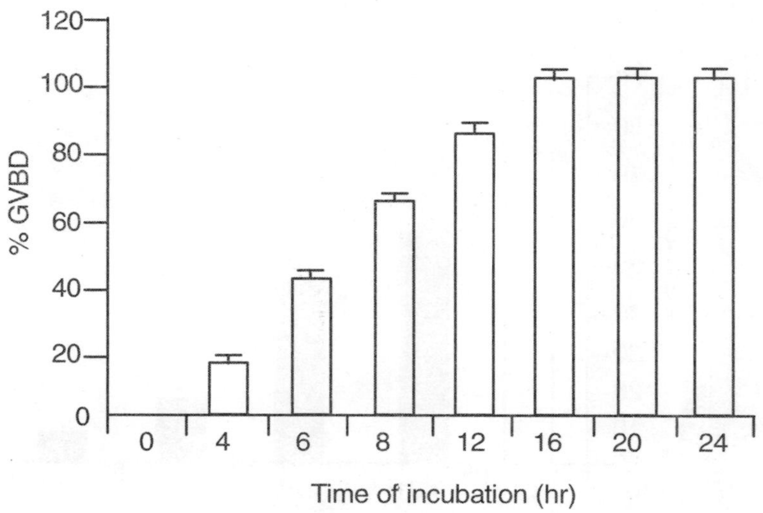

Fig. 2 : Time course study of oocyte maturation treated by SPE-PV in postvitellogenesic follicles of *Tenualosa ilisha.* Each bar indicates the mean (:i:SEM) values of five replicates from five donor fish.

Effects of Protein Synthesis Inhibitor on SPE-Induced GVBD

To observe the effects of transcription and translation inhibitors (actinomycin D and cycloheximide), on oocyte maturation, postvitellogenic follicles of *T ilisha* were incubated with SPE-PV (1/2 pit./ml) and various concentration (0.01,0.1, 1,2 Jlg/ ml) of actinomycin D (act-D) and cycloheximide (chx) for 24 hr. Cultures were pre-incubated with act-D and chx for 1 hr. prior to the addition of SPE-PV. After termination of incubation GVBD were scored. Figure 3 shows that act-D and chx in increasing concentration inhibited the SPE-induced oocyte maturation in a dose dependent manner. Both the inhibitors at concentration 0.01 μg/ml significantly reduced the percentage of oocyte undergoing GVBD.

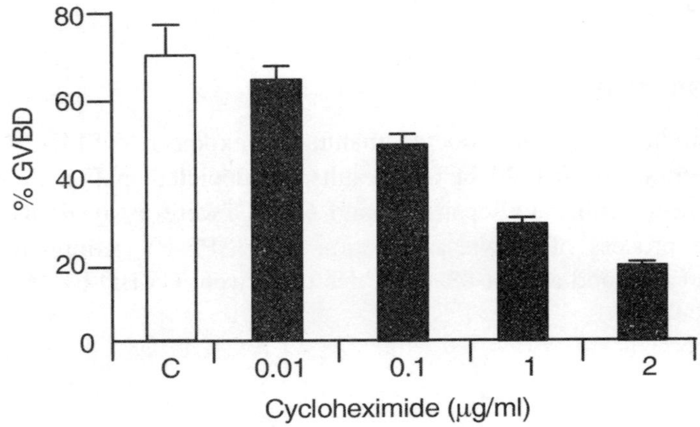

Fig. 3(A)

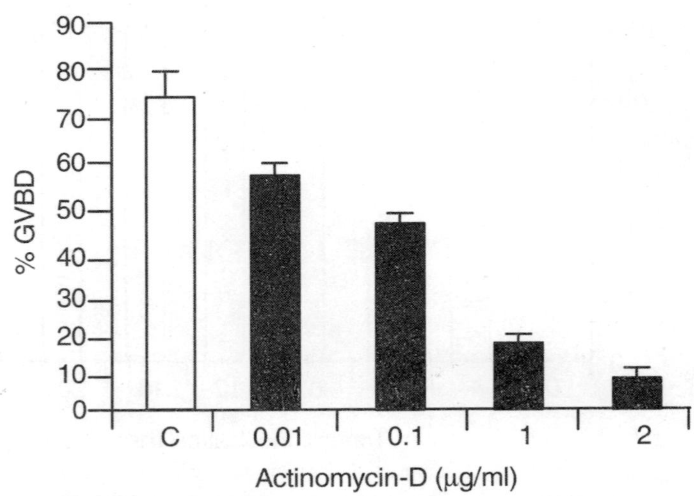

Fig. 3(B)

Fig. 3 : Effects of cycloheximide (A) and actinomycin-D (B) on SPE-PV-induced (112 pit.! ml) GVBD after 24 hr of incubation in postvitellogenic follicles of *Tenualosa ilisha.* Each bar indicates the mean (:i:SEM) values of five replicates from five donor fish.

DISCUSSION

In teleosts, follicles undergo maturation before ovulation. This final maturation has been linked to gonadotropic promotion of maturation involving steroid production. It is now known that fish gonadotropins, GTH-I and GTH-II are chemically and biologically similar to tetrapod FSH and LH respectively (Kawauchi *et at.* 1989; Swanson 1991). Like FSH, GTH-I is secreted during the period of gonadal growth (vitellogenesis) and function to stimulate gonadal growth and steroidogenesis at this stage. Like LH, plasma levels of GTH-II increase during the period of final oocyte maturation to alter steroidogenesis and to promote oocyte maturation and release (Swanson 1991). In the present study SPE-PV-induced oocyte maturation in postvitellogenic follicles of *T. ilisha* clearly indicates the involvement of GTH-II in this physiological process.

Present study demonstrates that vitellogenic follicles of *T. ilisha* were incapable to undergo final oocyte maturation (FOM) in response to SPE at their highest concentrations tested. This indicates that sensitivity of the shad oocyte to MIH at this stage has not developed. Current information indicates that GTH-II regulates oocyte maturation in two stages and that "oocyte maturation" can be appropriately described within the broader context of ovarian follicle maturation. During the first stage of maturation the follicle cells acquire the ability to produce MIH and the oocyte to respond to MIH (i.e., oocyte maturational competence, OMC) where as in the second stage the follicle cells produce MIH and consequently the oocyte is released from meiotic arrest (Patino *et at.,* 2001).

In agreement with the previous studies in fish (Bhattacharyya et. al. 2000) cycloheximide and actinomycin-D were shown to block SPE-induced oocyte maturation in *T. ilisha.* These results indicate the requirement of de novo protein translation for final oocyte maturation to proceed. It would be interesting to ascertain whether the novel protein is 20β-HSD enzyme required for MIH receptor (Patino and Thomas 1990a) or the maturation promoting factor induced by MIH (Nagahama *et al.* 1993).

REFERENCES

Chonder, S.L. 1999. *Biology of finfish and shellfish.* SCSC publishers, Howrah.

Kagawa, H., Kobayashi, M., Hasegawa, Y., and Aida, K. 1994. Insulin and Insulin-like growth factors I and II induce final oocyte maturation of oocytes of red seabream, *Pagrus major,* in vitro. *Gen Comp Endocrinol.* **95**, 293-300.

Kawauchi, H., Suzuki, K., Itoh, H., Swanson, P., Natio, N., Nagahama, Y., Nozaki, M., Naki, Y. and Itoh, S. 1989. The duality of teleost gonadotropins. *Fish Physiol. Biochem.* **7**, 29-38.

Nagaharna, Y. 1987. Gonadotropin action on gametogenesis and steroidogenesis in teleost gonads. *Zool. Sci.* **4,** 209-222.

Nagaharna, Y., Yoshikuni, M., Yamashita, M., Sakai, N. and Tanaka, M. 1993.

Molecular endocrinology of oocyte growth and maturation in fish. *Fish Physiol. Biochem.* **11,** 3-14.

Patino, R., and Thomas, R. P. 1990a. Characterization of membrane receptor activity for 17a, 20b, 21-trihydroxy-4-pregnen-3-one in ovaries of spotted seatrout *(Cynoscion nebulosus). Gen. Compo Endocrinol.* **78,** 204-219.

Patino, R., Yoshizaki, G., Thomas, P., and Kagawa, H. 2001. Gonadotropic control of ovarian follicle maturation: the two-stage concept and its mechanisms. *Compo Biochem. Physiol.* Part B. **129,** 427-439.

Paul, S., Pramanick, K., Kundu, S., Bandyopadhyay, S. and Mukherjee, D. 2009. Involvement of P 13 kinase and MAP kinase in IGF-I- and insulin-induced oocyte maturation in *Cyprinus carpio. Mol. Cell. Endocrinol.* 309,93-100.

Pillay, T.V.R. 1958. Biology of the hilsa, *Hilsa ilisha* (Hamilton), of the river Hoogly. *Ind. J. Fish.* **5(2),** 201-257.

Pillay, S.R. and Rosa, H.Jr. 1963. Synopsis of biological data on Hilsa, *Hilsa ilisha* (Hamilton, 1822). FAO *Fisheries Biology Synopsis* No. 25.

Sen, D., Mukherjee, D., Bhattacharyya, S. P, and Mukherjee, D., 2002. Seasonal changes in plasma steroid levels in Indian major carp *Labeo rohita:* influence of homologous pituitary extract on steroid production and development of oocyte maturational competence. *Gen. Compo Endocrinol.* **128,** 123-134.

Swanson, P. 1991. Salmon gonadotropins: reconciling old and new ideas. In : Reproductive Physiology of Fish. Edited by A.P. Scott, J. Sunipter, D.E. Kime and M.S. Rolfe, *Fish Symposium,* Sheffield, pp. 2-7.

Swanson, P. and Dittman, A. 1997. Pituitary gonadotropins and their receptors in fish. *In,* "Advances in Comparative Endocrinology", S. Kawashima and S. Kikuyama, eds., Monduzzi Editore, Bologna, Italy. pp. 841-846.

Yoshikuni M and Nagahama Y 1991 Endocrine regulation of gametogenesis in fish; *Bull, Ins!. Zool. Acad. Sinica Monograph.* **16,** 139-171

Bioresources for Food Security and Rural Livelihood, Pages 279–282
Edited by : G.K. Kulkarni and B.N. Pandey

FUNCTIONAL FOOD:
INCREASING THERAPEUTIC VALUE OF BUFFALO MILK THROUGH CONJUGATED LINOLEIC ACID AND ITS ANTI CARCINOGENIC POTENCY IN RATS

C. Kathirvelan and A.K. Tyagi

ABSTRACT

Conjugated linoleic acid (CLA) is milk fattyacid which has various health beneficial properties like anticarcinogenic, antiatherogenic, immunomodulatory and antidiabetic towards human being and whole milk contains an average 4.0 – 6.0 mg CLA / g milk fat. It was observed that feeding of mustard oil at 2 percent in concentrate diet to buffaloes resulted in 19.54 mg CLA / g milk fat compared to control 6.84 mg CLA / g milk fat. Feeding of CLA enhanced milk to chemically induced breast cancer rats resulted in 37 percent reduction in cancer incidence than control group and CLA not only inhibited benign tumour but malignant tumour as well.

Keywords: Conjugated linoleic acid, Milk Fat Cancer Mustard oil Human health

INTRODUCTION

Diet as a mean of controlling and reducing incidence of cancer has received considerable attention. Interest among the consumers about natural nutrients and nutraceutical present in foods having health benefits in human, is growing. One such natural nutrient is Conjugated Linoleic Acid (CLA). CLA occurs naturally in many foods; however the principle dietary sources are milk and dairy products (Chin *et al.*,1992).

Conjugated Linoleic Acid (CLA) is a collective term describing a mixture of positional and geometric isomers of linoleic acid (C 18 : 2). CLA is formed as an intermediate during biohydrogenation of linoleic acid by rumen bacteria or from tissue (mammary gland epithelial cell) synthesis of CLA by Δ^9 -desaturase conversion of *trans*- 11 fatty acids. Milk fat is the richest natural dietary source of CLA. The whole milk contains an average 4.5 mg CLA/g of fat (Kelly *et al.*, 1998). Recent studies have shown that the milk fat content of CLA can be markedly enhanced by several dietarymanipulation especially those involving dietary additions of plant oils which are high in unsaturated fatty acids (Griinari and Bauman, 1999).

CLA has received considerable attention as a chemo preventive agent in the past few years since being shown to in hibit rat mammary tumourigenesis, mouse fore stomach neoplasia, and mouse skin carcinogenesis (Ha *et al.,* 1990). Therefore, an experiment was carried out to study conjugated linoleic acid enriched ghee feeding on cancer incidence in 7, 12 dimethyl benz (a) anthracene induced breast cancer in rats.

METHODS AND MATERIALS

Lactation Trial

Fifteen multiparous Murrah buffaloes in early lactation were randomly grouped into three groups namely group I, group II and group III of five animals each. Buffaloes in group I, group II and group III were fed with three different types of concentrate mixture. The composition of concentrate mixture for group I animals has groundnut cake 27 parts, maize 50 parts and wheat bran 20 parts. For group II animals, mustard cake 39 parts, maize 41 parts and wheat bran 17 parts whereas for group III animals, the composition of concentrate mixture as similar as group II plus 2% mustard oil has added extra. Mineral mixture 2 parts and salt 1 part was added in all three groups. Lactation trial was conducted for a period of 90 days. After 30 days of adaptation period, milk samples (100 ml) were collected at fortnightly intervals and the fatty acid composition of milk including CLA was estimated.

Ghee Preparation

Low (6.92 mg/g fat) and high CLA (19.54 mg/g fat) ghee was prepared from milk of buffaloes which are maintained on groundnut oil cake and mustard oil (2%) based concentrates, respectively by creamery method.

Cancer Study

Female Wistar rats 21 days of age (32 to 38 g of body weight) were randomly divided into 3 groups of thirty animals each and group I animals *fed* with soybean based diet where as group II and group III animals were fed with low CLA and high CLA ghee based diet, respectively. The rat diets were prepared according to the AOAC (1995) guidelines and ghee was included 20 % in the rat diet. Feeding experiment was conducted for a period of 32 weeks after 7, 12 DMBA injection.

At 55 days of age, each animal was administered with 5 mg 7, 12 Dimethyl benz (a)anthrazene (DMBA) in soybean oil by oral intubation as single dose. At the end oftheexperimentation period, animals were sacrificed by cervical dislocation and tumour incidence and tumour weight were calculated. A portion of mammary gland was preserved in 10% formal saline for histopathological study. The data were subjected to analysis of variance and statistical significance was accepted at ($P < 0.05$).

RESULTS

The average total CLA content in groups I, II and III was 6.84, 12.12, and 19.50 mg/g fat, respectively (Table 1) and differences among the groups were significant ($P < 0.05$). Milk CLA was increased ($P < 0.05$) 185 percent in groups fed on mustard oil containing ration (19.50 mg/g fat) as compared to control group I (6.84 mg/g fat). The process of ghee making doesn't change the CLA content in ghee and the CLA content was similar with their corresponding milk.

Table 1 : Milk yield and milk composition of buffaloes fed with different type of concentrate mixture.

Attribute	Group I	Group II	Group III
Milk yield (kg/d)	$7.86^{a} \pm 0.1\ 0$	$8.28^{a} \pm 0.09$	$8.90^{b} \pm 0.11$
Fat (%)	$8.13^{a} \pm 0.03$	$7.90^{a} \pm 0.01$	$7.71^{b} \pm 0.05$
Total Solids (%)	16.75 ± 0.12	17.48 ± 0.04	17.88 ± 0.04
SNF (%)	8.62 ± 0.14	9.58 ± 0.05	10.17 ± 0.09
Total CLA (mg/g) of milk fat	$6.84^{a} \pm 0.23$	$12.12^{b} \pm 0.12$	$19.50^{c} \pm 0.28$

Values with different superscript in row differ at ($P < 0.05$)

Table 2 summarizes the data on tumour incidence and mean tumour weight from each cancer group. CLA ghee feeding produced statistically significant ($P < 0.05$) reduction in tumour incidence and it has noticed that 37% reduction in gross tumor incidence in high CLA ghee fed group. High CLA ghee fed rats have lowest ($P < 0.05$) mean tumour weight than the other two groups. Fibroma, adenoma, and fibroadenoma were the common types of tumours in all the three groups. However, malignant type tumour (adenocarcinoma) was detected only in those rats fed on soybean oil diet.

Table 2 : Effect of conjugated -linoleic acid enriched ghee feeding on mammary gland tumour incidence in 7, 12 DMBA administered rats

Groups	Percent incidence*	Percent Individual tumour incidence: Fibroma tumour	Adenoma	Fibro adenoma	Adenoma carcinoma	Mean tumour weight (g)
Soybean oil	83.33 (25/30)	36.00 (9/25)	24.00 (6/25)	28.00 (7/25)	12.00 (3/25)	$3.30^{a} \pm 1.39$
Low CLA ghee	63.33^{b} (19/30)	42.11 (8/19)	21.05 (4/19)	36.84 (7/19)	0.00	$2.42^{b} \pm 0.88$
High CLA ghee	46.70^{c} (14/30)	57.14 (8/14)	14.29 (2/14)	28.57 (4/14)	0.00	$1.29^{c} \pm 0.27$

Rats were killed 32 weeks after DMBA administration
Values with different superscript in coloum differ at ($P<0.05$)
* Includes both Gross - and histopathological examination

CONCLUSION

This study showed that CLA supplemented diets significantly lowered cancer incidence. CLA not only inhibited benign type tumour but malignant tumour as well. Prevention, rather than therapy, must be an important strategy for conquering cancer and CLA could well do that. Though CLA is not a cancer drug but it is something that would be useful in addition to other cancer treatments. The successful therapy of breast cancer will have to identify natural products that have significant therapeutic and preventive potential without toxic side effects of the common anticancer drugs.

REFERENCES

AOAC (1995) : Official methods of analysis, Association of Official Analytical Chemists. 16th Edn. Virginia, USA.

Chin, S.F., W. Liu, J.M. Storkson, H.A. Yi, M.W. Pariza (1992). Dietary sources of conjugated dienoic isomers of linoleic acid, a newly recognized class of anti carcinogens. *J Food Comp Analy.* **15:** 185-197.

Griinari, J.M and D.E. Bauman (1999). Biosynthesis of Conjugated Linoleic Acid and its composition, incorporation into meat and milk in ruminants. In: Advances in CLA research. AOCS Press, Champaign, 11. pp: 180-200.

Ha L, J. Strokson, and W. Pariza (1990): Inhibition of benzo Ca pyrene-induced mouse fonestomach neoplasia by conjugated dienoic derivatives of linoleic acid. *Cancer Res.* **50:** 1097-1101.

Kelly, M.L, E.S. Kolver, D.E. Bauman, M.E. Van Amburgh and Muller D (1998). Effect of Intake of pasture on concentrations of conjugated linoleic acid in milk of lactating cows. *Journal of Dairy Science* **81:** 1630-1636.

Bioresources for Food Security and Rural Livelihood, Pages 283–309
Edited by : G.K. Kulkarni and B.N. Pandey

DAIRYING FOR RURAL LIVELIHOOD AND NUTRITIONAL SECURITY

A.K. Srivastava

ABSTRACT

Agriculture and allied sectors contribute about 30% to the gross domestic product of India. Indian agriculture & livestock sectors support 17.5% of world population with 2.3% of global land and 4.2% of water resources. Livestock farming is an age-old tradition for millions of Indian rural households. Livestock play a pivotal role in the agrarian economy, as nearly 90 million farming families, cultivating 140 million hectare land area, rear 90 million milch animals. This sector alone provides regular employment to 18.4 million people constituting about 8.5% of the total work force. India has the largest population of livestock with 57% of the world buffalo population and 16% of the world cattle population. The population dynamics of dairy animals of year 2003 indicated that the total population of buffalo and crossbred cattle have increased by 7.45% and 12.6%, respectively, as compared to the data of year 1997, however, in this corresponding period, the population of indigenous animals has decreased by 7.85%.

Presently, the contribution of live stock sector to the national GDP and agricultural GDP are to the tune of about 5.59% and 31.7%, respectively. The global milk production is around 630 million tones with an annual growth rate of 1%. India is producing about 15% of total world milk yield (104.9 million tonnes) with a growth rate of 4.8%. The contribution of landless, marginal, small and medium farmers in dairy sector is loudable. Approximately 70% of milk is being produced by these groups of farmers. What is particularly remarkable about Indian dairying is that, it provides livelihood to millions of rural family including the landless farmers. "In India, buffalo rather than cow is the milk producing animal," was the opinion of the "Royal Commission on Agriculture" in 1928, and it still holds good. Indian buffalo accounts for nearly 38% of bovine population and contribute 55% of milk production while cow contribute 40%. Buffalo milk production is limited to about 18-20 countries of the world with India contributing a maximum of 66% of the total buffalo milk production. Buffalo appears to be the animal of future in increasing the milk production because of their adaptability to varied climactic conditions and easier management practices required to rear them. Buffalo withstand extremes of temperature that make them most preferred milch species under changing global climatic conditions. The feed conversion efficiency of buffalo is remarkable and can thrive well on all type of feed and fodder. Buffalo population registered a significant growth during 1992-2003 with concomitant growth rate of 50% in milk production. The average milk yield from buffalo is around 4.2 kg per day which is although lower than cross bred cattle but substantially higher than indigenous cows (1.9 kg/day). The zone wize contribution of milk to national milk pool is as follows: northern (38%), western (31%), southern (20%) and eastern (11%). Despite highest milk production, the per capita availability of milk in India is only 250g/day, which is below the world average of 270g/day. Another good feature of Indian dairy sector is that the cost of milk production is one of the lowest in the world. The Ex. Farm gate cost of milk production in India is US $ 21/ 100 kg as compared to US $ 28/ 100 kg in USA, US $ 68/ 100 kg in Japan, US $ 58/ 100 kg in Canada and US $ 57/ 100 kg in Switzerland. As far as processing and trading of milk is concerned, in India, 35% of the milk produced, is retained by the farmers and producers and only 65% is traded. The major chunk is still handled by unorganized sector. The share of milk handled by organized sector is very less i.e. only 16 – 17%. Another issue which requires to be addressed is the low productivity of Indian milch animals (987 Kg/year) as compared to their western counterparts (World Average 2200Kg/Year).

In addition to food security, milk and milk products are also playing very important role and are essential for National Nutritional Security. Milk and its products are preferred food from animal origin and the total expenditure on milk by rural and urban Indians are to the extent of 15.4 and 18.6, respectively, of the total expenditure on food. The supply of protein, micronutrients and vitamins through milk and its products are higher than their intake through other livestock products. Besides being a source of quality proteins and energy rich fat, it contains important micronutrients like calcium, potassium, sodium, magnesium and vitamins, which are vital for overall development of the human body. Also, several health attributes are associated with milk and its constituents, such as the role of calcium in bone strengthening, controlling hypertension and as colonic anticarcinogenic, protective roles of carotenenoids and conjugated Linoleic acid (CLA) against cancers. Butyric acid, the short chain fatty acid of milk fat has been shown to regulate cell growth and enhance the anti-tumor activities. Certain minor milk components either naturally occur or formed during processing have also been endowed with many unique health benefits. Examples include lactoferrin, lactulose, galacto-oligosaccharides (GOS), α-lactoglobulin, and bioactive peptides. Further milk may also serve as ideal vehicle for the supply of additional micronutrients and nutraceuticals to larger segment of population.

The strength of Indian dairy sector lies in the fact that despite limited investments, it has shown constant and sustainable growth, however, the conventional dairy commodities like liquid milk, milk powder & other don't appear to be sustainable in future. Finally, it is conclude that to get "sustainability of livestock sector as tool for rural livelihood and nutritional security", (i) The newer challenges in terms of value addition, health and functional foods and safety will have to be addressed in light of advanced developments, (ii) R&D efforts to modify conventional processes to be taken at larger scale to manufacture all dairy products from all milk and (iii) The Indian dairying has to become more competitive by adopting cost effective state-of-the art technologies to produce quality and safe products.

Keywords : Dairying, Milk and Milk Products, Indigenous Dairy Products.

INTRODUCTION

Milk is considered as Mother Nature's perfect food. Apart from providing valuable nutrients. Milk also possesses wide assay of life promoting components. Milk is known to contain about 10,000 compounds in varied concentration with the advent of nutraceuticals and functional foods. The difference between diet and medicine is slowly dimmishing. researcher across the world are looking for bio-molecule derived from natural resources and also possess disease preventing and combating ability.

The convergence of major consumer food and nutrition trends with significant scientific breakthroughs in biotechnology and medical sciences will have a profound effect on the therapeutics in 21st century. Consumers are beginning to understand the powerful influence of naturally occurring component particularly in diet on health and well being. Increasing scientific evidence confirms that specific components in diet might tend off certain chronic diseases. It has revitalized the interests not only in consumers, but also among researchers and processors to isolate bioactive components and utilize them to formulate products, which is "natural," "Therapeutic" and "nutritional." as well. In this regard a very high value is placed on milk component and since these molecules are safe and does not cause any toxicity these may become the ultimate choice for formulators. Ingredient suppliers are also keeping a close eye on trends impacting consumer purchase. Marketing ingredients for pharmaceuticals require equal emphasis on its physiological functionality apart from other advantage associated with it. Dairy milk ingredients including non fat dry milk, whey proteins and milk minerals, a variety of new product formulations meant for specialized group of peoples by adding nutritional value, providing functional benefits and improving sensory acceptability. New products meant for specialized group of people utilizing the advantage of dairy ingredients include popular sports drinks, meal replacement beverages, nutritional bars, infant formula. Consumers have a natural fondness for the taste of dairy products and consider dairy ingredients as good stuff. Minor milk constituents like whey proteins; peptides, lactose & lactose derivatives have been thoroughly investigated by researchers for their nutritional and therapeutic characteristics and promising ingredients for pharma industry as well as food industry. Some of the prospective promising milk components that have potential for formulation and designing of new age therapeutics have been discussed here under.

India's population is approximately 1.17 billion (estimate of July 2009) with 72.5% of people living in rural areas. With the ever progressive reduction in the man-land ratio, cultivable land per family is decreasing from generation to generation leading to increase in number of small holders. Presently, India has 577 million small holders comprising of landless, small and marginal farmers, of which 350 million (70 million rural households) keep dairy animals. Out of these households, 75 percent (52 million households) are small, marginal and landless milk producers with the average herd size of 2-8 animals (Banerjee 2007). These farmers contribute to more than 80 percent of the total milk produced in the country. This underscores that any positive development for these farmers in dairying would not only improve the country's milk production but also enhance the livelihood and nutritional security of these rural masses.

The importance of dairying has long been recognized for upliftment of rural poor and several attempts have been made in policy level to promote dairying as means to provide agricultural economics with the growth impetus required for broad based rural development. Dairy cattle/buffaloes have an immense contribution for sustainable rural development as unlike crop which is seasonal, dairying provides a stable, year-round income, which is an important economic incentive for the small farmer to take up dairying. Dairying directly enhance the household income by providing high value output from low value input besides acting as wealth for future investment. It also offers a buffer income to the family during crop-vulnerable calamities. Rural workforce involves and contributes significantly to the country's milk production leaving ample scope for employment generation through the profitable dairy venture. The need of the hour is to address these issues for reducing migration of people from rural to urban in need of employment. Further, if the technological competence of the rural people in dairying is substantially improved, it would not only improve the self employment and also enhance the rural economy and livelihood. This paper deals with the dairy production scenario in India, role of rural masses on contribution to national milk grid and the way forward for making dairy enterprise to enhancing livelihood of rural people.

Dairying in India

Dairying is an age old practice in India and is one of the most important means of providing livelihood and nutritional security to the rural masses. In India, the rural population is more in North and North Central region and it is positively correlated with the milk production as this region provides more than 50% of the total milk production. Since 2001, India is rated as the highest milk producer in the world. The livestock activities, among which dairy farming predominates, employ about 8.5 million workers i.e 8% of the total labour force in the country. Women constitute about 71% of the labour force in livestock sector where as in crop farming they constitute only 33% of the work forces. The traditional dairying aims at fulfilling mainly the household requirements and the excess milk if any is marketed in an unorganized way. Nevertheless, commercial dairying has gained momentum recently. Milk and milk products account for 9.2 and 12.4 percent of protein intake in rural and urban areas, respectively, which is higher than the protein intake through other animal proteins.

Milk production of the country has registered impressive growth during the post independence era which has jumped from 19.4 million tonnes in 1951 to more than 104 million tones at present. The increased production of milk has improved the per capita milk availability to 246 grams per day; which almost equals the recommended per capita requirement of 250g by ICMR. The rate of growth in milk production is also substantially higher (4.5%) than the world average of 1.5 per cent. The share of livestock sector in agriculture GDP is increasing mainly due to the enhanced rate of production in livestock sector especially for milk which accounts for 67 percent of the value of output from livestock sector. Dairy farmers, scientists, cooperative agencies and development departments have all played a pivotal role in this regard. The demand of milk and milk products in India is projected to increase to 142.9 million tonnes in 2015 and further to 191.3 million tonnes in 2020. At the existing rate of growth in milk production, in next ten years, supply is likely to fall short of the demand. Trade liberalization coupled with rising income in several developing countries has further opened up new vistas of international trade for the Indian dairy sector. However, for tapping the economic benefits of growing global demand of milk and milk products Indian dairy sector has to gear itself by increasing the milk production at the rate close to 7 percent as against 4 percent at present, increasing the production of value added dairy products and by providing complete quality assurance.

In India, the dairy production system is more complex and can broadly be divided into peri-urban dairying and rural dairying. Peri-urban dairies usually keep crossbred cows and or buffaloes whereas the rural dairy farming is small holding with average herd size of around 3 animals per household. More than 40% of Indian farming households are engaged in milk production as this is a livestock enterprise in which they can engage with relative ease to improve their livelihoods. An interesting feature of the Indian dairy farming is the production system where the dairy animals are largely fed on crop residues and high producing animals are supplemented with concentrate feed. As more than 40 million households in India at least partially depend on milk production, developments in the dairy sector will have important repercussions on their livelihoods and on rural poverty levels. Despite holding the number one position in milk production in the world, our milk productivity remains one of the lowest as compared to the many leading countries of the world. Country boasts of 17% of the bovines in the world comprising of 8.22 million crossbred cows, 28.37 million indigenous cows and 33.17 million buffaloes together producing 15% of the world's total milk production. On the other hand, USA is producing 12.5% of the total world milk from merely 6% of the world cattle production thereby indicating that we have a very large population of low and unproductive bovines to thrive on our scare fodder resources. Ever raising demand for milk and dairy products across the globe offers exciting market opportunities for India.

Regional Analysis and Prospective in Milk Production

Analysis of region wise milk production would help in evolving region specific strategies for increasing milk production. The estimates of major livestock products in 2007-08 by Department of Animal Husbandry and Dairying reveal that the North and North Central region contributes to 54.5 percent of the total milk production in the country followed by the South (21.05%), West (15.18%), East (12.94%) and Northeast (1.16%) region. This suggests need for prioritizing the dairy animal production in the low producing regions and to boost the production in the already performing areas. There are wide variations in respect to the contribution of different dairy animals to the total milk production in a region. The contribution of buffaloes was highest in North and North Central region (68.45%) where as in Northeastern region its contribution was merely 11.21%. The contribution of buffaloes is moderate in West (54.74%) and South (41.89%). Crossbred cattle contribute much to the total milk production in South (42.95%) compared to any other region suggesting that crossbred cattle are preferred by the population and cross breeding is taken up well. In Northeastern region, the contribution of Indigenous cattle is much higher (55.14%) than in any other region indicating that the crossbreeding has not percolated much in these areas. When the individual animal productivity is considered, across the country, on an average a crossbred cattle, indigenous cattle and buffalo yields 6.37, 2.20 and 3.90 liters/day. There is no much variations in respect to the per day milk yield of crossbred cattle in different regions except for South, in which the yield is higher (6.57 litres/day) compared to other regions (5.69 litres/day). The major concern is the individual animal productivity in case of Indigenous animals, which ranges from 1.06 liters/day in Northeast to 2.58 liters/day in North and North Central regions. It is pertinent to highlight here that the indigenous cattle accounts about 87% to the total cattle population in our country, leaving ample scope to intervene the productivity of these animals.

Swot Analysis: The SWOT analysis of the current dairy scenario of the country would reveal the following:

Strengths

- The vast livestock population of the country could prove to be a vital asset for the country and unlike many other natural resources which will deplete over the years, a sustainable livestock production system will continue to propel Indian economy.
- As the milk productivity of our animals is low, there is a vast scope for improvement of the milk production and consequently increased marketable surplus of milk for processing.
- Purchasing power of the consumers is on the upswing with growing economy & continually increasing population of middle class.
- Milk consumption in India is regular part of the dietary programme irrespective of the region and hence demand is likely to rise continuously.
- Indian dairy farming thrives largely on crop residues and agricultural byproducts keeping the input costs low. Labour cost is also fairly low making the industry fairly cost competitive.
- Industry continues to grow and the margins are still fairly reasonable. Indian dairy farmers do not receive any subsidy and when the world dairy market opens up post-WTO negotiations, our products could compete on the price front
- Large number of dairy plants in public and cooperative sectors besides several others in the private sector is coming up.
- Vast pool of highly trained and qualified technical manpower is available at all levels to support R&D as well as industry operations.

Weaknesses

- Though cross breeding programmes have significantly improved animal productivity, milk production system in many parts of the country is still largely dominated by low yielding animals.

- Poor condition of roads and erratic power supply remain a major challenge for procurement and supply of good quality raw milk. Furthermore, raw milk collection systems in certain parts of the country remain fairly underdeveloped.
- Maintenance of cold chain is still a major handicap. For organized marketing of milk, the milk produced is required to be transported to nearby processing plant which incurs cold storage and transportation costs which are quite high.
- Majority of producers is unaware about scientific dairy farming, clean milk production and value chain.
- Seasonal fluctuations in milk production pattern, regional imbalance of milk supply and species-wise variation (buffalo, cow, goat etc.) in milk quality received by milk plants continue to pose serious handicaps.
- Absence of comprehensive and reliable milk production data, impact assessment studies are almost non-existent, investments in dairy research is also not commensurate with returns and potential.

Opportunities

- Expanding market will see creation of enormous job and self employment opportunities.
- Economy is growing at the rate of nearly 8% of GDP. Consequently, the investment opportunities are also increasing continually.
- Demand for dairy products is income elastic. Continued rise in middle class population will see shift in the consumption pattern in favour of value added products besides the growth in demand for liquid milk.
- Greatly improved export potential for indigenous as well as western milk products.
- Opening of the world market offers opportunities for utilization of byproducts of the dairy industry for manufacturing value added products for import substitution.

Threats

- Excessive grazing pressure on marginal and small community lands has resulted in almost complete degradation of land.
- Indiscriminate crossbreeding for raising milk productivity could lead to disappearance of valuable indigenous breeds.
- Organized dairy industry handles only 15% of the milk produced. Cost effective technologies, mechanization, and quality control measures are seldom exercised in unorganized sector and remain key issues to be addressed.
- There is a gross lack of awareness among farmers about the quality parameters, including microbiological and chemical contaminants as well as residual antibiotics.
- Middlemen still control a very large proportion of the milk procurement. Serious efforts need to be taken to eliminate them from the supply chain.

- Export of quality feed ingredients viz., cakes, molasses etc. is making the domestic producers rely on low energy fodders.
- Entry of multinationals could result in a large portion of milk being diverted towards value added products which, though it augers well for the producers, is likely to affect the availability of liquid milk supply for mass consumption especially for the poor urban class.
- A parallel economy is thriving on adulterated liquid milk including synthetic milk in certain pockets which needs to nip in the bud.

Bringing Sustainability in Smallholder Dairy Production

Small scale milk producers face a number of challenges in realizing the opportunities offered by the growing demand for dairy products. Increasing milk output per animal is the first and foremost step to be taken to sustain productivity in smallholder production system. This will need interventions in breed improvement, improving the availability of feeds and fodder and control of diseases.

Germplasm improvement

The strategy for increasing milk production should focus on increasing animal productivity rather than animal population. The emergence of India as a premier dairy nation of the world could be attributed mainly to the intensive crossbreeding programmes implemented throughout the country over the last few decades. The population pressure and emerging global opportunities further necessitate that the efforts for enhancing animal productivity are accelerated. This calls for continuance of well-proven technologies such as crossbreeding with superior germplasm coupled with improvement in the productivity of vast population of generally low producing cattle and buffaloes at faster rate using emerging reproductive and molecular technologies including MOET, use of sexed semen and cloning for production and faster multiplication of superior germplasm of elite animals. For improvement of various breeds of cattle and buffaloes there is a need to form Breed Societies on the pattern of western countries. These societies could work in close cooperation with various research and developmental agencies including Non-Governmental Organisations (NGOs) for improving the livestock productivity. Interactive programmes involving field recording, progeny testing and data bank should be taken to supply information on availability of semen, bulls, females etc. for consistent genetic improvement of livestock. To effectively monitor these programmes, there is an urgent need for linking them with bio-informatics centres at state, national and international levels. There is also scope for identifying the unique genes specific to indigenous dairy animal genetic resources using the molecular techniques. It is expected that by the use of such technologies more number of superior animals per unit time can be obtained by reduction in generation interval which could further increase the pace of genetic advances by increasing the intensity of selection. Technologies for raising male buffalo calves economically should be developed as this potential source of income of farmers is almost going waste.

Improvement of Feed and Fodder Availability

Poor feeding practices also lead to poor reproduction, long calving intervals, and as a result insufficient replacement stock, which limits the opportunities for selection and breed improvement. New challenges in relation to forage production and feeding include climatic change and reduction of green house gasses. A major challenge for small scale farmers is the sustainable intensification of milk production to obtain higher milk production per animal, which will improve economic competiveness and may contribute to reduction of green house gasses. Availability of good quality feed (forage/roughage and concentrate feed) throughout the year is a major limitation for increased milk production, especially when better breeds are reared. The projected dry doffer, green fodder and concentrate demand for 2102-11 is 422,199 and 56 million tones on dry matter basis where as the availability stands at 382, 133 and 36 million tones leaving a short fall of 10, 33 and 35%, respectively. The situation is still worse in Arid Zones of the country where dry season feed shortages lead to seasonal fluctuations of milk supply. The local inhabitants in these zones have evolved suitable landuse and management systems of farming, pastoralism and animal husbandry. As time passed, these local survival systems have become inadequate to fulfill the ever increasing needs and resulted in over-exploitation of the resources causing rapid and widespread land degradation and decline in productivity. Two way approaches may be followed to improve the feed and fodder availability in these areas. Identification and development of suitable feed and fodder crops, grasses and top feed species for Arid zones and incorporation into water shed development areas and rational exploitation of the non-conventional feed resources on the other hand would go a long way in fulfilling, at least partly, the demands of the animals in this ecosystem. Sylvi-pastoral system can also be promoted in these areas.

In order to meet the nutritional requirements of animals there is a need to increase the bioavailability of the feeds and fodders using chemical, biological and biotechnological approaches. Also, it is necessary to improve the productivity of the land for meeting out the feed and fodder requirements from the limited area available for this purpose. There is also the need to encourage farmer's unions, NGOs and cooperatives to put common property lands under improved pasture and silvi pastoral systems. An exercise has to be conducted to develop an effective plan for improving the availability of feed resources of the country, as there is shortage of feeds, resulting in inadequate supply of nutrients especially protein, energy and minerals. Besides this, we need to explore new feed resources and evaluate them for livestock feeding and find out how much these can be useful in bridging the gap between supply and demand of the nutrients. For updating the information about feed resources, there is a need to generate information on the level of anti nutritional factors in the feeds. Presently, data on the levels of aflatoxins, pesticide residues, heavy metals in feeds and fodders is inadequate. It is therefore imperative to generate objective information on this vital aspect so that we are able to produce livestock feed of internationally acceptable quality.

Potential Dairy Products for Rural Entrepreneurs

A wide range of technological packages are available for the rural entrepreneurs and their adoption depends on several factors. It includes availability and type of milk, entrepreneurship skills, location and size of market and also capital requirement. Some of the dairy products that offer great potential are discussed here. Conversion of milk into indigenous milk products in and around milk production areas is least expensive and more remunerative because of simple technologies, use of low grade energy sources, inexpensive infrastructure and equipment and low operating and overhead costs as compared to European or western systems of dairy processing. These indigenous products have great social, religious, cultural, medicinal and economic importance and have been developed over a long period with the culinary skills of homemakers and *halwais*. In addition to preservation of milk solids for longer time at room temperature, manufacture of traditional dairy products add value to milk and also provide considerable employment opportunities to rural population. These products include curd, ghee, *khoa*, *chhana*, *paneer*, *shrikhand* and a variety of milk sweets, some of which are now produced by the organized diary industries as well. It is estimated that about 50% of total milk produced in India is converted into traditional milk products. The market for traditional dairy products in India is estimated to be US $ 10 billion, being the largest and fastest growing segment of the Indian dairy industry. This figure underlines the significance of Indian traditional products in the national economy.

1. **Khoa and Khoa Based Sweets:** Among the traditional dairy products, *khoa* – an intermediate concentrate that is base material for a wider range of sweetmeats such as *burfi*, *peda*, *gulabjamun*, milk-cake, *kalakand* and *kunda* has great significance in India. According to one estimate about 5.5% of total milk production is converted into *khoa* and on the basis of present milk production of about 96 million tones per annum, this is equivalent to about 14.7 million kg of *khoa* per day. *Khoa* is prepared by continuous boiling of milk with scrapping in a shallow pan (karahi). Milk thickens progressively as the evaporation of moisture takes place constantly. Vigorous stirring and desiccation continued until desired concentration (65 to 72% TS) is reached and the product become viscous and pasty. According to Prevention of Food Adulteration (PFA) (1955) rules, *khoa* sold by whatever variety or name such as *Pindi*, *Danedar*, *Dhap*, *Mawa*, or *Kava* means the product obtained from cow or buffalo (or goat or sheep) milk or milk solids or a combination thereof by rapid desiccation and having not less than 30 per cent milk fat on dry weight basis.. To achieve the PFA standard a minimum fat level of 5.5 in milk is essential. The quality of *khoa* is better when made from buffalo milk because *khoa* from cow milk is inferior due to its moist surface, salty taste and sticky and sandy texture which is not considered suitable for the preparation of sweetmeats. Also, buffalo milk results in higher yield of *khoa*. *Khoa* is manufactured primarily by *halwais* in jacketed kettles, which inherently suffers from several disadvantages particularly the poor and inconsistent quality of the product. It has a limited shelf life of about 5 days at 30°C. Rural entrepreneurs can manufacture *khoa* by adopting the latest technical know-how including the semi-

continuous processing equipments. A large quantity of *khoa* is being produced in various pockets around big cities. For example to meet the large demand of *khoa* in Delhi, it is being produced in the interior villages of western Uttar Pradesh, Haryana and Rajasthan, where milk is comparatively cheaper, demand is less and availability is high. *Khoa* based sweets can also be prepared and marketed without adding newer equipments and also the processing technologies are quite simple.

2. **Paneer : *Paneer,*** the indigenous variety of soft cheese, is obtained by the acid coagulation of heated milk. In the traditional process, buffalo milk is boiled in a vessel. To coagulate the milk while still hot, a suitable coagulant (lime/citric acid/ alum) is added, with a slow stirring. The technology of manufacturing *paneer* from buffalo milk has been standardized so as to obtain the most acceptable and safe product with maximum recovery of solids. Subsequently, process of preparing *paneer* of acceptable quality from cow milk has also been developed. Irrespective of the type, milk should be standardized to a fat and SNF ratio of 1:1.65 so that the final product conforms to PFA requirements. Good quality *paneer* is obtained by heating milk to about 90°C, acidifying the hot milk by adding citric acid solution at 70°C. Formation of clear whey is indicative of complete coagulation. After the formation of large lumps is complete, contents of vessel are pored over a muslin cloth to separate the coagulum from whey. The coagulum so obtained is highly pressed to facilitate formation of paneer blocks, followed by their immersion in chilled water to impart them distinctive texture. Chemical and physical changes in casein and whey proteins, brought about by the combined influence of heat and acid treatment, form the basis of *paneer* making. Conventionally, citric acid is used for coagulating hot milk for *paneer* making but certain non-conventional, low-cost coagulants have been suggested for manufacture of *paneer* without any loss of its yield and quality.

   The phenomenon of coagulation involves the formation of large structured aggregates of proteins in which fat and other colloidal and dissolved solids are entertained with whey. *Paneer* has granular structure consisting of protein particles having a core and lining ultrastructure irrespective of the type of milk used. Good quality p*aneer* is characterized by a white colour, sweetish, mildly acidic, nutty flavour, spongy body and closely knit texture. Buffalo milk *paneer* is of good quality as it has all these attributes. On the other hand, cow milk p*aneer* is of inferior in its textural attributes due to the lesser proportion of high melting triglycerides in its fat, lesser proportion of casein, smaller size of casein micelles and fat globules and lower content of calcium in it. Different factors, like type of coagulant, temperature and pH of coagulation, homogenization of milk, type and quality of milk, additives and storage conditions affect the quality of *paneer*. Processing makes *paneer* soft and spongy and reduces the textural differences in raw *paneer* from cow, buffalo and low fat buffalo milk. Acidic milk produced inferior quality product. Surface spoilage of paneer can be controlled by using antimicrobial agents like 5% brine, potassium sorbate, benzoic acid, sorbic acid etc. Normally, *paneer* blocks of required size/weight are packaged in polyethylene pouches, heat sealed and stored under refrigeration conditions. Alternatively, they are vacuum packaged in laminated or co-extruded films. The demand for the value added products with *paneer* as base material is growing in the urban areas. It is an excellent match for non vegetarian food. Moreover the cost of *paneer* production is low because of its significantly higher yield and short preparation time. This is likely to increase the returns on investment. There is a need to tap the market potential of *paneer* both for domestic consumption as well as export. Manufacture of *paneer*, therefore has great social and economical significance in our country.

3. **Chhana: *Chhana*** is an another heat and acid coagulated product which serves as a base product for a large variety of Indian delicacies, namely, *Rasogolla, Chhana gala, Sandesh, Cham cham, Rasmalai, Pantoha, Rajbhog, Chhana murki* etc. It differs from *paneer* as no pressure is applied to drain the whey and its pH is slightly higher. *Rasogolla* is undoubtedly the king of all Indian sweets. Its production is largely confined to the cottage and small scale industry. In the traditional production of *chhana*, a small portion of boiled milk is transferred to a small coagulating vessel. The required amount of coagulant (usually previous day sour *chhana* whey) is added to the hot milk and stirred with a wooden ladle till the coagulation is complete. The vessel contents are then poured over a clean muslin cloth held over another vessel in which the whey gets collected. The muslin cloth containing the curd mass is hung to further drain out whey and to cool the *chhana* simultaneously. Structure of *chhana* revealed a conglomerated compact protein matrix with numerous small uniformly distributed pores. A number of factors like type and quality of raw milk, type and strength of coagulant, temperature and pH of coagulation, homogenization of milk, method of straining and additives influence quality of *chhana.* Cow milk is better suited for *chhana* making because it produces *chhana* with soft body and smooth texture which is better for making sweets. On the other hand *chhana* prepared from buffalo milk is harder and less cohesive due to the higher concentration of casein more so in the micellar state with bigger size of the micelles along with large portion of high melting triglycerides with bigger size of fat globules and higher content of calcium in buffalo milk. However, good quality *chhana* can be prepared from buffalo milk by different treatments like addition of mono sodium phosphate, disodium phosphate or a combination of both, dilution of milk with water, homogenization of milk, addition of cow milk or skim milk, alteration in coagulation temperature and pH or method of straining. The temperatures of coagulation exert a significant influence on physico-chemical attributes. Coagulation of milk at 70°C produces a very soft *chhana* but higher temperature produces grainy texture and increased hardness, whereas coagulation at 35°C produces highly sticky product. Citric acid as coagulant yields more cohesive *chhana* and lesser losses of milk solids in whey.

4. **Ghee:** *Ghee* is heat-clarified butterfat derived solely from milk or curd or from *desi* butter (cooking butter) or from cream to which no colouring matter or preservative has been added. In India, *ghee* is considered as an excellent cooking or frying

medium. In addition, *ghee* is used for numerous religious rites by Hindus and it has also many medicinal uses. It is usually prepared from cow's milk, buffalo's milk or mixed milks. *Ghee* made from cow milk fat has a distinct golden yellow colour, attributable to carotene. On an average cow or buffalo ghee contains 99.0-99.5% fat and less than 0.5% moisture. Traditionally, *ghee* is produced by first converting milk into *dahi*, churning *dahi* to produce *makkhan*, and subjecting *makkhan* to heat clarification to yield the final product. *Ghee* production forms the largest segment of the milk consumption and utilization pattern in India. Hence for obvious reasons, many efforts have been made by various workers to mechanize the process of *ghee* production.

5. **Dahi and Other Fermented Milk Products:** Dahi is a well-known fermented milk product consumed by majority of population in India. It is used either as a part of diet or as a refreshing beverage. It is mildly sour product with a pleasant flavour formed by the combined action of acid producing and flavour producing lactic acid bacteria. Choice of majority of consumers of Dahi lies in a product with glossy to semi-glossy velvety appearance, medium firm body, moderately acidic and delicate flavour. Dahi is prepared by using starter cultures, growing at appropriate temperature depending upon the strains used. Production of the product varies from place to place. The types of organisms include *Lactococcus lactis* ssp. *lactis, Lactococcus lactis* ssp. *cremoris, Lactococcus lactis* ssp. *lactis* var. *diacetylactis, Leuconostoc mesenteroides* ssp. *cremoris*. Dahi is also prepared by using thermophilic starter, viz., *Streptococcus thermophilus, Lactobacillus delbrueckii* ssp. *bulgaricus* and other *Lactobacillus* species. Shrikhand is a semi solid, sweetish-sour fermented milk product, prepared from dahi (curd). Shrikhand is a popular dessert and forms a part of a meal on festive occasions, particularly in the states of Gujarat and Maharashtra. Like dahi, it is very refreshing, particularly during summer months. Whey is drained off from dahi to yield chakka, Sugar, flavour, colour and spices are thoroughly mixed into chakka to form a soft homogeneous mass that resembles sweetened quarg of Germany.

6. **Dairy Beverages:** Milk based beverages have emerged as one of the most promising segment among value added dairy products and milk processors must exploit the opportunities to keep pace with the changing pattern of product consumption. Apart from traditional milk beverages like lassi, flavoured milk etc. there are a wide range of fruit based dairy beverages that could be added to the product portfolio. Milk-fruit based beverages are another range of products that offer tremendous market potential for Indian dairy industry because of being nutritionally-rich. Similarly, minor cereals and millets based milk beverages seem to be lucrative products for school feeding programmes. Whey can be defined as the watery part of milk remaining after removal of protienacious mass that results from the coagulation of milk by acid or proteolytic enzymes. Whey is the largest by-product of the dairy industry both in terms of volume and milk solids as it contains 50% of original milk solids including whey proteins, lactose, minerals, water soluble vitamins and residual lipids. Whey is

obtained during the manufacture of cheeses, casein, paneer, chhanna, and shrikhand. In recent years the demand for cheeses, coagulated indigenous dairy products and casein has increased tremendously. Recent statistics suggests that around one million tone of whey is produced in India annually and it corresponds to approximately 70,000 tones of whey nutrients. Liquid whey has also been utilized for the manufacture of wide range of beverages and soups over the years and as result of which a number of such products are available to consumers.

The ready-to-serve (RTS) type beverage may be prepared by mixing an appropriate fruit juice or concentrate and minimally processed whey along with other minor additives to improve the sensory characteristics and thermally processed to make it shelf-stable. A process has been standardized to manufacture whey-fruit beverages using cheese or paneer whey with three different fruits viz. banana, lemon and pineapple. Among the beverages mango beverage scored maximum for all sensory attributes and it contained 15% mango pulp, 7% sugar, 78% whey and the pH of the beverage was kept at below 4.5. An attempt has been made at our Institute to develop sports beverage using whey. The hydrolyzed paneer whey was utilized for the preparation of lemon and pineapple based sports beverages. Acceptable lemon based sports drink can be made with 7.43% sugar, 3.1% lemon juice, 0.07% lemon flavour and 0.07% CMC. Addition of 0.35% salt enhanced the overall acceptability, but protein enrichment of the beverage using hydrolyzed whey protein resulted in lowering of product acceptability. The best formulation for pineapple sport beverage were 20% pineapple juice, 7.49% sugar, 0.15% stabilizer mix and 0.12% salt mix. The stored sports drink was acceptable up to 6 months.

7. **Whey Based Soups:** Soups are served as appetizers before meals as they stimulate the secretion of gastric enzymes that leads to feeling of hunger. In market a large number of ready-to-make soup mixes are available to suit the palate of consumers. Moreover apparently they do not seem to provide quality nutrients and utilization of whey for soup preparation is attractive possibility. The process for the manufacture of whey based soup involves blending of vegetables in whey and cooking of corn flour followed by heating. The time-temperature combination of cooking of vegetables, corn flour and seasoning is important for dispersion of vegetables, gelatinization of starch and flavour perception of soup respectively. The developed product could be stored for a week under refrigeration and UHT treatment can be adopted to improve the shelf-stability. Paneer and cheese whey were utilized for the potato-carrot-tomato and spinach soups. Cheese whey was preferred for the manufacture of vegetable soups than paneer whey. The reason could be the low pH of paneer whey that resulted in acidic product not usually compatible with most vegetables. Whey based soups have been reported to be more viscous as compared to water based most probably gelation of whey proteins on heating. Whey based soups require less amount of salt, thickener and fat and technology for manufacture of retort processed low fat tomato-whey soup has been developed recently at out Institute. These products can be marketed through retail outlets, hotels & restaurants and others.

Critical Issues in Rural Dairy Entrepreneurs

1. **Safety of Processed Dairy foods :** Despite all impressive figures the presence of Indian dairy products in international market is quite negligible, mainly due to poor quality of these products. To compete in international market Indian dairy industry has to gear up to meet emerging challenges. In this context, food safety issues associated with milk and milk products now has become a matter of great concern to the industry, regulatory bodies and the consumers. Assurance of both the safety and quality of food is the most important aspect of food production because of increased public awareness towards health care and changing food habits. Since perishable foods like dairy products are extremely vulnerable to microbial contamination during their processing, especially under tropical, humid and unhygienic conditions prevalent in third world countries like India, their microbiological safety against high risk food pathogens has always been a matter of considerable public health concern. In addition, the health hazardous effects are also due to mixing of non-potable water, neutralizer, detergents, preservatives, urea and presence of residues of pesticide, heavy metals, antibiotics, aflatoxins and other veterinary drugs. Improper handling of milk can also serve as a potential vehicle for transmission of many diseases like tuberculosis, brucellosis, diphtheria, anthrax, foot and mouth disease, hepatitis, Q-fever, salmonellosis, shigellosis, streptococcal infections, staphylococcal poisoning, *E. coli* poisoning and botulism. Further, now a day's public has also become more concerned about synthetic milk. Therefore, full assurance for the supply of safe milk has added an edge on quality.
2. **Poor Shelf-life of Dairy Products :** During storage and distribution, dairy products are exposed to a wide range of environmental conditions. Environmental factors such as temperature, humidity, oxygen and light can trigger several reaction mechanisms that may lead to their degradation. As a consequence, these products may be altered to such an extent that they are either rejected by consumers, or they may become harmful to the person consuming them. It is, therefore, imperative that a good understanding of different underlying reactions that cause deterioration is gained prior to setting up of any processing unit. Knowledge of influence of processing on shelf-life, kind of packaging required and also the maintenance of cold chain across the production-to-consumption continuum is essential not only for shelf-life enhancement but to ensure consumer safety as well.
3. **Training of Rural Entrepreneurs :** Most of the trade of Indian milk products is with the *halwais* and the small-scale operators. Most of them have art and skill of manufacturing varieties of dairy products. However, no attention is paid by them on quality of milk, hygienic handling, proper packaging and storage due to ignorance. Knowledge regarding physico-chemical attributes of these dairy products and effect of different factors influencing the same may also contribute a great deal to enable them to produce these products under hygienic environment with better quality and longer shelf-life. Capacity building through innovative and science based training and

orientation programmes should be given major emphasis for promoting entrepreneurship in dairy processing. The training of operators in this sector in hygienic handling and quality control aspects will go a long way in improving the quality of these products. The regional Agricultural Universities and Krishi Vigyan Kendras (KVK) will have to play an active role in training of small entrepreneurs. In a significant move to support and nurture the dairy entrepreneurs, a Technology Business Incubator (TBI) has been started by Society for Innovation and Entrepreneurship in Dairying (SINED) at NDRI, Karnal. It helps technology-based dairy enterprise with all the necessary support that the needs to evolve and grow into a mature business.

Whey Proteins as a Protein Source

Whey may be defined, broadly, as the serum or watery part of milk remaining after separation of the curd which results from the coagulation of milk by acid or proteolytic enzymes. Whey is the largest by-product of the dairy industry. It is obtained during the manufacture of cheese, casein, paneer, channa, and shrikhand. Whey contains proteins, lactose, mineral matters and water-soluble vitamins. Proteins can be separated from whey using ultrafiltration or diafiltration technologies and during this process low molecular weight compounds (lactose, non-protein nitrogen, vitamins and minerals) are removed from whey into the permeate. The remaining proteins are, in turn, concentrated in the retentate. Whey protein is a combined word for diverse proteins that are present in whey. In general, whey proteins are globular, smaller in size, and heat denaturable. Whey proteins can be used for food applications in various forms such as concentrated whey; lactose hydrolyzed dried whey, UF-retentate, WPC and WPI. In the form of whey protein concentrate (WPC) and whey protein isolate (WPI), whey proteins have high nutritional and functional properties and are capable of fulfilling the diverse attributes to satisfy different forms of utilization (de Wit, 1998). Infact at present the pharmaceutical industry is capitalizing the muscle growth enhancing ability of whey proteins in formulation of specialized preparation for sports and athalete.

Composition and Quality of Whey Proteins

Liquid whey contains approximately 20 % of the original proteins of milk ranging from 4 to 7g/lt of which 3.7g is β – Lactoglobulin, 0.6g is α – Lactalbumin, 0.3g is Bovine Serum Albumin, and 1.4g is proteose – peptone fractions (Marshall, 1982). In addition, it contains other proteins such as lactoferrin, immunoglobulins, ceruloplasmin, and milk enzymes such as lysozyme, lipase, and xanthine oxidase, which are present in low concentrations. The nutritional quality of dietary protein is essentially related to its amino acid composition as well as to the availability of these amino acids. In this respect it is well known that milk proteins have a high content of essential amino acids. Whey proteins are also known to have high nutritional value due to their varied amino acid composition and good digestibility (de Wit, 1998). Whey proteins are rich source of all essential amino acids (Regester *et al.*, 1996). The BV (104) and PDCAAS (1.0) values of whey protein are high and therefore whey proteins are superior to other dietary proteins. The proportion of sulphur containing amino acids i.e. cysteine and methionine is also reported to be on the higher than meat, soy and casein. Whey proteins provide more than 100% of the requirement for sulphur amino acids in the growing human being, whereas soy protein is limiting in these amino acids (Sanwar *et al.*, 1985). Tryptophan, which acts as building block for niacin, is present in higher amount in whey proteins. Therefore, whey protein based pharmafoods like powders, biscuits can be developed for different groups.

Therapeutic Value of Whey Proteins

Dietary whey proteins have a number of putative, biological effects when ingested (Horton, 1995). The ability of whey proteins to increase the level of natural anti-oxidants within the body and possibly in stabilizing DNA during cell division is emerging as premier contribution to population health (Bonous *et al.*, 1989). Growth factors present in the cheese whey have attracted increasing interest since Howarth and co-workers (1996) showed that oral administration of this extracted growth factors from cheese whey could reduce small bowl damage in methotrexate-treated rats. Some of the growth factors that have been identified in whey include: transforming growth factor (beta and basic fibroblast growth factors) and an inhibitor (mammary derived growth inhibitor) with an anticarcinogenic activity at extremely low concentrations (Steijns, 2001). These compounds may survive digestion in sufficient quantities to generate a physiological response. Experimental animals fed with four different proteins namely whey, soybean, casein and red meat, as sources of protein, were administered with injection of the carcinogen, dimethylhydrazine. Whey protein-fed animals showed the lowest incidence of colon cancer (McIntosh *et al.*, 1995). Experiments in rodents indicate that the antitumor activity of the dairy products lies with protein fraction and more specifically in the whey protein component of milk. Possible modes of action maybe biochemical, including levels of sulphur containing peptide, glulathione and the influence of protein on fat metabolites generated in gut, or immunological or a combination of both (Regester *et al.*, 1996). The anticarcinogenic properties of whey proteins are related to compounds rich in sulphur containing amino acids, methionine and cysteine. They contain ã glutamyl-cysteine residue, which makes cysteine readily available for synthesis of glutathion, (ã ??glutamyl- cysteinyl- glycine) a strong xenobiotic deactivating and anti-neoplastic agent (Parodi, 1998). Methionine is utilized for glutathion synthesis in times of cysteine deficiency and it also acts as methyl donor. Hypomethylation of a DNA is an important risk factor for cancer at number of sites. Glutathion is, believed to act as an antioxidant, anticarcinogenic and in stabilisation and repair of DNA. Evidences indicate that its level increased after receiving the whey protein diet (Bonous *et al.*, 1989; Bonous and Gold, 1991). The whey proteins act as precursor glutathion, this at least in part, explains the ability of whey protein diet to enhance humoral and cell-mediated immune responses in laboratory animals.

Another explanation that has been put forward for anticarcinogenic activity of whey proteins is its ability to induce bio-synthesis of folic acid, vitamin $B_{12,}$ riboflavin, retinol and vitamin D (Parodi, 1998). Binding of iron by lactoferrin makes this potential pro-carcinogenic unavailable for intestinal damage. Binding of vitamin B to proteins makes them more bioavailable and protects them for being utilized by intestinal microorganisms. Probiotic cultured milk products containing condensed whey and lactose-hydrolyzed condensed whey have been demonstrated to lower the total and LDL cholesterol in experimental animals (Beena and Prasad, 1997). In another experiment, the serum total cholesterol level for the group of rat fed on whey protein concentrate containing milk fermented with both *Lacto bacillus casei* TMC 1543, was significantly lower than that of control group (Kawase *et al.*, 2000). They also conducted similar long-term intake (4weeks) trials with twenty healthy adult men and found a significant rise in HDL cholesterol level and decrease in triglyceride level. The a-La contains 2-3 times more tryptophan than an average protein. In body, tryptophan is converted into 5-hydroxytryptophan and then to 3-hydroxytryptamine (sorotonin). Inadequate 1evel of serotonin in the brain has been linked to depression, obesity, insomnia and chronic headache (Walzem, 2001). Whey protein isolates (WPI) has been used to treat HIV patients because immunoglobulins and bovine serum albumins present in it, may stave off this disease (Horton, 1995; Walzem, 2001).

Whey Protein in Health Foods

Functional or health foods are foods and food components that provide a health benefit beyond basic nutrition (for the intended population). Examples may include conventional foods; fortified, enriched, or enhanced foods; and dietary supplements. In recent years, scientific evidence has revealed that bioactive dietary components may benefit health in ways that extend beyond meeting basic nutrition. Some components, when consumed often enough and in sufficient quantities, may help reduce risk of chronic diseases such as heart disease, cancer, diabetes, or obesity. Human nutrition science has moved from a focus on the prevention of nutrient deficiencies to an emphasis on health – maintenance and reduced risk of chronic diseases. In this context, the food industry has come to realize that milk proteins in general and whey proteins in particular have potential to improve the nutritional and therapeutic quality of food products. Some of the food applications of whey protein include infant food, plasma protein-enriched foods, nutrition bars etc.

Lactose and Lactose Derivatives

Lactose, the principal milk sugar, is slowly metabolized and therefore considerable portion of it passes in to the large intestine where it promotes the growth of lactic acid producing bacteria. It is one of the major milk constituent that is used in pharmaceutical industry as binder. Lactose is an important constituent of majority of tablets and medicines mainly used as filler material. In recent years lactose as well lactose derivatives are becoming popular because of their health promoting effects.

Latulose : Lactulose (4-o- β -D- galactosylpyranosyl – D - fructose) is an isomer of lactose. Lactulose is formed during heating of milk in small amounts. Pasteurized milk does not contain any detectable amount of lactulose, highly pasteurized milk (< 50 mg / lt), UHT milk (100 –400 mg/litre), and sterilized milk (600-1200 mg/litre). International dairy federation (IDF) has set the upper limit of lactulose content in UHT milk as 600 gm / liter.

Lactulose is white crystalline odourless powder with melting point of 169°C. Lactulose is water-soluble and solubility increases with increases with temperature, exhibiting 86% solubility at 90°C. The degree of sweetness ranged between 0.48 – 0.62 compared to 1.0 of sucrose, however with increase in concentration the difference in sweetness decreased, lactulose is 1.5 times sweeter than lactose.

Lactulose as "Bifidus factor" : Lactulose plays a role in proliferation of Bifidobacterium spp. that has a very close relationship with human health. Lactulose is generally not metabolized in small intestine and passed directly in large intestine, where it is broken down by the microbes, most specifically by *bifidobacterium* and acidophilus bacteria with lactic acid as end product. Bifidobacteria has been thoroughly investigated as therapeutic and prophylactic agents for human and animal health. *Bifidobacterium* is a predominant microorganism in the intestine of breast feeding infants because of lower protein and buffering capacity of human milk. Investigation regarding the effect of incorporation of lactulose in infant formula on the intestinal *bifidobacterial* flora in rats indicated that 0.5 – 1.0% lactulose content in formula had no adverse effect on the absorption and retention of nitrogen, calcium, phosphorus and iron from the formula by *bifidobacterial* flora. It also did not affect the growth, blood composition or histopathology of liver and shows no adverse effect on continuous consumption duering growth period. Incorporation of 0.5% lactulose in formulation was considered adequate to stimulate bifidobacterial flora and prevent laxative effect of it. Borculo- Domo ingredients, a major Europian dairy manufacturer, have developed a prebiotc ingredient named "Elix'or". The product is both in powder or syrup forms and contains 60% galactooligosaccharides, 20% lactose, 19% glucose and 1% galactose. As per manufacturer's claim the product in pH and thermostable, has a bifidogenic effect and increase intestinal absorption of minerals. Health benefit associated with lactulose among elderly is its ability to act as mild purgative, thus it helps in reducing the growth of ammonia producing organism. This can be attributed to Lactobacilli stimulation ability of lactulose. This particular property of lactulose has successfully utilized by medical practitioners in the treatment of portal systemic encephalopathy and chronic constipation. Lactulose also has been gaining attention for being prospective prebiotic substance.

Galactooligosaccharides : Human milk contains various types of oligosaccharides and most predominant among them is galactooligosaccharides (GOS). The presence of GOS is breast milk is linked with higher *bifidobacterial* count in infants. They are produced commercially from lactose using the glycosyl transferase activity of enzyme β-galactosidase EC 3.2.1.23. GOSs consist of a number of β- (1→6) linked galactopyranosyl unit linked to as terminal glucopyranosyl unit via a α- (1→4) glycosidic bond. These galactooligosaccharides were earlier considered as unwanted products but now they are considered as prebiotics because they function as bifidobacteria growth promoting factors, reduce risk of colon cancer, prevent bone loss and lower serum cholesterol concentration

Physiological Functions of Galactooligosaccharides : *In-vitro* studies revealed that GOS are not hydrolyzed by human digestive secretion, but many beneficial intestinal microflora including *bifidobacteria* and lactic acid bacteria can effectively metabolize them. Literature also showed that GOSs could be used more readily and selectively by the bifidobacteria than other oligosachharides. Clinical trials with GOS consumption with animal and human subjects indicated its positive role improving constipation, reducing blood ammonia level, cancer and osteoporosis prevention. The constipation may be improved by an increase in the osmosis because of the production of higher concentration of SCFAs in the intestinal as result of GOSs consumption. GOS consumption along with *Bifidobacterium breve* supplementation decreased the blood ammonia levels in patients with hyper ammonia associated liver cirrhosis. *Bifidobactria* not only suppress growth of ammonia producing organisms like *baceteroidaceae* but utilize ammonia as sole source of nitrogen. Thus help in maintaining level of nitrogenous compounds.

Bioactive Peptides as Therapeutic Components : Dietary proteins possess nutritional, functional and biological properties, and the technological processes used in food manufacture and processing often affect these properties. The role of proteins as physiologically active components in the diet has been increasingly acknowledged in recent years. Such proteins or their precursors may occur naturally in raw food materials, exerting their physiological action directly or upon enzymatic hydrolysis *in vitro* or *in vivo*. Several dietary proteins, can act as a source of biologically active peptides. These peptides inactive within remain the parent protein, and released during gastrointestinal digestion or food processing. Once liberated, the bioactive peptides may provide different functions *in vitro* or *in vivo*. Bioactive peptides have to be released from the parent protein by enzymatic hydrolysis. This can be achieved by the use of isolated enzymes, as well by microbial fermentation .The process is influenced by the type of enzyme, by the processing conditions used (pH, temperature, incubation time enzyme / substrate ratio), but also the species from which the milk is derived, the length and the nature of the bioactive peptides, as well as by the presence of other peptides in the medium. Biologically active peptides are of particular interest for pharma industry because they have been shown to play different physiological roles, including opioid like activity, antimicrobial, immunomodulatory and antihypertensive. Such peptides can be released during hydrolysis by digestive or microbial enzymes. Microbial enzymes from lactic acid bacteria have demonstrated to be able to liberate theses peptides from milk proteins, in various fermented milk products. Upon oral administration bioactive peptides may affect the major body systems- namely the cardiovascular, digestive, immune and nervous systems. For this reason, the potential of certain peptides sequences to reduce the risk of chronic diseases or boost natural immune protection has aroused a lot of scientific interest over the past few years. These beneficial health effects may be attributed to known peptide sequences exhibiting, e.g., antimicrobial, antioxidative, antithrombotic, antihypertensive and immunomodulatory activities. The activity of peptides is based on their inherent amino acid and composition and sequence. The size of active sequences may vary from two to twenty amino acid residues, and many peptides are known to possess multi- functional properties. Milk proteins are considered the most important source of bioactive peptides and an increasing number of bioactive peptides have been identified in milk protein hydrolysates and fermented dairy products. Enzymatic hydrolysis has been the most common route to produce functional peptides – and pancreatic enzymes, especially trypsin, have been associated with efforts toward production, characterization and identification of many known peptides .The release of various bioactive peptides from milk proteins through microbial proteolysis has been reported

BIOACTIVE PEPTIDES POTENTIAL FROM MILK

The peptides with potential physiological and bio-functional activities from milk such as opioid peptides, antihypertensive peptides, anti-thrombotic peptides, immunomodulating peptides, anti-microbial peptides, mineral binding peptides etc. are detailed as follows.

Peptides with opioid activity: Opiates are drugs containing opium, with basic substance morphine in it. They have been used since ancient times in medicine to relieve pain and induce sleep. Opioid peptides, such as *enkephalins*, are defined as peptides having both an affinity for an opiate receptor and opiate like effects inhibited by naloxone. The Opioid peptides originate from three precursor proteins: proopiomelanocortin (endorphins), proenkephalin (enkephalin) and prodynorphin (dynorphins). These peptides have the same N terminal sequence, Tyr – Gly – Gly – Phe. Opioid peptides that are derived from a variety of precursor proteins are called "atypical", since they carry various amino acids sequence at their N terminal regions; only the N terminal tyrosine is conserved. The N terminal sequence of "atypical" Opioid peptides is Tyr – X – Phe or Tyr – X_1 – X_2 – Phe. The tyrosine residue at the N terminal and the presence of another aromatic amino acid at the third or fourth position form an important structural motif that fits into the binding site of the Opioid receptors. The major exogenous opioid peptides, b - casomorphins, are fragments of the b - casein sequence 60 – 70 (Tyr – Pro – Phe – Pro – Gly – Pro – Ile – Pro – Asn – Ser – Leu) (Chiba and Yoshikawa, 1986). Whey proteins contain opioid – like sequences, namely a - Lactalbumin (both bovine and human) f (50 - 53) and b - Lactoglobulin (bovine) f (102 – 105), in their primary structure. These peptides have been termed a - and b - Lactorphins. Proteolysis of a - Lactalbumin with pepsin produces a - Lactorphin, and while that digestion of b - Lactoglobulin with pepsin and then with trypsin, or with trypsin and chymotrypsin, yields b - Lactorphin. Bovine blood serum albumin f (399 - 404) named serrophin, displays opioid activity. Peptides with an affinity for opioid receptors have been observed in an artificially methyl – esterified peptic digest of human Lactoferrin. These peptides are named Lactoferroxins and behave like opioid antagonist peptides, such as Casoxins.

Peptides with angiotensin – I converting enzyme inhibition activity : Angiotensin, a blood polypeptide exists in two forms, the physiologically inactive angiotensin – I and the active angiotensin – II. The inactive form is converted into active by angiotensin – I converting enzyme (ACE), which is a key enzyme in the regulation of peripheral blood pressure. ACE plays a major physiological role in the regulation of local levels of several endogenous bioactive peptides. Various ACE inhibitors have been isolated from enzymatic hydrolysates and related synthetic peptides of food proteins. These food proteins include such as bovine and human casein and whey, zein, gelatin, yeast and corn.

Casokinins : Casokinin sequences have been found in all casein fractions, but a_{s1} – and b - caseins, in particular, are rich in ACE inhibitory sequences. Tyrpsin hydrolysis of casein produces ACE inhibitors. Yamamoto, et.al (1994) showed that peptides produced by Lactobacillus proteases had ACE inhibitory activity. Furthermore, two ACE – inhibitory tri peptides, Val – Pro – Pro and Ile – Pro – Pro, were obtained during the fermentation of milk proteins with *Lactobacillus helveticus* and *Saccharomyces cerevisiae* (Nakamura, et.al. 1995).

Lactokinins : The first ACE inhibitory peptides derived from whey proteins were synthetic peptides corresponding to the known bioactive sequence of b - Lactoglobulin (b - Lactorphin and b - Lactotensin) and a - Lactalbumin (a - Lactorphin). The tripeptide Tyr – Gly – Leu (a - Lactalbumin f (50 - 52)) has ACE inhibitory activity at about the same range as the a - Lactorphin (Pihlanto-Leppala, et.al. 2000). Several peptides have been isolated from whey protein digested with proteinase – k, among them is a peptide corresponding to b - Lactoglobulin f (78 - 80) (b - Lactosin), which showed the highest ACE inhibitory activity (Abubakar, et.al 1998).

Peptides with anti-thrombotic activity : Thrombosis is defined as the formation or presence of a blood clot within a blood vessel (Gerdes, et.al. 2001). Plasma protein called as fibrinogen is produced in the liver and is converted into fibrin necessary for platelet aggregation. Milk peptides are known to inhibit this platelet fixation. Hydrolysis of bovine k - casein by chymosin constitutes the first stage of milk clotting. In this reaction, one bond (Phe_{105} – Met_{106}) of k - casein is rapidly hydrolyzed, leading to the release of an insoluble N – terminal fragment (para - k - casein; residues 1 - 105) and a soluble C – terminal fragment (caseinomacropeptide; residues 106 - 169) from which a series of tryptic peptides active in platelet function has been characterized. These peptides are called as Casoplatelins. Jolles, et.al. (1986) reported that a dodecapeptide, inhibits ADP – induced platelet aggregation and combines with the fibrinogen – binding with blood platelets, in a concentration – dependent manner. Three peptides, which completely inhibit thrombin- induced aggregation, have been derived from an enzymatic Hydrolysate of caseinoglycopeptide .(Qion,et.al.1995).

Peptides with immuno modulating activities : The immune system plays a central role in protection against bacterial, viral, parasitic, and fungal infection and also cancers. Deficiencies in any aspect of immune system can pre dispose an individual to a greater risk of infection and may enhance the severity of disease (Bajaj and Sangawan, 2002). Milk protein – derived peptides are known to have an effect on the cells of the immune system, as well as on downstream immunological responses and cellular functions. The immune system employs non – specific and specific responses to confer protection against the disease. Non-specific components of the host defense. Specific immune responses are mediated by antibodies (Ig G, Ig A, Ig M, Ig E, and Ig D) produced by b - lymphocytes (humoral immunity), while T – lymphocytes give rise to T – helper, T – successor, and cytotoxic lymphocytes (cell mediated immunity) (Bounous and Kongshavan, 1985).

Bovine milk Derived Immuno-peptides : Casein hydrolysates modulate the immune function, with different modulatory effects attributable to varying enzyme digestion regimes. Pancreatin and trypsin digests of as2 – and b - casein have been shown to significantly inhibit the proliferative responses of murine splenic lymphocytes and rabbit peyer's patch cells, whereas digests derived from pepsin and chymotrypsin treatment have no effect when added to *in vitro* culture with mitogen – stimulated cells (Otani and Hata, 1995). Tryptic hydrolyzates stimulate murine peritoneal macrophages at quite low doses. Two peptides (Val – Glu – Pro – Ile – Pro – Tyr and Gly – Leu - Phe) from human caseins have been shown to increase the phagocytosis of human and murine macrophages and protect mice against *klebsiella pneumoniae* infection (Fiat, et.al. 1993). Two synthetic peptides mimicking milk protein derived peptides have been shown to enhance the proliferation of human peripheral blood lymphocytes. These peptides correspond to Tyr – Gly and Tyr – Gly – Gly which, correspond to fragments of bovine k casein and a - Lactalbumin.

Antimicrobial Peptides : The natural antimicrobial activity of milk is mainly associated with whey proteins, primarily Lactoferrin. Early report associating lactoferrin with prevention of microbial growth attributed this function to its ability to bind and sequester iron, and to deprive microorganisms of this essential nutrient (Sanchez, et.al. 1992). However, hydrolysis of lactoferrin by pepsin produces hydrolysates in which the antimicrobial potency is higher than in undigested lactoferrin. The iron binding capacity of the hydrolysates is lost, but the antimicrobial activity is not affected by the addition of iron. These results, thus, indicate that the antibacterial activity of these lactoferrin hydrolysates is not dependent on iron. A bactericidal domain has now been isolated and identified. All the identified antimicrobial peptides, called Lactoferricins, are cationic and originate from the N – terminal of the molecule (Bellamy, et.al. 1992). These peptides have antimicrobial activity against various gram-positive and gram-negative bacteria, yeast and filamentous fungi. An antibacterial peptide has also been purified from bovine milk hydrolyzed by serine proteases. This peptide was identified as a_{s2} – casein f (165 - 203), casocidin – I, and it inhibits the growth of *Escherichia coli* and *Staphylococcus carnosus* (Zucht, et.al. 1995). The polypeptides were mostly active against gram positive while gram-negative bacteria were only poorly susceptible to the bactericidal action of the polypeptides. *Bacillus subtilis* was the most susceptible bacterial strain to the action of these peptides.

Mineral Binding Peptides : It is well known that casein – derived phospharylated proteins enhance vitamin D – independent bone calcification in rachitic infants. The extent of phospharylation is dependent on the casein type. These confer to the proteins the ability to chelate calcium, which is related to their level of phospharylation; thus αs2> βs1> κ>. Further, studies on the primary structure have indicated that the phospharylated residues are not evenly spread throughout the protein chain, but are often clustered with three or more common cluster sequences such as Ser (P) – Ser (P) – Ser (P) – Glu – Glu (Kitts, 2005). These phospharylated fragments are believed to play a crucial role in protecting the milk gland against calcification by controlling the calcium phosphate precipitation. Enzymatic hydrolysis of casein using trypsin and alcalase, results in formation of several caseinophosphopeptides (CPPs). The calcium chelating activity of CPP – fragments *in vitro* has been attributed to the role of component phosphoserine residue in stabilizing the colloidal calcium phosphate of casein micelles. Dephosphorylated peptides don't bind minerals. The proportion of phosphopeptides interacting with colloidal calcium phosphate correlates with their relative content of phosphoserine residues. Other side – chains such as glutamic and asparatic acids, may also contribute to metal binding. In addition, peptides binding minerals have been found in whey proteins, i.e., from β – lactoglobulin and β – lactalbumin and lactoferrin. Since these proteins are not phospharylated, the minerals seem to bind through other binding sites than caseins. β-Lactoglobulin and β-Lactalbumin peptides show a higher affinity for iron than the native proteins (Vegarud, et.al. 2000).

Whey Minerals : Whey contains most of minerals present in milk. Whey-based products are an excellent and cost efficient source of calcium (500-2000 mg/100g). These preparations are neutral taste and bland odour, hence can be utilized effectively in formulation of calcium-enriched foods. Age-related osteoporosis a common bone disease is characterized by reduced bone density resulting in increased bone fragility and susceptibility to fracture calcium is widely recognized as important for bone health. This calcium promotes the bone health because 99% of the body calcium resides in the skeleton. Calcium alone is not enough for healthy bones. Recent researchers have shown that poor calcium nutritional status is related to certain diseases like hypertension. Calcium blocks the tumor promoting effects of dietary fat by chelating unconjugated bile acids and or fat ionized fatty acids in the colonic lumen to form relatively insoluble calcium complexes, which are less damaging to the colorectal mucosa. Many studies confirmed that increased calcium intake definitely reduced the risk of colon cancer. The bioavailability of calcium from milk or whey products is more as compare to other relatively rich sources. Mineral concentrated whey functions to improve texture, flavour, solubility and nutritional profiles in food formulations. The foods containing mineral-concentrated whey will have a higher nutritional density than comparable other products. These minerals from whey can be utilized to develop calcium-enriched foods products like candies, sauces, seasonings, frozen desserts, nutritional drinks.

CONCLUSION

In recent year ingredients market in India is witnessing an explosive growth, which was never seen earlier. Many companies are entering with wide range of ingredients that are natural, healthy and functional as well. Whey components may be preferred by consumers and help manufactures design products with a "clean label" and "natural image. However there is lot of scope in technological front for the preparation, isolation and purification of these components in cost effective manner. Well-documented therapeutic role established through intensive scientific investigations certainly make whey derivatives as food ingredient of 21st century.

REFERENCES

Abubakar A, Saito T, Kitazawa H, Kawai Y, and Itoh T (1998) Structural analysis of new antihypertensive peptides derived from cheese whey protein by proteinase K digestion. *J. Dairy Sci.* Vol. **81:** 3131–3138.

Bajaj R K and Sangawan R B (2002) Health enhancing potential of whey proteins – a review. *Indian Jr. Dairy Sci.* Vol. 55 **(5):** 253–260.

Beena, A. and Prasad, V. (1997). Effect of yogurt and bifidus yoghurt fortified with skim milk powder, condensed whey and lactose-hydrolyzed condensed whey on serum cholesterol and triacylglycerol levels in rats. *J. Dairy Res.* Vol. **64**: 453–457.

Bellamy W R, Takase M, Yamauchi K, Wakabayashi H, Kawase K and Tomita M (1992) Identification of the bactericidal domain of lactoferrin. *Biochim. Biophys. Acta.* Vol. **1121:** 130–136.

Bounous G and Kongshavan P A L (1985) Differential effect on protein type on the B – Cell and T – Cell immune responses in mice. Jr. Nutrition. Vol. **115:** 1403–1408.

Bounous, G. and Gold, P. (1991). The biological activity of undenatured dietary whey protein on tissue glutation and the disease of ageing. *Clin. Invest. Med.* Vol. **14**: 296–309.

Bounous, G., Batist, G. and Gold, P. (1989). Immunoenhancing property of dietary whey proteins in mice: role of glutathione. *Clin. Invest. Med.* Vol. **12**: 154–161.

Chiba H and Yoshikawa M (1986) Biologically functional peptides from food proteins: new opioid peptides from milk proteins. In: Protein tailoring for food and medical uses (Ed: R. E. Feeney and J. R. Whitaker). Pp: 123–153. Marcel Dekker, New York.

de Wit, J. N. (1998). Nutritional and functional characteristics of whey proteins in food products. *J. Dairy Sci.* Vol. **81**: 597–608.

Fiat A M, Migliore – Samour D, Jolles P, Drouet L, Collier C and Caen J (1993) Biologically active peptides from milk proteins with emphasis on two examples concerning antithrombotic and immunomodulating activities. *J. Dairy Sci.* Vol. **76:** 301–310.

Gerdes S K, Harper J W and Miller G (2001). Bioactive compounds of whey and cardiovascular health. Pp: 1–8. Applications Monograph. USDEC, USA.

Horton, B. S. (1995). Commercial utilization of minor milk components in the health and food industries. *J.Dairy Sci.* Vol. **78**: 2584–2589.

Jolles P, Levy-Toledano S, Fiat A M, Soria G, Gillessen D, Thomaidis A, Dunn F W and Caen J P (1986) Analogy between fibrinogen and casein. Effect of an undecapeptide isolated from k-casein on platelet function. *Eur. J. Pharmacol.* Vol. **158:** 379–382.

Kawase, M., Hastimoto, H., Hasoda, M., Morita, H. and Hosono, A. (2000). Effect of administration of fermented milk containing whey protein concentration to rats and healthy men on serum lipids and blood pressure. *J.Dairy Sci.* Vol. **83 (2):** 255–263.

Kitts D D (2005) Antioxidant properties of casein-phosphopeptides. *Trends in Food Science and Technology*. Vol. **16:** 549–554.

Marshall, K. R. (1982). Industrial isolation of milk proteins: whey proteins. In, Developments in Dairy Chemistry – 1: Proteins (ed Fox P F) Applied Science Publishers, London. pp 339–373.

McIntosh, G. H., Regester, G. O., Leu, R. K. Le, Royale, P. J. and Smithers, G. W. (1995). Dairy proteins protect against dimethylhydrazine-induced intestinal cancers in rats. *J. Nutrition*. Vol.**125 (4):** 809–816.

Nakamura, Y., Yamamoto, M., Sakai, K., Okubo, A., Yamazaki, S., and Takano, T. (1995). Purification and Characterization of angiotensin I converting enzyme inhibitors from sour milk. *J. Dairy Sci.* Vol. **78**: 777–783.

Otani H and Hata I. (1995) Inhibition of proliferative responses of mouse spleen lymphocytes and rabbit Peyer's patch cells by bovine milk caseins and their digests. *J. Dairy Sci.* Vol. **62:** 339–348.

Parodi, P. W. (1998). A role for milk proteins in cancer prevention. *Aust. J. Dairy Technol*. Vol. **53 (1):** 37–47.

Pihlanto-Leppala A, Koskinen P, Piilola K, Tupasela T and Korhonen H (2000) Angiotensin I converting enzyme inhibitory properties of whey protein digests: concentration and characterization of active peptides. *J. Dairy Sci.* Vol. **67:** 53–64.

Qian Z Y, Jolles P, Miglore-Samour D and Fiat A M. (1995) Isolation and characterization of sheep lactoferrin, an inhibitor of platelet aggregation and comparison with human lactoferrin. *Biochim. Biophys. Acta.* Vol. **1243:** 25–32.

Regester, G. O., McIntosh, G. H., Lee, V. W. K. and Smithers, G. W. (1996). Whey proteins as nutritional and functional food ingredients. *Food Australia.* Vol. **48 (3):** 123–127.

Sanchez L, Calvo M and Brock J H (1992) Biological role of lactoferrin. *Arch. Dis. Child.* Vol. **67:** 657–661.

Sanwar, G., Peace, R. W. and Botting, H. G. (1985). Corrective relative net protein ratio (CRNPR) method based on differences in rat and human requirements for sulfur amino acids. *J.Assoc.Off.Anal.Chem.* Vol. **68**: 689.

Steijns, J. M. (2001). Milk ingredients as nutraceuticals. *Int. J. Dairy Technol*. Vol. **54 (3):** 81–88.

Vegarud G E, Langsrud T and Svenning C (2000) Mineral-binding milk proteins and peptides: occurrence, biochemical and technological characteristics. *British Journal of Nutrition.* Vol. **84:** 91–98.

Walzem, R. L. (2001). Health enhancing properties of whey proteins and whey fractions. Applications monograph. US Dairy Exports Council, USA. (www.udec.org). pp: 1–7.

Yamamoto M, Akino A, and Takano T (1994) Antihypertensive effect of the peptide derived from casein by an extracellular proteinase from *Lactobacillus helveticus* CP790. *J. Dairy Sci*. Vol. **77:** 917–922.

Zucht H, Raida M, Adermann K, Magert H J and Forssmann W (1995) Casocidin – I: a casein a_{s2} derived peptide exhibits antibacterial activity. *FEBS Lett.* Vol. **372:** 185–188.

Bioresources for Food Security and Rural Livelihood, Pages 311–318
Edited by : G.K. Kulkarni and B.N. Pandey

MARINE LIVING RESOURCES OF THE EEZ OF INDIA

L. Kannan, T. Thangaradjou and K. Sivakumar

ABSTRACT

India has a coastline of about 8,040 km, an Exclusive Economic Zone (EEZ - 2.02 million sq km) adjoining the continental regions, offshore islands and a very wide range of coastal ecosystems such as estuaries, lagoons, coral reefs, mangroves, seagrasses, backwaters, salt marshes, rocky coasts and sandy stretches, which are characterized by unique biotic and abiotic properties and processes. As of today, the bioresources of the EEZ of India are underutilized as compared to their real potential. However, there are certain groups which are under threat due to target fishing and certain other groups are less targeted. In addition to this, we have to take into account the growing concerns of the global climate change and its consequences on the marine fisheries, which may have the ability to change trends of fisheries in a negative line. In this paper we attempt to provide a detailed account on the coastal resources of India with special reference to their sustainable exploitation and the threats they are facing.

Keywords : Coastline, Eez, Bioresources, Sustainability threats.

INTRODUCTION

India is one among the 17 mega-biodiversity countries, encompassing highly diverse habitats ranging from high altitude cold deserts in Himalayas to hot deserts, tropical rain forests, east and west coasts etc. with different ecosystems. Among the Asian countries, Indian coast is one of the highly explored coasts for its biological wealth and India is perhaps the only nation that has a long record of inventories of coastal and marine biodiversity, dating back to at least two centuries.

In terms of the marine environment, India has a coastline of about 8,040 km, an Exclusive Economic Zone (EEZ) adjoining the continental regions, offshore islands and a very wide range of coastal ecosystems such as estuaries, lagoons, mangroves, backwaters, salt marshes, rocky coasts, sandy stretches and coral reefs, which are characterized by unique biotic and abiotic properties and processes (Venkataraman, 2003).

The EEZ starts at the seaward edge of the territorial sea and extends out into the sea to a distance of 200 nautical miles (370 km) from the baseline (Article 55). This rule will not be applicable if the EEZ of two countries overlaps. If overlap occurs, it is upto the countries to resolve the boundary bilaterally. However, countries have the right over the seabed of the continental shelf up to 350 nautical miles from the coast and it is not their EEZ. Countries have exclusive right to explore, use and generate income by using the biological and non biological resources available within their EEZ. In India, the entire EEZ measuring 2.02 million sq km is divided into four regions viz. Arabian Sea (0.698×10^6 sq km), Bay of Bengal (0.525×10^6 sq km), Lakshadweep Sea (0.230×10^6 sq km) and Andaman and Nicobar Seas (0.467×10^6 sq km) (Goswami, 2004).

MARINE RESOURCES

Seaweeds

Seaweeds are marine macro-algae that form an important renewable resource in the marine environment and have been a part of human civilization from time immemorial. They synthesize a wide range of chemicals, some of which stand as the only natural resource, e.g. agar, carrageenan and alginate. The seaweed flora of India are highly diversified and comprise mostly tropical species. In all, 271 genera and 1,153 species of marine algae, including different forms and varieties have been enumerated till date from the Indian waters (Krishnamurthy, 2005). Survey on seaweed resources of the country was carried out at different periods in different parts of the Indian coast. A compilation of all these data reveals a standing stock of approximately 6,63,371 tons of fresh seaweed along the Indian coast. The estimated standing crop of the seaweeds is higher than that of the real exploitation of the seaweeds for agar industries which is around 40,000 t per year.

Standing stock of seaweeds in the different coastal states of India (from Rao and Mantri, 2006)

State	Tons (FW)
Gujarat	105931
Maharashtra	20315
Goa	2255
Kerala	1000
Tamilnadu	219280
Andhra Pradesh	7500
Orissa	269705
Lakshadweep	10077
Andaman	26048
Drift seaweeds	1260
	6,63,371

Seagrasses

Seagrasses form an important coastal ecosystem, surpassing the tropical rain forests in productivity and provide with an excellent habitat for rare and endangered flora and fauna. Most important and striking feature of the seagrass ecosystem is its habitat heterogeneity and its use as a nursery and breeding ground for commercially very important shellfish and finfish. In India, luxuriant seagrass beds are occurring along the coasts of Tamilnadu, Lakshadweep and Andaman and Nicobar islands, in addition to sparse seagrass patches distributed along the entire Indian coast . A total of 14 species of segrasses belonging to 6 genera have been recorded from here. Data on actual coverage of seagrasses in the Indian EEZ are largely not available but sporadic data from different regions are available in literature. Jagtap *et al.* (2003) have provided with a list of floral and faunal groups associated with the seagrasses of India.

Seagrass distribution in selected sites of the Indian coast.

Region	Area (Sq. km)	Reference
Gulf of Mannar	55.16	Thangaradjou *et al.*, 2008
Palk Bay	26.58	Sridhar, 2006
Lakshadweep	10.9	DOD and SAC, 1997
Andaman & Nicobar	8.3	Jagtap *et al.*, 2003

Associated biota recorded from the seagrass beds of India (Jagtap *et al.*, 2003).

Group	Number of Species
FAUNA	
Bait fishes	21
Ornamental fishes	138
Fin fishes	33
Crustaceans	150
Molluscs	143
Echinoderms	77
Turtles	4
Mammals	1
FLORA	
Marine algae	100
Phytoplankton	13
Fungi	9

Coral Reefs

In India, major coral reefs are distributed in Gujarat, Tamilnadu, Andaman and Nicobar and Lakshadweep islands. Of these, reef flat cover is more in Lakshadweep followed by Gujarat, Andaman and Nicobar and Tamilnadu regions. A total of 199 species belonging to 37 genera of coral species have been reported from Indian waters.

All the coral reef areas have been identified as biodiversity rich areas. Coral reefs serve as a habitat for a variety of fishes and other marine organisms. Though the exact number of different species associated with the coral reefs of India is still to be obtained, more than 1,367 species of fishes in Indian Ocean region have been reported. Lakshadweep islands have a total of 603 species of fishes (Jones and Kumaran, 1980), Andaman and Nicobar islands have about 1,000 species and the Gulf of Mannar supports over 538 species of fishes (Venkataraman and Wafar, 2005).

Coral reef categories of India (area in sq. km) (DOD and SAC, 1997)

Category	Gujarat	Tamilnadu	Andaman and Nicobar	Lakshadweep
Reef flat	148.4	64.9	136.5	795.7
Sand over reef	11.8	12.0	7.3	73.3
Mud over reef	117.1	—	—	8.4
Coralline shelf	—	—	230.9	45
Live coral platform	—	—	43.3	—

Marine fisheries

India remained to be a coastal fishing nation until the Marirtime Zones of India Act came into force in 1981. Thereafter, it promoted and regulated the offshore fishery in the Indian EEZ. In 1991, the Act was amended with the main objectives of establishment of abundance of fishery resources in the offshore waters of the Indian EEZ, identification of suitable crafts and gears for fishing operations, enlarging the offshore fishing fleets, transfer of technology of offshore fishing to the local fishermen and elaborating the domestic and overseas market for non conventional fish species. This enabled the Indian offshore fishing industry to get a boost and as a result of which, the Indian annual marine fish production increased constantly.

Marine fish production in India which was only 0.5 mt in 1950 consistently increased through the time scale and reached 2.7 m t in 1997 but has remained between 2.7 m t and 2.9 m t since then. The substantial increase during 1950-1997 is largely due to the introduction of mechanized fishing vessels and synthetic gear materials, expansion of trawl fleets, motorization of traditional fishing crafts, substantial growth in the number and efficiency of trawlers and motorized crafts, and development of infrastructure for fish preservation, processing and storage (Devaraj and Vivekanandan, 1999). This modernization in fishing industry also increased the pressure on marine resources of the EEZ especially the biological resources within the 100 m depth. This necessitated the search for alternate fishing grounds and shifted the momentum towards the deep sea fishing. Recent reports of the Working Group for validating the potential of fishery resources suggest that the estimated potential yield of fish stocks in the Indian EEZ is 3.92 million tonnes (m t) (Anon, 2000).

Of the current annual yield of 2.7 to 2.9 m t, about 2.2 to 2.4 m t is mainly from 387,500 km^2 of inshore waters (within 100 m depth). The additional annual harvestable yield of 1.22 m t is available in the extensive offshore (beyond 100m depth) area of 1.63 million km^2 in the EEZ (Vivekanandan *et al.*., 2008). The FSI has estimated the potential of marine fisheries as 3.92 million tonnes in the Indian EEZ. About 58 % of the resources are available at a depth of 0-50 m, 35 % at 50-200 m and 7 % from beyond 200 m depth (Boopendranath, 2007). However, the marine fish landings of India during the year 2007 have been estimated as 2.88 million tonnes showing an increase of about 1.7 lakh tonnes (6.3%) against the estimate of the previous year. The pelagic finfishes constitute 57%, demersal fishes, 25%, crustaceans, 14% and mollusks, 4% of the total landings (CMFRI, 2008).

Estimated potential (m t) of fishery resources in the Indian EEZ (Anon, 1991, 2001; Somvanshi, 2001)

Resource	Depth (m)			Total (mt)
	1 – 50	51 – 500	501 to EEZ	
Demersal	1.28	0.65	–	1.93
Pelagic	1.00	0.74	–	1.74
Oceanic	–	–	0.25	0.25
Total	2.28	1.39	0.25	3.92

India's production of tunas and tuna like fishes during 2006 was about 1,14,269 tonnes. The coastal fishery contributed 1,13,145 tonnes whereas, the contribution of deep sea / oceanic tunas was 1124 tonnes. The total catch of neritic tunas was 64,056 tonnes, billfishes, 301 tonnes and seer-fishes ,49,089 tonnes (Somvanshi, 2007). FSI surveys clearly indicate that the estimated potential of tunas in the Indian waters will be around 209,400 – 242,000 t (Somvanshi, 2001) and among all the tunas, the yellow fin tunas are considered to be the dominant forms.

Similarly, FSI has estimated that the potential of the shark fishery in the Indian EEZ is around 45,064 tons and beyond the shelf area, it has an additional potential of 26,200 t (Anon, 1991, 2001) i.e. the estimated potential of the sharks in the EEZ is 71,264 t. However, considering the current annual average catch of 36,021 t per year, it appears that there is scope for increased shark fishery in the Indian waters (Vivekanandan, *et al.*., 2008).

Shellfish

According to the FSI's marine fish landings of the 2007-08 data, the shellfish landing during this period was 4,44,368 t. This indicates that there is a vast potential for shellfish resources in the Indian EEZ; however, a clear standing stock assessment is required before going for large scale exploitation of the resources. In India, so far, a total of 5,070 species of mollusks have been recorded of which 3,370 species are from the marine environment. Similarly, of the 2,934 species of Crustacea reported, 94.85% is from the marine environment. This includes 139 species of stomatopods, 26 species of lobsters, 162 species of hermit crabs, 705 species of brachyuran crabs, 84 species of shrimps and prawns and 159 species of Caridea. Besides these, 540 species of copepods, 104 species of cirripeds and 120 species of ostrocods have also been reported (Venkataraman and Wafar, 2005).

Shellfish landing during 2007-08 (FSI, website)

Group	Total catch (in tonnes)
Penaeid shrimp	2,89,771
Non-Penaeid shrimp	22,668
Deep see shrimp	0
Lobsters	2,473
Deep sea lobster	813
Crabs	56,824
Bivalves / gastropods	330
Squid and cuttle fish	70,530
Octopus	959
Total	4,44,368

Squids

The Arabian Sea is considered to be one of the richest regions for the oceanic squids in the Indian Ocean (Zuyev *et al..*, 2002). The purpleback flying squid *Sthenoteuthis* (*Symplectoteuthis*) *oualaniensis,* was available in plenty in the Arabian sea in the Indian ocean region (Zuyev *et al..*, 2002). Silas (1985) reported that 25,000 to 50,000 t of squids are available in the Indian Ocean regime. However, Zuyev *et al..* (2002) estimated that the squid stock in the Arabian Sea alone varies between 0.9 and 1.6 m t. This indicates that the Indian waters especially, Arabian Sea has vast potential of quid resource which can be explored for the human use.

CONCLUSION

As of today, the bioresources of the EEZ of India are underutilized as compared to their real potential. However, there are certain groups which are under threat due to target fishing and certain other groups are less targeted. In addition to this, we have to take into account the growing concerns of the global climate change and its consequences on the marine fisheries. There are also other sources of threats to the living resources of the EEZ, specifically destruction of coastal habitats, unregulated and illegal fishing activities, over exploitation of the resources especially undersized fishes, pollution due to land based and sea based activities, coastal developmental activities, ghost fishing etc. which need to be addressed properly for the sustainable development of the fishery resources of the EEZ of India.

REFERENCES

Anon. 1991. Report of the Working Group on Revalidation of the Potential Marine Fisheries Resources of Exclusive Economic Zone of India. *Ministry of Agriculture, Govt of India,* 66 pp.

Anon. 1993. Global Ocean Ecosystem Dynamics Implementation Plan. *Workshop Report of US GLOBEC studies in the Arabian Sea,* **9:** 105 pp.

Anon. 2001. Report of the Working Group for Revalidating the Potential of Fishery Resources in the Indian EEZ. *Ministry of Agriculture, Govt of India,* 81 pp.

Article 55. 1982 UN Convention on the Law of The Sea.

Boopendranath, M.R., 2007. Fishing Capacity Management. IFP Souvenir, 7-14.

Devaraj, M. and E. Vivekanandan. 1999. Marine capture fisheries of India: challenges and opportunities. *Curr. Sci.,* **76:** 314 –332.

DOD & SAC, 1997. Coral reef maps of India. Department of Ocean Development and Space Application Centre, Ahmedabad, India.

Goswami, S.C., 2004. Biological productivity and potential resources of the exclusive economic zone (EEZ) of India. In: Large marine ecosystems: Exploration and exploitation for sustainable development and conservation on fish stocks, (ed.) V.S. Somvanshi, 43-53.

Jagtap, T.G., D.S. Komarpant and R.S. Rodrigues, 2003. Status of seagrass ecosystems of India, *Wetlands,* **23 (1):** 161-170.

Jagtap, T.G., Komarpant, D.S., Rodrigues, R., 2003. The seagrasses of India. In *World Atlas of Seagrasses*, 113–120.

Jones, S. and M. Kuraman, 1980. Fishes of Laccadive archipelago. The Nature Conservation and Aquatic Sciences Service, Trivandram, Kerala, 760 pp.

Krishnamurthy, V., 2005. Seaweeds: Wonder Plants of the Sea, Aquaculture Foundation of India, Chennai, p. 30.

Rao, P.V.S. and V. A. Mantri, 2006. Indian seaweed resources and sustainable utilization: Scenario at the dawn of a new century. Curr. Sci., **91(2):** 164-174.

Silas, E.G (1985). Cephalopod bionomics, fisheries and resources of the EEZ of India. Bull. Cent. Mar. Fish. Res. Inst**. 37:** 195p.

Somvanshi, V.S. 2001. Problems and prospects of deep sea and far sea fishing. In: T.J. Pandian (ed.), *Sustainable Indian Fisheries, NAAS, New Delhi,* 71-87.

Somvanshil, V.S., S. Varghese and NGK Pillai, 2007. National report on tuna fisheries and its development in India. IOTC-2007-SC-INF **19:** 1-4.

Subarao, 1998. Mollusca. In: Faunal diversity of India. Zoological Survey of India, Culcutta, 104-117.

Venkataraman, K. 2003. Natural aquatic ecosystems of India, Thematic Working Group, The National Biodiversity Strategy Action Plan, India, 1-275.

Venkataraman, K. and M. Wafar, 2005. Coastal and marine biodiversity of India. Indian J. mar. Sci., **34(1):** 57-75.

Vivekanandan, E., K.S. Mohamed, N.G.K. Pillai and Prathibha Rohit, 2008. Offshore fisheries Resources of India. FAO, APFIC.

Zuyev, G., Nigmatullin, C., Chesalin, M., Nesis, K.N. 2002. Main results of long-term worldwide studies on tropical nektonic oceanic squid genus *Sthenoteuthis*: an overview of the Soviet investigations. Bull. mar. Sci., **71(2):** 1019–1060.

Bioresources for Food Security and Rural Livelihood, Pages 319–338
Edited by : G.K. Kulkarni and B.N. Pandey

ADVANCES IN EXPLORATION OF INDIAN SEAS FOR BIORESOURCES PROSPECTING BIODIVERSITY, BIOTECHNOLOGY AND ASSOCIATED CHALLENGES

V.S. Somvanshi

INTRODUCTION

Human beings have relied on marine resources for food, jobs, recreation, etc from time immemorial. Just as they were curious about exploring the universe around, earlier humans were equally fascinated by the sometimes mysterious cycle of life in marine environments, and how it impacts their life. Because of this curiosity, humans were driven to discover as much as possible about natural world and how we interact with its resources. Marine environments hold vast wealth of biodiversity as a consequence of millions of years of evolutionary history. The oceans, which cover 70 percent of the earth's surface, are home to 34 of the 36 phyla of life, while, in contrast, only 17 phyla reside on land. Furthermore, 15 phyla are exclusively marine. Consequently, marine organisms display a much larger phyletic diversity than those on land (Ray, 1988). Ocean creatures contain a diversity of survival strategies not found on land and therefore, they may exhibit a higher functional diversity. Therefore, it is possible that the marine environment has not only a high phyletic diversity but also a great functional diversity. Marine biodiversity is undoubtedly the most diverse, not only in terms of number of diverse kinds of organisms but also in terms of unique habitats and ecosystems. Compared to terrestrial biodiversity, diversity of the marine organisms attracts more importance due to its fragile nature. Regardless of the species counts, marine animals are more diverse than land animals at the higher, phyletic levels of evolutionary and taxonomic differentiation (Venkataraman, 2007). Marine environment encompass an impressive diversity of ecosystem and habitats. Coral reefs, which are among the planet's largest and oldest structures created by living organisms are home to dense concentration of species and complex webs of interspecies interactions. Some coastal systems such as marshes, mangrove forests, and sea grass beds are characterised by high biological production. They are important both to other marine ecosystems and to human development because of the fisheries and other resources and services they provide (De Fontaubert *et al.*., 1996).

In the face of harsh conditions like salinity, hydrostatic pressure, low temperature etc., the physiological adaptations, as offensive and defensive capabilities of marine organisms, are in the form of bioactive metabolites (Martin and Padilla, 1973; Scheuer, 1973 and Hashimoto, 1979). Bioactive molecules are substances that have an effect on the cellular function of an organism or living tissue. Bioactive molecules from biological resources are a potential source of pharmaceuticals and nutraceuticals. In addition, food bioactives such as phenolics, carotenoids, alkaloids and terpenoids are the potential chemopreventive agents against cancer (Pan, *et al.*., 2008). In addition, investigations on the mechanisms of action of toxins and poisons in causing harmful effects on cells are another important area of research in bioactives. As the marine environment is a rich source of both biological and chemical diversity and thus is of enormous importance to present day humans as a resource for food, pharmaceuticals and ecosystem services and it has been classified as the largest remaining reservoir of natural molecules to be evaluated for drug activity (Gerwick, 1987).

Since the later part of the twentieth century, researchers and entrepreneurs realised the untapped potential of marine biodiversity, thereby marking the beginning of an era of marine exploration, extraction, experimentation and commercialisation. This has lead to a sudden surge in the investment made by industrialists and academicians in marine resources based research (Demunshi and Chugh, 2009). However, potential of marine bioresources is largely unexplored, mainly due to the difficulty in access and expense involved. A majority of marine organisms are yet to be identified and those which are identified are not that studied, preventing their commercial exploitation. Bioresources prospecting or bioprospecting — the exploration and sampling of biodiversity for commercial and industrial purposes — holds substantial promise for the development of new compounds for consumer goods, public health, and energy and environmental uses. Life science corporations and academic institutes frequently engage in bioprospecting to explore and discover biological molecules for commercial applications. Exploratory surveys leading to documentation and quantification of different taxa are the first important step in biodiversity studies. The rich biodiversity of India attracts much interest in the exploration of its bioresources for new bioactive molecules, in the attempts of discovering new drugs.

MARINE BIODIVERSITY OF INDIA

The marine fauna of India is rich and varied. The Table 1 summarises the number of species of marine biodiversity of India as per two recent reviews (Venkataraman and Wafar, (2005) and Venkataraman, (2007).

Among the coastal wetland estuaries, mangroves and coastal lagoons are biodiversity-rich areas, whereas the other brackish habitats have only a few specialized species. It is generally believed that there is a reduction in the species number in estuaries compared to adjacent seas and in-flowing river system. However, as far as Indian estuaries are concerned the statement is partly true. The upper riverine ecosystem does not harbor as many species as its estuary. It has been observed that as the distance increases from the sea the number of species decrease. Salinity becomes an influencing regulatory factor.

The marine fauna of India is rich and varied. The coastline encompasses almost all types of inter-tidal habitat, from hyper-saline and brackish lagoons, estuaries, and coastal marsh and mudflats, to sandy and rocky shores with every degree of exposure and widely varying profile. Sub-tidal habitats are equally diverse. Each local habitat reflects the prevailing environmental factors and is further characterized by its biota. Thus, the marine fauna itself demonstrates gradients of change throughout the Indian coasts.

Much remains to be known about the number and types of the species of flora and fauna, which live in India. The algae, fungi, ferns, bryophytes and lichens are less known than the flowering plants. For the flowering plants, the location and population size are known even for the most threatened species. Invertebrates are generally less well known than the vertebrates. For birds, and now increasingly for reptiles, amphibians and mammals, most species have estimates.

Out of the total 32 animal phyla, 15 are represented by the taxa in the marine ecosystem. They may constitute either migratory or resident species. The former includes pelagic crustaceans, coelenterates (medusae), Cephalopod molluscs, fishes, reptiles, birds and mammals. Amphibians are generally absent in estuaries. The benthic macro fauna comprises resident species of polychaetes, molluscs, sipunculas and mud- burrowing fishes. Among invertebrate, sponges, phoronids and echinoderms generally do not prefer estuarine ecosystem. In India estuarine species diversity seems to be maximum in Molluscs. About 245 species belonging to 76 genera under 54 families were cataloged. Another important taxa, polychaeta are represented by about 167 species belonging to 97 genera under 38 families. Maximum diversity has been observed in the much-studied Hoogly-Matlah estuary (West Bengal). Macro organisms and meofauna of Indian estuaries are not properly investigated. Estuarine mud may contain rich variety of bacteria, flagellates, ciliates nematodes, ostracodes, harpacticoid copepods, rotifers, gastrotriches, arachnids and tardigrades.

Free swimmers or nekton are important components of marine biodiversity and constitute important fisheries of the world. The dominant taxa in the nekton are fish, others being crustaceans, molluscs, reptiles and mammals. Out of a total 22,000 species, it has estimated that about 4,000 species occur in the Indian Ocean of which 1,800 species are from the Indian seas. Majority of the nektonic species is found in the coastal waters. It is estimated that 40 species of sharks and 250 bony species represented the oceanic fishes.

Among reptiles, sea snakes and turtles are important and represented worldwide by 50 and seven species respectively. These are generally oceanic forms but majority of

Table 1 : Diversity of marine flora and fauna of India based on recent compilations

Name of the group	No. of species	
	Venkataraman and Wafar, 2005	Venkataraman, 2007
Plantae		
Diatoms	200+	
Dinoflagellates	90+	
Algae	844	425
Rhodophyta	434	
Phaeophyta	191	
Xanthophyta	3	
Chlorophyta	216	
Sea grasses	14	
Protista		
Protozoa (Sarcomastigophora and Ciliophora)	532+	750
Animalia		
Porifera	486+	500
Cnidaria	842+	842
Ctenophora	12	100
Platyhelminthes		350
Gastrotricha		98
Kinorhyncha		10
Annelida	270	440
Mollusca	3370	3370
Bryozoa	200+	170
Entoprocta		8
Phoronida		3
Brachiopods		3
Arthropoda		2448
Crustacea	2154	
Malacostraca	422+	
Decapoda	922+	
Sipuncula	35	35
Echiura	33	43
Tardigrada	10+	5
Chaetognatha	30+	30
Echinodermata	765	765
Hemichordata		12
Chordata		
Hemi chordata	12	
Proto chordata	119+	116
Pisces	2546	1800
Amphibia		3 *
Reptiles	35	31
Aves		145
Mammalia	25	30
TOTAL	**13,000+**	**12,462**

these often swim near to the shore and visit the shore at some part of their life. About 26 species of sea snakes belonging to one family, hydrophiidae and five species of sea turtles were reported from seas around India. Oceanic islands seem to harbor more reptiles in their marine environment. All the sea snakes and four species of turtles in their marine environment are known from islands of Andaman and Nicobar. Nesting sites of an amphibious snake were reported from the shores of north Andaman Islands (Whitaker, 1985). Turtles visit the shore during the breeding season to lay their eggs. The shore visit of these turtles especially the olive ridley is a spectacular sight on the sandy beaches at Gaginnatha near Bitharkanika in Orissa. The Andaman and Nicobar islands have best nesting beaches for the leatherback, the hawksbill and the green turtle and also the olive ridley. The seashore offers a variable feeding and breeding ground for a number of birds. It is difficult to define precisely the avian component of marine biodiversity. There are some special species, which are exclusively dependent on marine ecosystem, while a few others are generalists without much dependence on it. From the available data, it has been inferred that there are 12 families, 38 genera and 145 species, which occur in the coastal ecosystem.

Marine mammals belong to three different orders - Sirenia, Cetacea and Carnivora. About 120 species are estimated to occur in all the seas and of these 30 are reported from seas around India. But majority of these is oceanic forms and occasionally a few individuals may get stranded on the shore. Sea cow occurs in near shore waters.

MARINE BIOTECHNOLOGY

Marine Biotechnology is defined as the application of science and engineering in the direct or indirect use of marine organisms or parts or products of living marine organisms in their natural or modified forms. The oceans encompass about 71% of the surface of our planet, but over 99% of the biosphere (since organisms are found throughout the water column), and they represent the greatest extremes of temperature, light, and pressure encountered by life. Adaptation to these harsh environments has led to a rich marine bio- and genetic-diversity with potential biotechnological applications related to drug discovery, environmental remediation, increasing seafood supply and safety, and developing new resources and industrial processes as pharmaceuticals, cosmetics, nutritional supplements, molecular probes, fine chemicals and agrochemicals. Sessile marine invertebrates such as sponges, bryozoans, tunicates, mostly lacking morphological defence structures have developed the largest number of marine-derived secondary metabolites including some of the most interesting drug candidates (Newman and Cragg, 2004; Sipkema *et al.*., 2005; Mayer and Gustafson, 2006; Mayer *et al.*., 2007; Tan, 2007). In recent years, a significant number of novel metabolites with potent pharmacological properties have been discovered from the marine organisms. Although there are only few marine-derived products currently in the market, several marine natural products are now in the clinical pipeline, with more undergoing development (Haefner, 2003; Rawat *et al.*., 2006).

Marine biotechnology includes several components: aquaculture biotechnology (fish health and brood stock optimization); aquatic bioprocessing (obtaining valuable compounds from marine organisms); and aquatic bioremediation (use of microorganisms to degrade toxic chemicals in the aquatic environment). The short-term opportunities are primarily in aquaculture biotechnology while long term opportunities are expected in the form of marine pharmaceuticals, aquatic bioremediation, bioinformatics and biosensors. Biotechnology has the potential to help aquaculture enhance a cultured organism's growth rate, reproductive potential, disease resistance, and ability to endure adverse environmental conditions, such as warm or cold water. Biotechnology also holds the promise of increasing food supplies from the ocean; this is critical as overfishing and increased demand has taxed the limits of marine sustainable stocks. Additionally, species- and serotype-specific vaccines are being developed using the techniques of modern biotechnology against the disease agents and parasites that commonly affect marine organisms. Aquaculture techniques have been enhanced by the discovery of natural inducers for spawning and larval settlement, as well as remediation of disease within culture facilities. Genome studies of different commercially important fish and the genetic variability of their populations is essential for effective management of fish populations. These studies demonstrate the promise of molecular techniques for addressing questions that were previously difficult to answer. The studies regarding the genetic variability of commercially harvested fishes like salmon and tuna have demonstrated that new mitochondrial and nuclear DNA technologies will permit a definition of population structure, identification of stocks and quantification of gene flow (Tahkur *et al..*, 2008). More recently, biotechnology has been used to assess seafood safety through the development of novel test kits that react to the presence of human pathogens in the food. These "biosensor" techniques are currently being employed as early warning systems for environmental health. For example, harmful algal blooms and/or anthropogenic toxins have the potential to cause great harm to marine habitats. Unique biosensors are being developed to inform scientists of increasing levels of specific biomarkers. Moreover, specific marine bacteria have been developed that can aid in the clean-up effort if a toxic spill should occur; thus biotechnology not only helps society, but it helps the environment as well. Numbers of molecular and quantitative genetic approaches have been combined to investigate and protect endangered marine species (Tahkur *et al..*, 2008). Molecular genetic characterization of endangered and threatened species including cetaceans and sea turtles has allowed better definition of population structures, genetic diversity and variability. Marine bio-invasion is the introduction of marine organisms alien to local ecosystem through ship hulls and ballast water. This process has serious consequences to native biota, fishery and general coastal ecosystem (Anil *et al..*, 2002). Biotechnolgy approaches can be applied in the investigation of bio-invasion as well as in the finding economical and environmental viable solutions for this problem. The research in this area involves the development and testing of molecular probes for the detection of alien organisms in ballast water. For the reliable monitoring of ships ballast, specific nested PCR-based tests for detection of the harmful species in environmental and ballast water samples have been developed recently (Deagle *et al.*., 2003). Bioremediation shows great promise for addressing problems in marine environments and in aquaculture. These problems include catastrophic spills of oil in harbors and shipping lanes and around oil platforms; movement of toxic chemicals from land, through estuaries, into the coastal oceans; disposal of sewage sludge, bilge waste, and chemical process wastes, etc. During the last decades, major advances have been made in understanding the mechanisms of interactions between microorganisms and pollutants and in the application of specialized microorganisms for the in situ and ex situ treatment of organic xenobiotics, heavy metal and oil. Molecular biological methods for the identification and characterization of bacteria and their specific properties have been used to assess the composition and activity of microbial communities at contaminated sites (Geets *et al.*., 2003). Molecular and microbial monitoring tools, combined with pilot studies have proved to be effective in bioremediation strategy.

In recent years marine microorganisms have become important in the study of novel microbial products exhibiting antimicrobial, antiviral, antitumor as well as anticoagulant and cardioactive properties (Marderosian, 1969; Molinski, 1993; Austin, B., 1989). These active compounds may serve as model systems in the discovery of new drugs (Bernan, 1997; Fenical, 1997; Gochnauer, *et al.*., 1975; Yoshida, *et al.*., 1991; Ruan, *et al.*., 1994; Kates, *et al.*., 1987). To date, researchers have isolated approximately 7000 marine natural products, 25 percent of which are from algae, 33 percent from sponges, 18 percent from coelenterates (sea whips, sea fans and soft corals), and 24 percent from representatives of other invertebrate phyla such as ascidians (also called tunicates), opisthobranch molluscs (nudibranchs, sea hares etc), echinoderms (starfish, sea cucumbers etc) and bryozoans (moss animals) (Kijjoa and Sawangwong, 2004). Various sponges and crustaceans and sea weeds and microbes produce a variety of bioactive compounds that are useful in pharmaceuticals. Marine Pseudomonas species produce some bioactive compounds having antibacterial activities. There are many arthropods which produce antibacterial compounds such as penaeidins, defensins etc. Horse shoe crab (*Limulus polyphemus*) is another example whose haemolymp is of great commercial value (Destoumieux, *et al.* (1997), being used in homeopathic therapies. Horse shoe crab extract is use to treat mental exhaustion, gastro-enteric problems, cholera, and drowsiness after sea bathing. The soft coral, *Plexaura homomalla* was one of the earliest resources identified for extraction of prostaglandins, which is a rich source of 15-*epi*-PGA_2 and its acetate, methyl ester derivative. These can be converted into PGE_2 and PGF_{2a}, which at the time were the compounds in demand for prostaglandin research (Schneider *et al.*., 1972).

The following is an account of the marine organisms which are potential sources of drugs as reviewed by Blunden (2001) and others.

Antimicrobial Compounds

Cephalothin sodium, widely used as antibiotic drug is a semi-synthetic derivative of cephalosporin C, synthesised from the marine fungus, *Cephalosporium acremonium*.

Istamycins, produced by the marine actinomycete *Streptomyces tenjimariensis* SS-939 are reported to have in vitro activity against both Gram-negative and Gram-positive bacteria, including those with known resistance to the aminoglycoside antibiotics (Okami *et al..*, 1979). During the course of screening for antibiotics from marine actinomycetes, Kokare *et al..*, (2004) isolated a halophilic *Actinpolyspora* species from Alibag, a coastal region in the west coast of India, which is proved to have antimicrobial activities.

Antiviral Compounds

The most important antiviral compound reported to have significant therapeutic activity is ara-A, which is a semi-synthetic substance based on the arabinosyl nucleosides isolated from the sponge *Tethya crypta*. Other compounds reported to have antiviral properties include the didemnins, which are cyclic depsipeptides isolated from *Trididemnum* species (tunicates). Didemnin B, as well as being antiviral, also shows pronounced antitumour activity. Other antiviral compounds include avarol and avarone, isolated from a sponge, *Disidea avara*. These compounds inhibit the immunodeficiency virus, have high therapeutic indices and the ability to cross the blood–brain barrier (Sarin *et al..*, 1987). HIV-inhibitory peptides from sponges include cyclic depsipeptides Papuamides A–D (Ford *et al..*, 1999) isolated from *Theonella* sp. Another anti-HIV candidate from the sponge *Sidonops microspinosa* is the Microspinosamide (Rashid *et al..*, 2001). Patellazole B, isolated from the tunicate, *Lissoclinum patella* (Zabriskie *et al..*, 1988), has very potent in vitro activity against herpes simplex viruses. Another example of an antiviral compound isolated from a tunicate is eudistomin A, which, along with related β-carbolines, was first isolated from *Eudistoma olivaceum* (Rinehart *et al..*, 1993).

Cytotoxic Compounds

Many of the compounds isolated from marine organisms have been tested for cytotoxicity in the search for drugs active against cancer. Probably the best known of the compounds with potential as anticancer drugs are the macrolides known as bryostatins, isolated primarily from the bryozoan, *Bugula neritina*, although some have been extracted from sponges and tunicates. A family of cyclic and linear peptides and depsipeptides known as dolastatins have been isolated from the sea hare, *Dolabella auricularia* Which are highly cytotoxic (Sone *et al..*, 1996). Also cytotoxic are the polypropionates, auripyrone-A and –B, which have also been extracted from *D. auricularia* (Suenaga *et al..*, 1996). Ecteinascidins extracted from the tunicate *Ecteinascidia turbinata* has undergone phase 1 clinical trials as an anticancer agent (Faulkner, 1998). Sponges have been a rich source of cytotoxic compounds. As well as the bryostatins, many of the other active compounds are macrolides. Examples of other cytotoxic compounds reported for sponges are the antineoplastic alkaloid niphatesine D, which was extracted from a *Niphates* species (Kobayashi *et al..*, 1990); the epinardins, isolated from an unidentified species (D'Ambrosio *et al..*, 1996); Kahalide F discovered in *Elysia rufescens*; Discodermins A-H and the structural related polydiscamide A, obtained from *Discodermia* sp.

(Aneiros and Garateix, 2004); Geodiamolides A–G initially isolated from the Caribbean sponge *Geodia* sp.; Arenastatin A (Kobayashi *et al.*., 1995) a cyclic depsipeptide from *Dysidia arenaria;* Halicylindramides A–C (Li *et al.*., 1995) and D and E (Li *et al.*., 1996) with antifungal and cytotoxic properties obtained from *Halichondria cylindrata.* and the globostellatic acids A to D, which are isomalabaracane triterpene constituents of *Stelletta globostellata* (Ryu *et al.*., 1996). Cytotoxic compounds isolated from coelenterates include the sesquiterpene, suberosenone, extracted from *Subergorgia suberosa* (Bokesch *et al.*., 1996); the cembranoids from *Sinularia gibberosa*; and the diterpene, sarcoglane, from *Sarcophyton glaucum* (Fridkovsky *et al.*., 1996).

Examples of active compounds from tunicates are the cytotoxic alkaloids clavepictine-A and –B, reported for *Clavelina picta* (Raub *et al.*., 1991) and the cytotoxic dimeric disulphide alkaloid polycarpine, isolated from *Polycarpa clavata* (Kang and Fenical, 1996).

Anti-inflammatory Compounds

From the marine cyanobacterium, *Rivularia firma*, a series of bi-indoles was isolated, having anti-inflammatory qualities (Baker, 1984). Other examples of anti-inflammatory compounds extracted from marine organisms include the sesterterpene, palaulol, extracted from the sponge *Fascaplysinopsis* sp. (Schmidt and Faulkner, 1996) and a sesquiterpene furan from the coelenterate, *Sinularia* sp. (Williams and Faulkner, 1996).

Antiparasitic Compounds

Digenia simplex, a red alga, has been used as a vermifuge for very many years. Its active component, α-kainic acid, is marketed for the treatment of parasitic round worm, whip worm and tape worm. Marine animals have been tested as sources of antiparasitic compounds including bengamide F, which has anthelmintic properties and cucumechinoside F, from a sea cucumber (Miyamoto *et al.*., 1990) which has antiprotozoal activity. Microsclerodermins C–E, isolated from two species of sponges, *Theonella* sp.; *Microscleroderma* sp. (Schmidt, and Faulkner, 1998) and Aciculitins A–C (Bewley, *et al.*., 1996) and the Theonegramide (Bewley and Faulkner, 1994) are reported to have antifungal qualities.

MARINE BIOTECHNOLOGY IN INDIA

The vast area of coastal and sea waters with rich marine biodiversity together with strength in pharmaceuticals and research capabilities offers India the promise of exciting scientific and commercial opportunities. Various CSIR laboratories and Institutes including National Institute of Oceanography, Central Drug Research Institute, Central Salt and Marine Chemicals Research Institute, ICAR Institutes like Central Marine Fisheries Research Institute National Bureau of Fish Genetic Resources and Government Agencies/ Ministries like Ministry of Earth Sciences and various Universities and Departments including Centre of Excellence in 'Marine Microbiology' at Goa University, Goa, National Facility for Marine Cyanobacteria, Bharatidasan University, Tiruchurapalli, Annamalai

University, University of Kerala, Cochin University of Science and Technology, University of Mumbai are some of the Institutes conducting active research in this field. Private entrepreneurs also have interest in marine biotechnology and there are three exclusive marine biotechnology companies registered in India, while a number of other biotechnology and pharmaceutical companies also are active in marine biotechnology research and product development. The increasing importance of marine biotechnology in India is indicated by patenting trend in the country (Demunshi and Chugh (2009). There is a steady increase in the number of patents filed by Indian companies/research Institutes, claiming various pharmaceutically, cosmeceutically useful molecules isolated from marine resources. Based on the data collected from the Indian patent office, Demunshi and Chugh (2009) found that snails and sea weeds are one of the most useful and popular source for many such potentially valuable products. Corals, cyanobacteria, sponges are also gaining momentum as a potential research target. The overall patenting trend was observed to be encouraging as it strongly indicated rapid growth of the nascent Indian marine biotechnology industry.

EXPLORATION OF INDIAN MARINE BIODIVERSITY AND BIOTECHNOGY

India is one among the twelve mega-biodiversity countries and twenty five hotspots of the richest and highly endangered eco-regions of the world (Myers *et al*.., 2000). She is one among the 12 mega biodiversity countries (Other countries being Brazil, Colombia, Ecuador, Peru, Mexico, Madagascar, Zaire, Australia, China, Indonesia and Malaysia). About 60-70% of the world's recorded biodiversity are from these mega biodiversity countries. India has a long coastline of 8118 km, an Exclusive Economic Zone of 2.02 million km^2 and continental shelf area of 0.506×106 km^2. The two island ecosystems of India – Lakshadweep and Andaman and Nicobar Archipelagos add to the ecosystem diversity of this country. Gulf of Mannar, Gulf of Kachch and the two island ecosystems have rich coral reefs harbouring valuable marine biodiversity (Venkataraman, 2007).

It is said that even before 1000 B.C., the Phoenicians who were among the first mariners of the world, visited the Indian seas also and attempted to gather valuable information about some parts of our seas (Seshappa (1991). The early (of about 600 BC) navigators of the Indian Ocean were probably ancient Egyptians. Arabs and Chinese followed them later. Study of marine fauna in India has drawn greater attention from the 18th century onwards. Some English members of the Royal Asiatic Society of Bengal who were engaged in Zoological researches took the initiative in this direction, and as a result of their efforts, an expert committee was appointed by the Government of India in the year 1871 A.D. for going into the question of conducting researches on the deep sea organisms of the Indian seas. The Government started the Marine Survey Department in the year 1874; Owing to insufficient facilities however, this department could not undertake any detailed work until 1881, but the scientists were still able to collect some samples of animals from the sea in depths down to 100 fathoms by means of dredges. In 1872 Wood-Mason was able to collect deep sea animals from the Andaman and Nicobars areas on behalf of the Indian Museum of Calcutta. The Marine Survey Department was able to collect samples of about 24 per year down to about 2000 fathoms between the years 1884 and 1894 A.D. using the Research Vessel "Investigator" for the purpose. Samples were collected during this period (by this vessel) of water, animals and also other materials from the Persian Gulf, Ceylon, Andamans, Burma and other regions of our seas. The results of most of these researches are described in the publications of the Indian Museum brought out from the year 1898 onwards periodically. Francis Day (1829-1889) was Inspector-General of Fisheries in India (from circa 1871) and Burma and an ichthyologist. He wrote a monograph on fishes between 1875-1878 "The Fishes of India" with a supplement in 1888 and two volumes on "Fishes" in the Fauna of British India series in which he described over 1400 species. Col. Seymour Sewell's work on the copepods (a group of animals belonging to the Arthropods and that are very useful as food of many kinds of fishes in the seas) is even today an authoritative reference book on these animals in this areas of the world. The other great oceanographic expeditions in the Indian Ocean included the British Challenger Expedition (1872-76); the German Expedition Valdivia (1898-99), the Dutch Expedition Siboga (1899-1900); the John Murrey Expedition (1930-34); Dana (1928-30); Galathea (1845-47 and 1950-52); Albatross (1950-52) that brought us to the time of the International Indian Ocean Expedition (I.I.O.E., 1959-65), with the participation of numerous ships (Vema, Argo, Horizon, Pioneer, Chain, Vega, Anton Brunn, Discovery, Challenger II, Vityaz, Meteor, Diamantina etc.,) belonging to several countries (Venkataraman, K., 2007). The International Indian Ocean Expedition was unprecedented in many respects, and several countries outside our region also participating in the work. About forty research ships from twenty countries were involved in this gigantic effort at the scientific surveying of the Indian Ocean; regions of the ocean never visited before were also covered and samples and data were collected of the organisms of the sea bottom, plankton, seawater, mudsamples and various other materials occurring in the different regions of the great ocean. Most of the samples and the data have since been worked out and the results published. The knowledge available from these results provides the much-needed basic material and foundation for further work and for the building up of the edifice of marine biodiversity of our area on very sound and dependable lines, for the future scientists of our country and of the other countries around the Indian Ocean. It need hardly be emphasized that the earlier workers did not have the advantage of such a comprehensive collected fund of knowledge of such a large area of our seas.

From the beginning of this century the fishery scientists of the Government of Madras and the specialists attached to the Connemara Museum in Madras also conducted some researches on the animals and plants of the sea, especially of the shore. Thurston and Herdman conducted researches independently in the Gulf of Mannar on the sough-east coast of India; Gravely (1927) & 1941) examined the shells and other animal remains from the beach; the littoral fauna of Krusadai Island in the Gulf of Mannar were worked out by various persons including Gravely, Fauvel, Burton and others; these results are detailed in Vol.1, No. 1 of the Madras Government Museum Bulletins (1927 to 1937). P. Fauvel's contributions (1932 & reprint 193) to our polychaete fauna are very important also. Among other works from the Madras Museum must be specially mentioned those of Burton (1937) on the sponges, Hickson on the Alcyonaria (1931), M.G.K. Menon on the Scyphomedusae (1930 and 1932), Krishna Menon on the Decapod larvae and life histories (1933, 1937 and 1940) and of C.C. John on the Chaetognaths of Madras (1933). Another expert namely, J.S. Gardiner (1903) studied the animals of the Lakshadweep group of islands and published his results in book form. At the beginning of this century, a Fisheries Departmental Research establishment was opened at Calicut on the west coast of India. Research Units were also set up in due course at Krusadai near Mandapam and at Ennur near Madras to carry out some limited investigations; these units were also helpful in supplying marine specimens to workers all over the country for routine study as well as research.

Various fishing voyages in the sea have also helped in the progress of marine biodiversity researches in India in some way or the other from time to time. From the year 1928 onwards, the Madras Fisheries Department conducted several exploratory surveys on both the east and west coasts for spotting regions of good fish stocks. In 1928-30 the Lady Goschen did the first extensive exploration of this kind on both the coasts, Similarly but even earlier (Hefford 1923) the Bombay fisheries Department conducted exploratory trawling surveys off Bombay and the north west coast with the vessel William Carrick. The Government of Bengal had also surveyed the deep sea fishing potentialities using the vessel Golden Crown even much earlier (1908-1911). From the experience of these surveys, not only were considerable data made available of the fishery wealth in the respective regions of the sea around India at that time, but some knowledge also became available about the oceanographic conditions and general marine biology in the areas investigated, though the trips of all these vessels were essentially fishing voyages only. In recent years exploratory and fishing voyages of large vessels in the offshore regions of our seas are very common, both Governmental organizations and private business interests being represented among these, Research workers frequently get facilities to accompany the fishing crew in such voyages not only to observe the fishing activity but also to collect samples and data of different kinds for study.

There has been an all-round increase in the facilities for the study of marine biology in our country after the attainment of Independence. The Government of India had established several Institutes for the survey of Indian seas for studying and monitoring the biodiversity, resources potential of our seas and the physical and biological processes driving them. The prime Institutes among them include Zoological Survey of India, National Institute of Oceanography, Central Marine Fisheries Research Institute, Centre for Marine Living Resources and Ecology, NIFPHAT and Fishery Survey of India. The Fishery Survey of India is the main Agency mandated for survey and management of the fisheries and related resources of the Indian Seas. Taking into consideration of the need for identifying marine fish diversity and density in the Indian waters, the FSI has been collecting base line data on biodiversity of finfishes, crustaceans and cephalopod resources in the Indian EEZ. This was driven by the need to acquire accurate assessments of fish stocks for sustainable fisheries, and for the study of biological features of marine organisms of commercial or scientific interest. The FSI, equipped with 12 survey/research vessels will also be important instruments in the cultivation of and study of marine organisms for biotechnological purposes. Marine Living Resources (MLR) programmes of the Ministry of Earth Sciences, utilising the facilities available onboard the FORV Sagar Sampada also had generated considerable data on the marine biodiversity of Indian seas. The facilities onboard various other research vessels of the Ministry of Earth Sciences also had been instrumental in the survey of marine biodiversity and bioprospecting of Indian marine biodiversity.

In addition to the Central Institutions, the maritime States of the Country also have some research activities administered through their forest and fishery departments, the problems of investigation being mainly of local industrial application; the more academic and comprehensive problems of an all-India nature are tackled by the Central Research Institutions. The various Universities in the maritime States of the country i.e. situated in the States of Gujarat, Maharashtra, Goa, Karnataka, Kerala, Andhra Pradesh, Tamil Nadu and Orissa as well as West Bengal, also conducts limited survey and studies on the biodiversity and its applications.

THE LEGAL FRAMEWORK OF BIODIVERSITY CONSERVATION

The steep decline in marine biodiversity prompted in enacting various international/national legal frameworks and conventions. The most important international convention is the convention on marine biodiversity (1992), in which over 150 countries signed documents declaring their commitment to the conservation of marine biodiversity (Convention on Biological Diversity). The Convention on Biological Diversity (CBD) is a landmark in the environment and development field, as it takes for the first time a comprehensive rather than a sectoral approach to the conservation of Earth's biodiversity and sustainable use of biological resources. It was in the year 1984 that the need to have in place a global convention on biological diversity started gaining momentum. In response, the United Nations Environment Programme (UNEP) in the year 1987 recognised the need to streamline international efforts to protect biodiversity. The Convention on Biological Diversity (CBD) was negotiated and signed by nations at the UNCED Earth Summit at Rio de Janeiro in Brazil in June 1992. The Convention came into force on December 29, 1993. India became a Party to the Convention in 1994. At present, there are 175 Parties to this Convention. The main objectives of the Convention are Conservation of biological diversity; Sustainable use of the components of biodiversity; Fair and equitable sharing of benefits arising out of the utilisation of genetic resources. India became a signatory to the Convention on Biological Diversity (CBD) in June 1992.

Apart from the Biodiversity Convention itself, the UN's Convention on the Law of the Sea (UNCLOS III 1982), which came into force in November 1994, is of major significance in relation to biodiversity (1997). IUCN has recently produced a comprehensive analysis of the Law of the Sea and other legal issues relating to marine conservation. UNCLOS establishes a comprehensive framework for use of the ocean and its resources. Other international agreements that relate to fishing and conservation of marine resources, such as conventions on whaling, marine mammal conservation, regional seas, Antarctic resources, transboundary fisheries (e.g. salmon and tuna) etc.

Other important international conventions include the 1973 Convention of International Trade in Endangered Species of Wild Fauna and Flora (CITES) and the 1979 Convention on the Conservation of Migratory Species of Wild Animal (CMS).

The Indian Acts/laws dealing directly with the biodiversity includes Indian Forest Act, (1927), Wildlife (Protection) Act, (1972), Forest (Conservation) Act, (1980) and Biological Diversity Act, (2002).

The Government of India has brought Biological Diversity Act, 2002 with the following salient features :

1. To regulate access to biological resources of the country with the purpose of securing equitable share in benefits arising out of the use of biological resources; and associated knowledge relating to biological resources;
2. To conserve and sustainably use biological diversity;
3. To respect and protect knowledge of local communities related to biodiversity;
4. To secure sharing of benefits with local people as conservers of biological resources and holders of knowledge and information relating to the use of biological resources;
5. Conservation and development of areas of importance from the standpoint of biological diversity by declaring them as biological diversity heritage sites;
6. Protection and rehabilitation of threatened species;
7. Involvement of institutions of state governments in the broad scheme of the implementation of the Biological Diversity Act through constitution of committees.

BIODIVERSITY AUTHORITY

Under the Biological Diversity Act, a National Biodiversity Authority has been set up at Chennai vide Gazette Notification dated 1st October, 2003. The Act also provides for establishment of State level Boards and Local level Bio-diversity Management Committees to deal with any matter concerning conservation of Biological Diversity, its sustainable use and fair and equitable sharing of benefits arising out of the use of Biological resources and associated knowledge. Detailed rules under the Act have been notified for implementation of the various Provisions of the Act. The National Biodiversity Authority is in process of developing norms for Access and Benefit Sharing as related to genetic Resources and Associated knowledge.

CONCLUSION

India, blessed with rich marine biodiversity, has great scope for developing marine biotechnology by bioprospecting our marine biodiversity utilising the infrastructure facilities and well trained man power available with government Institutes and universities. Development involving varied applications with concerted effort synergised through Government developmental agencies and individual entrepreneurs/corporate houses would help us to enhance contributions to food security, medicines and cosmetics in near future.

REFERENCES

Aneiros, A. and Garateix, A. 2004. Bioactive peptides from marine sources: pharmacological properties and isolation procedures. Journal of Chromatography B, 803 (2004) 41–53

Anil AC, Venkat K, Sawant SS, Dileep Kumar M, Dhargalkar VK, Ramaiah N, 2002. Marine bioinvasion: Concern for ecology and shipping. Curr Sci **83:** 214-8.

Arderosian, A. D., 1969. Marine pharmaceuticals. J. Pharm. Sci., **58:** 1–30

Austin, B. 1989. Novel pharmaceutical compounds from marine bacteria. J. Appl. Bacteriol. **67:** 461-470

Baker J.T. 1984 Seaweeds in pharmaceutical studies and applications. Hydrobiologia **116/117:** 29-40

Bernan, V. S., Greenstein, M. and Maiese, W. M., 1997. Marine microorganisms as a source of new natural products. Adv. Appl. Microbiol., **43:** 57-90.

Bewley, C.A. He, H. Williams, D.H. and Faulkner, D.J., 1996. J. Am. Chem. Soc. 118 4314.

Bewley, C.A. and D.J. Faulkner, 1994. J. Org. Chem. **59:** 4849.

Blunden, G. 2001. Biologically Active Compounds from Marine Organisms. Phytother. Res. **15:** 89–94 (2001).

Bokesch, H.R., McKee, T.C., Cardellina, II J.H. and Bord M.R., 1996 Suberosenone, a new cytotoxin from Suberorgia suberosa. Tetrahedron Lett **37:** 3259-3262.

Burton, M. 1937. Supplement to the littoral fauna of Krusadai Island in the Gulf of Manaar: Porifera. Bull. Madras Govt. Mus., n.s., Nat. Hist. Sect. **1 (no 2, part 4):** 1-58

Camacho FC, Rodriguez JG and Miron AS, 2007. Biotechnological significance of toxic marine dinoflagellates. Biotech Adv. **25:** 176-94.

Chisholm, S.W., Olsen, R.J., Zettler, E.R., Goericke, R., Waterbury, J.B. and Welschmeyer, N.A. 1988. A novel free living prochlorophyte abundant in the oceanic euphotic zone. Nature. **334:** 340-343.

Chu, W. and Radhakrishnan, A.K., 2008. Research On Bioactive Molecules: Achievements And The Way Forward. IeJSME: **2 (Suppl 1):** S21-S24.

Collette, B.B., Russo, J.L., and Zavala-Camin, L.A. 1978. Scomberomorus brasiliensis, a new species of Spanish mackerel from the western Atlantic. Fish. Bull. **76:** 273-280.

D. F. Martin and G. M. Padilla (eds), 1973. Marine Pharmacognosy. Action of Marine Biotoxins at the Cellular Level. Academic Press, NY

D'Ambrosio R, McKhann GM, Janigro D (1996) Whole-cell recording from hippocampal astrocytes during orthodromic stimulation. Soc Neurosci Abstr **22:** 128.16.)

de Fontaubert, C. A., Downes, D.R. & Agardy, T.S. 1996. Biodiversity in the seas: Implementing the convention on biological diversity in marine and coastal habitats. IUCN Environmental Policy and Law Paper No. 32. A Marine Conservation and Development Report. 82 pp.

Deagle, BE, Bax, N, Hewitt, CL, Patil, JG .2003 Development and evaluation of a PCR-based test for detection of Asterias (Echinodermata : Asteroidea) larvae in Australian plankton samples from ballast water Marine and Freshwater Research **54:** 709-719

Demunshi, Y. and Chugh, A., 2009. Patenting trends in marine bioprospecting based pharmaceutical sector. Journal of Intellectual Property Rights, **14:** 122-130.

Destoumieux D *et al.*, 1997. Penaeidins, a new family of antimicrobial peptides isolated from the shrimp Penaeus vannamei (Decapoda). Journal of Biological Chemistry **272(45):** 28398-28406.

Faulkner, 1998. Marine Natural Products. Nat. Prod Rep (R Chem Soc) **16:** 113-158.

Fenical, W., 1997. New pharmaceuticals from marine organisms. Trends Biotechnol., **15 (9):** 339–341

Ford, P.W., Gustafson, K.R., McKee, T.C., *et al.*., 1999. Papuamides A–D, HIV-inhibitory and cytotoxic depsipeptides from the sponges Theonella mirabilis and Theonella swinhoei collected in Papua New Guinea. J. Am. Chem. Soc. **121:** 5899–5909.

Fridkovsky, E., Rudi, A., Benayahu, Y., & Kashman, Y. 1996. Sarcoglane, a new cytotoxic diterpene from Sarcophyton glaucum. Tetrahedron Letters, **37:** 6909-6910.

Gardiner, J.S. (1903) The Fauna and Geography of The Maldive and Laccadive Archipelagoes, being the account of the work carried out and collections made by an expedition during the years 1899 and 1900. Volumes 1 and 2. Cambridge University Press.

Geets J, Vangronsveld J, Diels L, and van der Lelie D., 2003. The Application of Molecular Tools to Follow up Bioremediation. J Soil Sediments; **3:** 251.

Gerwick, W.H. 1987. Drugs from the Sea: The Search Continues. J. Pharm. Tech. **3:** 136-141.

Gochnauer, M. B., Leppard, G. G., Komaratat, P., Kates, M., Novitsky, T. and Kushner, D. J. 1975. Isolation and characterization of Actinopolyspora halophila, gen. et sp. nov., an extremely halophilic actinomycete, Can J Microbiol **21:** 1500-1511.

Gochnauer, M. B., Leppard, G. G., Komaratat, P., Kates, M., Novitsky, T. and Kushner, D. J., 1975. Isolation and characterization of Actinopolyspora halophila, gen. et. sp. nov. an extremely halophilic actinomycete. Can. J. Microbiol., **21:** 1500–1511

Gravely, FH., 1927. The littoral fauna of Krusadai Island in the Gulf of Mannar. Bulletin Madras Government. Museum Natural History. **1(1):** 163-173.

Gravely, FH., 1941. Shells and other animal remains found on the Madras beach. Bulletin Madras Government, Museum Natural History. **5:** 86-90.

Haefner, B., 2003. Drugs from the deep: Marine natural products as drug candidates, Drug Discovery Today, **8(12):** 536-544.;

Hashimoto, Y. 1979. Marine toxins and other bioactive marine metabolites. Japan Scientific Societies Press, Tokyo.

Hefford, A.E. (1923) Report of the work of the steam trawler, William Carrick May 1921 to February 1922. Government Central Press: Bombay, India.

Hickson, S. J., 1931. Three species of Alcyonaria from Gulf of Mannar. Bulletin Madras Government, Museum Natural History **1(2):** 79-89.

John, C. C., 1933. SagItta of the Madras Coast. Bull. Madras Gov. Mus., (N.S.), Nat. Hist. Sect., **3:** 1-10

Kang H, Jensen PR and Fenical W, 1996. Isolation of microbial meta- bolites from a marine ascidian of the genus Didemnum. J. Org. Chem. **61:** 1543–1546)

Kates, M., Porter, S. and Kushner, D. J., 1987. Actinopolyspora halophila does not contain mycolic acids. Can. J. Microbiol., 1987, **33:** 822–823

Kijjoa, A. and Sawangwong, P. 2004: Drugs and Cosmetics from the Sea. Mar. Drugs 2004, **2:** 73-82.

Kobayashi, M., Aoki, S., Gato, K., Matsunami, K., Kurosu, M., Kitagawa, I., 1994. Marine natural products XXXIV. Trisindoline, a new antibiotic indole trimer, produced by a bacterium of Vibrio sp. separated from the marine sponge Hyrtios altum. Chemical and pharmaceutical bulletin **42(12):** 2449-2451

Kobayashi, T., Nakanishi, K., Kajio, H., Morinaga, S., Sugimoto, T., Murase, T. and Kosaka, K, 1990. Pancreatic cytokeratin: an antigen of pancreatic exocrine cell autoantibodies in type 1 (insulin-dependent) diabetes mellitus. Diabetologia; **33:** 363-370

Kokare, C. R., K. R. Mahadik, S. S. Kadam and B. A. Chopade, 2004. Isolation, characterization and antimicrobial activity of marine halophilic Actinopolyspora species AH1 from the west coast of India Cuur. Sci. 86, **4:** 25 593-597)

Kokare, C.R., Mahadik, K. R. and Kadam, S.S., 2004. Isolation of bioactive marine actinomycetes from sediments isolated from Goa and Maharashtra coastlines (West coast of India). Ind. J. Marine Sci., **33:** 248-256.

Krishna Menon (1933, 1937 and 1940)

Li H. Y., Matsunaga, S and Fusetani, N., 1995. Halicylindramides A-C, antifungal and cytotoxic depsipeptides from the marine sponge Halichondria cylindrata. Journal of medicinal chemistry: **38(2):** 338-43.

Li H., Matsunaga S. and Fusetani N., 1996. Halicylindramides D and E, antifungal peptides from the marine sponge Halichondria cylindrata. J. Nat. Prod. **59:** 163.

Li, H., Matsunaga, S. and Fusetani N. 1996. Halicylindramides D and E, Antifungal Peptides from the Marine Sponge Halichondria cylindrata J. Nat. Prod., **59(2):** 163–166

Li, H., S. Matsunaga y N. Fusetani. 1995. Halicylindramides A-C, antifungal and cytotoxic depsipeptides from the marine sponge Halichondria cylindrata. J. Med. Chem., **38:** 338-343.

Lobel, P.B., Belkhode, S.P., Jackson, S.E. and Longerich, H.P. 1990. Recent taxonomic discoveries concerning the mussel Mytilus: Implications for biomonitoring. Arch. Environ. Contamin. Toxicol. **19:** 508-512.

M G K Menon. 1930. The scyphomedusae of Madras and the neighbouring coast. Madras Government Museum Bulletin. New Series, Natural History Section : Vol 3 No 1

M G K Menon. 1932 The hydromedusae of Madras. Madras Government Museum Bulletin. New Series, Natural History Section : Vol 3 No 2

M Krishna Menon. 1933 The life histories of decapod crustacea from Madras. Madras Government Museum Bulletin. New Series, Natural History Section : Vol 3 No 3

Mayer AMS and Gustafson KR 2006. Marine pharmacology in 2003-2004: Anti-tumour and cytotoxic compounds. Eur J Cancer 42: 2241-2270.;

Mayer, A.M.S., A.D. Rodriguez, R.G.S. Berlink and M.T, Hamann 2007. Marine pharmacology in 2003-2004: Marine compounds with anthelmintic, antibacterial, anticoagulant, antifungal, anti-inflammatory, antimalarial, antiplatelet, antiprotozoal, antituberculosis, and antiviral activities; affecting the cardiovascular, immune and nervous systems, and other miscellaneous mechanisms of action. Comparative Biochemistry and Physiology, Part C 145. **61:** 553-581.;

McDonald, J.H., Seed, R. & Koehn, R.K. 1992. Allozymes and morphometric characters of three specis of Mytilus in the Northern and Southern Hemispheres. Mar. Biol. **111:** 323-333.

Menon, M. K., 1937. Decapod larvae. from the Madras plankton. Mad. Covt. Mus,, Bull., N.S., Nat. Hist. Sect. **3 (5):**

Miyamoto T, Togawa K, Higuchi R, Komori T, Sasaki T. 1990. Six newly identified biologically active triterpenoid glyco- side sulfates from the sea cucumber Cucumeria echinata. Leibigs Ann Chem: 453-460.

Molinski, T. F., 1993. Developments in marine natural products- Receptor-specific bioactive compounds. J. Nat. Prod., 1993, **56:** 1–8

Myers, N., Mittermeier R. A., Mittermeier S. G., da Fonsec G.A.B. nd Kent J., Nature, 403 (2000) 853-858.

Narsinh L. Thakur, Roopesh Jain, Filipe Natalio, Bojan Hamer, Archana N. Thakur, Werner E.G. Müller, 2008. Marine molecular biology: An emerging field of biological sciences. Biotechnology Advances 26 (2008) 233–245.

National Research Council. 1995. Understanding marine biodiversity: A research agenda for a nation. National Academy Press, Washington D.C. 114 pp.

Newman, D.J. and Cragg, G.M., 2004. Advanced preclinical and clinical trials of natural products and related compounds from marine sources. Curr Med Chem. **11(13):** 1693-713.

Norse, E.A. 1993. Global marine biological diversity: A strategy for building conservation into decision making. Island Press, Washington D.C.383 pp.

Norse, E.A. 1993. Global marine biological diversity: A strategy for building conservation into decision making. Island Press, Washington D.C.383 pp.

Okani, Y., K. Hotta, M. Yoshida, D. Ikeda, S. Kondo, and H. Umezawa. 1979. New aminoglycoside antibiotics, istamicins A and B. J. Antibiot. **32:** 964-966)

Olsen, R.J., Chisholm, S.W., Zettler, E.R., Altabet, M. and Dusenberry, J. 1990. Spatial and temporal distributions of prochlorophyte picoplankton in the North Atlantic Ocean. Deep-Sea Research. **37:** 1033-1051.

P. Fauvel's (1932 and reprint 193)

P.W. Ford, K.R. Gustafson, T.C. McKee, N. Shigematsu, L.K. Maurizi, L.K. Pannell, D.E. Williams, E.D. De-Silva, P. Lassota, T.M. Allen, R. Van-Soest, R.J. Andersen, M.R. Boyd, J. Am. Chem. Soc. 121 (1999)

Pan MH, Ghai G, Ho CT. Food bioactives, apoptosis, and cancer. Mol Nutr Food Res 2008; **52:** 43-52.

Rashid, M.A., K.R. Gustafson, L.K. Cartner, N. Shigematsu, L.K. Pannell and M.R. Boyd, 2001. Microspinosamide, a new HIV-inhibitory cyclic depsipeptide from the marine sponge Sidonops microspinosa, J. Nat. Prod., **64:** 117-121.)

Rashid, M.A., K.R. Gustafson, L.K. Cartner, N. Shigematsu, L.K. Pannell and M.R. Boyd, 2001. Anti-HIV candidate from the sponge Sidenops microspinosa. J. Nat. Prod., 64: 117., J. Nat. Prod. 64 (2001)

Raub, M.F., Cardellina, J.H., Choudhary, M.I., Chao-Zhou, N., Clardy, J., and Alley, M.C. 1991. Clavepictines A and B: Cytotoxic Quinolizidines from the tunicate Clavelina picta. J. Am. Chem. Soc. **113:** 1378-1380)

Rawat DS, Joshi MC, Joshi P Atheaya H. Marine peptides and related compounds in clinical trial. Anti–Cancer Agents – Med Chem 2006; **6:** 1–8.)

Ray, G.C. 1988. Ecological diversity in coastal zones and oceans. In E.O. Willson (ed), Biodiversity. National Academy Press, Washington, D.C. Pp 36-50

Ray, G.C. 1988. Ecological diversity in coastal zones and oceans. In E.O. Willson (ed), Biodiversity. National Academy Press, Washington, D.C. Pp 36-50.

Rinehart KL, Shield LS, Cohen-Parsons M. 1993. Antiviral substances. In. Marine Biotechnology, Vol. 1. Pharmaceutical and bioactive natural products. Attaway DH, Zaborsky, OR (eds) Plenum Press: New York and London; 309-342.)

Ruan, J., Mahmoud, A., Zhou, Z. and Liang, H., 1994. Actinopolyspora iraqiensis sp. nov., a new halophilic actinomycete isolated from soil. Int. J. Syst. Bacteriol., 1994, **44:** 759–763

Ryu *et al.*., 1996 Ryu, G., Matsunaga, S. and Fusetani, N., 1996. Globostellatic Acids A-D, new cytotoxic isomalabaricane triterpenes from the marine sponge Stellata globostellata. J Nat Prod **59:** 512-514.

Sarin PS, Sun D, Thornton A, Müller WE. Inhibition of replication of the etiologic agent of acquired immune deficiency syndrome (human T-lymphotropic retrovirus/lymphadenopathy-associated virus) by avarol and avarone. J Natl Cancer Inst. 1987 Apr; **78(4):** 663–666.)

Scheuer, PJ 1973. Chemistry of Marine Natural Products. Academic Press. New York and London., 1973

Schmidt, E.W and Faulkner, D.J. 1996. Palauolol, a new anti-inflammatory sesterterpene from the sponge Fascaplysinopsis sp. From Palau. Tetrahedron Lett. **37:** 3951-3954. 54 (1998) 3043

Schmidt, E.W.; Faulkner, D.J. Absolute configuration of methyl (2Z,6R,8R,9E)-3,6-epoxy-4,6,8-triethyl-2,4,9-dodecatrienoate from the sponge Plakortis halichondrioides. Tetrahedron Letts. 1996, **37:** 6681-6684

Schmidt, E.W.; Faulkner, D.J. Microsclerodermins C-E, antifungal cyclic peptides from the lithistid sponges Theonella sp. and Microscleroderma sp. Tetrahedron, 1998, **54:** 3043-3056.

Schneider WP, Hamilton RD, Rhuland LE. Occurrence of esters of (15S)-prostaglandin A 2 and E 2 in coral. J Am Chem Soc. 1972 Mar 22; **94(6):** 2122–2123)

Seshappa, 1991. Indian Marine Biology. Daya Publishing House, Delhi. 5-10

Sipkema D, Franssen MC, Osinga R, Tramper J, (2005). Marine sponges as pharmacy. Mar. Biotechnol. **7:** 142-162.;

Sone H, Shibata T, Fujita T, Ojika M, Yamada K (1996) Dolastatin. H and isodolastatin H, potent cytotoxic peptides from the sea hare Dolabella auricularia Tetrahedron **53(24):** 8149-8154)

Suenaga, K., H. Kigoshi and K. Yamada, Isolation and stereostructure of aurilide, a novel cyclodepsipeptide from the Japanese sea hare Dolabella auricularia. Tetrahedron Lett. **37:** 6771-6774)

Tan, L.T. 2007 Bioactive natural products from marine cyanobacteria for drug discovery, Phytochemistry 68 (2007), pp. 954–979

Thakur, N. L., Roopesh Jain, Filipe Natalio, Bojan Hamer, Archana N. Thakur, Werner E.G. Müller, 2008. Marine molecular biology: An emerging field of biological sciences. Biotechnology Advances 26 (2008) 233–245.

Venkataraman K. and Wafar, M.V.M. 2005. Coastal and marine biodiversity of India Indian J. Mar. Sci. **34 (1) :** 57-75., (2005)

Venkataraman, K., 2007, Marine Ecosystems of India. 7-26. Indian Journal of Environmental Education, Vol. **7:** 7-26.

Whitaker, R. 1985. Rational use of estuarine and marine reptiles. In: Proc. Symp. Endangered Mar. Animals and Mar. Parks. 1: 298-303. Cent. Mar. Fish Res. Inst. Spec. Pub. **18:** CMFRI, Cochin.

Williams, D.H. and Faulkner, D. J., 1996. Two practical syntheses of an anti-inflammatory sesquiterpene furoic acid from Sinularia spp. Tetrahedron, 52, **(12):** 4245-4256)

Yoshida, M., Matsubara, K., Kudo, T. and Horikoshi, K., 1991. Actinopolyspora mortivallis sp. nov., a moderately halophilic actinomycete. Int. J. Syst. Bacteriol., 1991, **41:** 15–20

Zabriskie TM, Mayne CL, Ireland CM. Patellazole C. A novel cytotoxic macrolide from Lissoclinum patella. J Am Chem Soc 1988; **110:** 7919-7920)

Bioresources for Food Security and Rural Livelihood, Pages 339–346
Edited by : G.K. Kulkarni and B.N. Pandey

A CASE STUDY ON AN ECONOMIC ASPECT OF COMMON PROPERTY RESOURCES AND SHARE IN THAR DESERT OF WESTERN RAJASTHAN

G.R. Parihar

ABSTRACT

The arid zones exhibit a spectacular and vivid biotic diversity and an interesting rural environment, dominated by man as everywhere. The desert dwellers human beings, animals and plants all are generally well adapted to face problems of paucity of water, food and shelter. Sixty percent of the Thar desert is located in northwestern part of Rajasthan state and it is the most populated desert of the world (i.e. 64 persons per sq.km.). The livestock animals maintained by man are outnumber (about 150 per sq.km.) the carrying capacity of this arid land. The economy of the region is almost livestock oriented, which are extremely important component of the ecology of most protected areas and should be censored accurately. Live stock compete directly with wild herbivores for limited fodder. There is a management problem but also a major socio-economic support to rural people in and around the protected areas which should not be ignored. The Thar desert, part of the great Indian desert constitute such a unique assemblage of animals and plants most of which are not found elsewhere. The nature selected biota of the desert are hardy and drought resistant to survive in low water and high temperature. This feature of desert have influenced the total social fabric of the native population and economy of the region. The biodiversity of Thar has been maintained and replenished over the centuries by the people themselves.

Keywords: Common Property Resources, Economy, Thar Desert, Rajasthan

INTRODUCTION

The economy of the region is alm ost totally livestock oriented. Livestock are extremely important component of the ecology of most protected areas and should be censuses. What effects do these organisms have an soil formation? How do animals help or obstruct in the germination of seeds subsequent development of seedlings and ultimately the vegetation and animal life in the desert? How the human interference is responsible for the survival and cutting of tree species in desert? How many local genetic races of 'sevan' or 'dhaman' and other palatable grasses are found in the Indian desert? Why this desert represent such a great concentration of black bucks, chinkaras, nilgais and thousands of migratory birds? There are many uncertainties involved in answering these questions but we are sure that it is only the intimate knowledge about the life and ecology of the plants and animals of the desert which help us to preserve disappearing species and depleting habitat in the arid zones.

Thar desert of Rajasthan is situated between 22^0 30 N and 32^0 05' N and between 68^0 05' E and 75^0 45' E. It is the eastern extension of the great Sahara desert and covers 2,85,680 km^2 (Krishnan, 1968). The western Rajasthan which houses major part of Thar is comprised of malani igneous smite with tuffs, welded tuffs, rhyolites, rhyolite porphyry and mafics (ca 747 my). The late proterozoic to early palaeozoic is represented by sandstones and limestone coupled with denuded and rugged terrain of malanis. These rock sequences form the Marwar supergroup. Dolomite and minor shale overlie the malani basement rocks at places. A great deal of mesozoics are represented by the sedimentary sequences of Jaisalmer, Barmer and Palana-Ganganagar shelf (Pareek, 1981).

Like the other deserts of the world, the Thar desert in Rajasthan is characterized by sparse and highly variable rainfall, extreme variation in diurnal and annual temperatures and high evaporations. One of the interesting feature of this desert is its high relative humidity (Krishnan, 1977). In this paper, the economy and the present biodiversity scenario from selected areas of the desert has been dealt.

During the study period i.e. 2004-05, to scan the biodiversity of desert, a mixed approach of understanding and recording the biodiversity and its economy was employed. This method of screening the biodiversity was not confined to the list of flora and fauna about which fairly good information exists (Bhandari, 1991; Prakash, 1994b; Ghosh et.al. 1996). The approach therefore was to understand the social fabric, peoples science, and physiographical situation, landform, soil geology and climate of some selected areas of the desert to understand the biodiversity and its influence in giving rise to a sustainable ecosystem. For centuries the biodiversity makeup of desert is the basis of socioeconomy of the area. To attain these objectives, this study was pursued in several stages and the data so generated are knit together to record the represent biodiversity scenario from the selected areas of the desert.

Field Surveys

Random field surveys during January through March 2004 were carried out in Barmer, Jaisalmer and Jodhpur districts of western Rajasthan to find out the representative areas where intensive studies could be confined. These surveys were meant to assess the physical features, status of biodiversity and extent of boitic and aboitic interventions. The visual observations during field surveys helped considerably to identify issues and people perception of biodiversity. All the selected 29 villages in Barmer, Jaisalmer and Jodhpur districts were surveyed. Based on field surveys of these villages, 5 villages - Kanasar and Bhinyad in Barmer, Rasla in Jaisalmer and Maluna and Nevra in Jodhpur districts were selected as study villages for intensive study.

Interviews

The field surveys were supplemented by developing simple questionnaire to elicit information through people's knowledge about the area like land cover, institutional forms, biodiversity niches, subsistence practices, and so on. These interviews provided basic biodiversity issues of the Thar and with more insight of past, present and future scenario of the biodiversity.

Surveys and interviews helped considerably to demarcate variety of villages subhapatis or institutional areas each representing a subecosystem in itself. Studies were attempted in these villages institutions to understand the biodiversity and variations. These village institutions are :

1. **Gaucher**: Defined as gram panchayat controlled grazing lands.
2. **Oran**: Defined as area where tree-cutting is banned by community consent and divine control and where such restraint is actively observed.
3. **Fallow Lands**: Defined as Khatedari (private agricultural) lands with usually 'kharip' crops. At least 2-3 years in fallow, managed as closed fodder fields during the growing season and thereafter as community-wise open grazing.
4. **Plough Fields**: These are of two types -
   i) **Traditional**: Those in which ploughing is done by animals to avoid destruction of useful species in the field during tilling.
   ii) **Disc**: In which mechanical methods are used wherein massacre of species takes place.
5. **Sand dunes**: Denied as natural sand masses either stationary or stabilized or shifting. Usually with good moisture.
6. **Forest Enclosure**: Defined as land under the present control of the Forest Department for plantation.
7. **Gravel Lands**: Defined as low productive area with little soil cover and moisture.
8. **Wastelands**: Defined as areas provided with mixture of characteristics with low productivity.

Plants and Animal Sampling

Sampling were done in September and early October in 2004-05 in the high growth period. 34 sampling sites in villages Kanasar, Rasla and Nevra were selected for plant cover and 8 sites in Kanasar and Rasla each were selected for animal study. At each site, a central point was selected where the institutional area was larger than 4 hectares (40000 sq. meters), a point was selected at random. In the case of extremely large institutional areas (for eg. the Rasla Oran), sample points were spatially stratified across a loose grid. From the sample point, a compass direction was taken at random and a 100 meter transact was laid from the point. A line transact method was employed to a major coverage by species intercept and 10 quadrates were laid alternately on either side of the transact line to measure frequency. The basal intercepts of each species was recorded along the transact to derive cover percentage and the number of individuals appearing in each quadrate was recorded to derive frequency.

There is a strong relationship between the presence of certain tree species and animals, between tree species with understory, herbs and grasses. This relationship is important for village biodiversity. By and large, the flora and fauna, biocommunities and their interrelationships are the same in the study area as found in the desert countryside.

Biological Diversity

Ecologically, the vegetation of the study area fall under the category of 'thorn forest' type. Some changes are seen due to cultivation and changes in the soil-climate condition. The natural vegetation has a substantial contribution to the productivity of the trees like, Khejri (*Prosopis cineraria*) which are highly valued and conscientiously maintained. Besides this, a lot of shrubs, and grasses are found growing in fallow as well as crop lands and of course in grazing lands. These are deep rooted, tenacious enough to survive expanded droughts and yet efficient in putting up a good biomass. A variety of animals are found in the study area. The area exhibits a spectacular biodiversity because of its evolutionary history and geographical location.

Flora

The vegetation of this tract is quite sparse with limited number of species. Its vegetation can be described into following types based on land forms or habitats. The plant communities growing on rocky habitats varies from dwarf semiscrub desert to mixed xerophytic thorn forest types in which the principal plant communities are *Euphorbia caducifolia - Acacia senegal*; *Euphorbia caducifolia - Grevia tenax*; *Euphorbia caducifolia - Capparis decidua*, etc.

The associated trees and shrubs which are widely scattered and sparse predominates with *Salvadera oleoides*, *Acacia senegal*, *Maytenus emerginata*, *Grewia tenax*, *Zizypus numelaria*, *Lycium barbarum*, *Capparis decidua* and *Euphorbia caducifola*. The last named species, in particular, supports large number of climbers such as *Coculus peridulus*, *Asparagus racemosus*, *Ephedra foiata*, *Rivea hypocrateri*, *Rhynchosia minima* and *Coccina grandis*. The plants of the ground floor vary at different places but the most common plants are *Oropetium thomeum*, *Tephrosia uniflora*, *Encapagon brachystachya*, *Pupalia lappaca*, *Lepidagathis bandraensis*, *Barleria acanthoides* and *Diptera canthus*.

Capparis decida and *Zizyphus nummularia* are the common plants of the gravely plain ecosystem. These are found in association with *Lycium barbarum* and *Leptadenia pyrotechnica*. The undergrowth in this habitat consists of *Boerhavia diffusa*, *Blepharis sindica*, *Heliotropium strigosum*, *Molluge cerviana* and *M. nudicaulis*, *Salvia aegyptiaca*, *Indigofera triogonelloides*, *Anticharis senegalensis*, *Pactyloctenium sindicum* and *Aristida mutabibis*. Alluvial plain represent mixed xenomorphic wooded desert. The dominant tree species of this habitat are *Prosopis cineraria*, *Salvadora oleoides* and *Tecomella undulata*. The common associate shrubs of these trees are *Calotropis procera*, *Laptademia pyrotechnica*, *Mimosa hamata*, *Acacia jacquemontii*, *Convolvulus microphyllus*, *Crotalaria burhia*, *Cenchrs biflorus*, *C. pricurio*, *Eragrotis pilosa*, *Cyperus rotundus*, *Indigofera cordifolia*, *Tribullus terrestris*, *Aevra persica* and *Sericostemma pauciflorum*.

Sand dunes in these study villages are either parabolic or barchans. The parabolic dunes which dominate the study are mostly have *Calligonum polygonoides*, *Panicum turgidum*, *Lasiures sindicus*, and *Crotolaria burhia*. *Indigofera cordifolia*, *Cyperus arenarius* and *Cenchrus biflorus* are fairly common, which *Dipterygium glaucum* and *Forsetia hamiltonii* are some time abundant. Barchans dunes which are also found in the study villages near settlements, are mostly devoid of vegetation.

Fauna

The study are represent fairly large number of animals from insects to mammals. For clarity the commonly met texa in the study area have been included. Here, most animals are exposed only occasionally to the extremes of desert climate. They may dwell in caves, crevices, beneath stones, in burrows, under the shade of plants, on the underside of leaves and thus live in relative comfort and safety in their own microclimate of desert is therefore essential to understand the actual physical conditions in which a desert animal lives. Studies were confined to macrotaxa encountered during field surveys and information supplemented by local people. These macrotaxa and microtaxa were found in two principal habitats viz. terrestrial and aquatic. The majority of vertebrates occupy the terrestrial habitat, while the aquatic habitat is confined to most of the amphibians, all fishes, and several invertebrate groups such as some protozoan, sponges, many insects, crustaceans, molluscan, etc.

For the invertebrates, the terrestrial habitats as under stones, under leaves, under cow dung's, on trees and bushes, etc. are met with. A special types of habitat is underground burrow - small ones (used by rats, bandicoots, gerbils) and long-narrow burrows (used by termites and ants). The aquatic habitat in deserts is limited in extent. Rivers here are ephemeral, but there are a few rivulets, tanks, 'talab' and 'nadis', where rain water is available year found in normal monsoon years. In addition, there are ephemeral pools and puddles formed during the rainy season where water may last from a few weeks to a few months.

The rich insect fauna which is a characteristics role as the micro-consumer in the existing ecosystem. Although most elements of this fauna are found to be present almost throughout the year. They exhibit a high population build-up particularly during the rainy season (i.e. July to September) and cause much loss to the growing vegetation especially in this season. Among the termites, the most destructive species are *Psammotermes rajasthanicus*, *Anacanthotermes macrocephalus*, *Microtermes spp.* and *Odontermes spp.* Locusts and grasshoppers breed in the region, especially in the rain season when sufficient moisture is available in the soft-sandy soil. These pests do much loss to the growing vegetation. Swarms of the desert locurt, *Schistocerca gregaria* visit occasionally, whenever suitable environmental conditions are formed. The predatory habit of some insects like mantids (*Schizocephala bicornis*, *Statilia spp.*), Coleopteran (*Monochilis sexmaculatus*), ant (*Camponotus gigas*) and spiders are quite marked in the rainy season.

A strong relationship between some species and their understory has been established under test-farm conditions. Ahuja (1980) and Kaul & Gangule (1963) recorded levels of forage, production under Thar desert trees. In particular, 'Bordi' (*Zizyphus nummularia*) and 'Khejari' (*Prosopis cineraria*) cover were found to be associated with increased forage production and increased biodiversity. Similarly, some studies revealed a negative association between cover of the exogenous 'angrezi babul' (*Prosopis juliflora*) and forage cover, especially valuable perennial grasses which suppress the biodiversity (Agarwal et.al., 1976; Ghosh et.al. 1998). In studies of soil, 'Khejri' and deshi babul (*Acacia nilotica*) wee found to increase organic matter, nitrogen, phosphorus and calcium to lower pH levels (Singh and Lal, 1969) and increase soil moisture (Gupta and Saxena, 1978).

Faunal variables may explain some of the variation as well. Browsing species, especially goats and sheep, were observed spending longer periods of time beneath these trees that out in the open, especially under 'bordi' and 'ker' (*Caparis decidua*) trees, for which there was a browsing preference. In this case, browsers may plan a maintaining role in the grazing system, rather than simply a destructive one. It was not possible to test this hypothesis during the study period. The nutrient inputs, disturbances in the soil, and cutting of perennial grass tops, encourage fast nutrient cycling and rhizomatic underground growth, which result in high productivity and stable soils (Clark et.al. 1980; Collins, 1987; Kanodiya and Patil, 1982).

Different animal species prefer different plants and different plants survive and adapt to grazing pressure differently. Some grazing and browsing increases available forage by "opening up" thatch and cover for other species and encouraging growth. The grazing and browsing have different impacts and effects on plant structure and reproductive capacity. The quantity (in terms of coverage) and quality (in terms of available nutrients) of different species varies throughout the year; perennials and annuals are valuable at varying times and conditions.

There are a large number of species consumed by small stock that are not consumed by cattle. These include browse species such as the thorn/brush trees 'bordi' and 'ker', but also include a large number of 'nuisance' species like 'dhamasa' (*Fagonia cretica*), 'siniya' (*Crotolaria burhia*) and 'biyoni' (*Tephrosia purpuria*) which are common on the 'degraded' pastures. Different species are feeding from different levels of the pasture/ scrub. Therefore, an increase in the overall number of animals may not cause immediate and uniform degradation if there is little overlap in diet (Parihar, 1997). It has been suggested that this stratification in grazing/browsing patterns may actually work in the reverse of the succession/degradation model by helping to maintain the 'canopy' balance, decreasing over-growth of shrub species and so maintaining high under-growth graze, production of perennial grasses and ephemerals for other, non-browsing species (Huss, 1972).

The seasonality of available nutrition in desert pasture species is also an important factor in determining grazing and browsing pressure patterns. Chakrawarty et.al. (1970) reported in their study of crude protein content in pasture species that while perennial species like 'sevan' (*Lasiurus sindicus*) and 'dhaman' (*Cenchrus ciliaris*) maintained nutrient value throughout the year. They are surpassed in available protein by marginal annual species like 'bekar' (*Indigofera cordifolia*), 'sinter' (*Concolvulus microphyllus*) and 'lamp' (*Aristida spp.*) during the months just following the monsoon. This means that the absence of particular perennial species from a landscape may not represent a loss of productivity from that landscape and that "degraded" pastures, low on the successional track, may provide a wealth of nutrition during part of the year.

In summary, several factors determine the kinds of faunal pressure exerted on community lands. The types of animals present will determine the range of affected flora. The ratio of grazers to browsers will determine the balance of pressures on different species. The yearly strategy of producers managing animals in the region will determine where and when pressure on the ecology is greatest.

ACKNOWLEDGEMENTS

The author is grateful to Prof. S.M. Mohnot, Emeritus Professor of Zoology, former Head, Department of Zoology, J.N.V. University, Jodhpur for regular guidance and encouragement. To Principal, Govt. College, Barmer (Rajasthan) for providing the logistic support during this field study. Thanks are also due to Dr. L.S. Rajpurohit, Associate Professor-Zoology, J.N.V. University, Jodhpur for improving the manuscript.

REFERENCES

Agarwal, R; Gupta, J.P; Saxena, S.K. and Muthana, K.D. 1976. Studies on Soil Physiochemical and Ecological Changes Under Twelve Years Old Five Desert Species of Western Rajasthan. *Indian Forester*, **102:** 863-72.

Bhandari, M.M. 1991. Evolution of Desert Vegetation in Thar (eds. Dhir et.al.). *Geol. Soc. India*: 73-78.

Chakravarty, A.K; Ratan, R. and Singh, K.C. 1970. Grazing Studies in the Arid and Semiarid Zones of Rajasthan: Utilization of Vegetation Cover, Grazing Behaviour of Sheep and Seasonal Variation of Crude Protein Content of Plants in Different Pasture. *Ann. Arid Zone*, **9:** 10-18.

Clark, F.E; Cole, C.V. and Bowman, R.A. 1980. Nutrient Cycling Grasslands, Systems Analysis and Man. (ed. by A.I. Breymeyer and G.M. Van Dyne), pp. 659-712. Cambridge Univ. Press, Cambridge.

Collins, S.L. 1987. Interaction of Disturbances in Tallgrass Prairie: A Field Experiment. *Ecology*, **68:** 1243-1250.

Ghosh, P.K; Baqari, Q.H. and Prakash, I. (eds.) 1996. Faunal Diversity in the Thar Desert: Gaps in Researches. Scientific Publisher, Jodhpur.

Gupta, J.P. and Saxena, S.K. 1978. Studies of the Monitoring of the Dynamics of Moisture in the Soil and the Performance of Ground Flora Under Desertic Communities of Trees. *Indian J. of Ecol.*, **5:** 30-36.

Huss, D.L. 1972. Goats Responses to the Use of Shrubs as Forage. Wildlands Shrubs, Their Biology and Utilization. USDA Forest Service. General Technical Report INT-I: International Forest and Range Experimentation Station.

Kanodiya, K.C. and Patil, B.D. 1982. Pasture Development In: *Desert Resources and Technology* (Ed. by A. Singh), pp. 263-305. Scientific Publishers, Jodhpur.

Krishnan, A. 1968. Delineation of Different Climatic Zones in Rajasthan and Their Variability. *Int. J. Geology*: 33-40.

Krishnan, A. 1977. A Climatic Analysis of the Arid Zone of North West India. In Desertification Control, ICAR, pp. 42-57.

Parihar, G.R. 1997. Studies on the Biodiversity of Some Selected Areas in the Desert. Ph.D. Thesis, J.N.V. University, Jodhpur (Rajasthan).

Pareek, H.S. 1984. Pre-Quaternary Geology and Mineral Resources of North-Western Rajasthan. *Geo. Sur. India Mem.*, vol. 115.

Prakash, I. 1994b. Biodiversity Conservation in the Thar Desert. Indian Forester, **120:** 873-879.

Singh, K.S. and P. Lal. 1969. Effects of KHEJARI (*Prosopis cineraria*) and Babool (*Acacia arabica*) Trees on Soil Fertility and Profile Characteristics. Ann. Arid Zone: 33-36.

Bioresources for Food Security and Rural Livelihood, Pages 347–353
Edited by : G.K. Kulkarni and B.N. Pandey

BIO-RESOURCES AS A TOOL FOR FOOD SECURITY AND RURAL LIVELIHOOD:

CHALLENGES OF 21st CENTURY

Sadhana Deshpande

INTRODUCTION

Nutrition has far reaching consequences as it has direct link to the physical ability of the people to work, earn livelihood and maintain the family up to a satisfactory level. It also has relation to the mental condition of the individuals. It is very much essential that the masses (particularly rural) understand what the importance of nutrition is in the overall well-being of the society.

Health is one of the fundamental human rights. The importance of physical and mental health can never be under estimated. This essential fact has to be constantly pointed out for the benefit of the general public. The international organizations like World Health Organization of UN have endorsed this principle. Unfortunately health cannot be given or distributed but has to be actively acquired. In order to promote the health, it is necessary to understand the fundamentals of health and diseases. Health is defined as a state of complete physical, mental and social well being. One can promote health only by understanding what it is dependent on and then applying this knowledge in every day life. The care of the body regarding food, hygiene, exercise, rest, measures for protection against diseases etc. has to be known to the people.

Food is one of the basic necessities of life. Good nutrition is the foundation of good health. Food is required to supply the body building materials necessary for an infant to grow into an adult or for an adult to replace the daily wear and tear in tissues occurring in the course of normal living. Diet is nothing but a combination of different foods as they are available in nature. Diet should not only be adequate but also be balanced. The components of food which fulfill all the basic function are known as proteins, fats, carbohydrates, minerals, and vitamins. These are collectively known as nutrients.

Among these, proteins form the prime substance of our muscles and organs and are important body building material. Fats, carbohydrates and proteins constitute the fuels that are burned inside the body and supply energy. Vitamins are necessary to regulate the proper utilization and assimilation of proteins, fats, carbohydrates and minerals to attain proper growth and also to prevent certain diseases. Water forms nearly 2/3rd of our body weight and is an important constituent of body tissues. It also acts as a vehicle to transport the other nutrients.

Following table shows some commonly consumed commodities across the globe and its production in terms of million MT,

Crop	Yield (mmt) (1986)
Wheat	552
Rice	469
Maize	449
Potatoes	290
Barley and Oat	178
Cane and Beet Sugar	100
Pulses	96
Vegetable Oil	47
Vegetables and Fruits	347
Meat and Milk	140
Fish and Sea Food	70

WORLD SCENARIO

Food and water are out barest necessities to remain alive. World food supply has been rising at spectacular rate during last fifty years. There is enough food to supply everyone in the world with more than the daily food requirements, but food is inequitably distributed. About 750 million people are under nourished or mal-nourished and 20 million children die each year from diseases related to mal-nutrition, others suffer from stunted growth mental retardation and developmental disorders. Even the richest countries like U.S.A. have under nourished population. For the ever-increasing population we need sustainable form of agriculture to feed future generation. Following chart shows the world food production statistics. While the production in absolute terms seems to be increasing, the increased quantities may mot be adequate for the increased population.

NEEDS FOR NUTRITION

Our body requires a constant supply of energy and raw material for the processes involved in day to day living. The amount of energy each of us needs to remain strong and healthy depends on body weight, climate, stress level and basic metabolism. The Food and Agriculture Organization (FAO) of the United Nations estimates that the average minimum daily calorie intake over the world is about 2500 calories per day. People who receive 90% of there minimum dietary intake on long term basis are considered under nourished. They are more susceptible to infectious diseases. People who receive less than 80% of there minimum dietary intake are seriously under nourished. This cycle extends

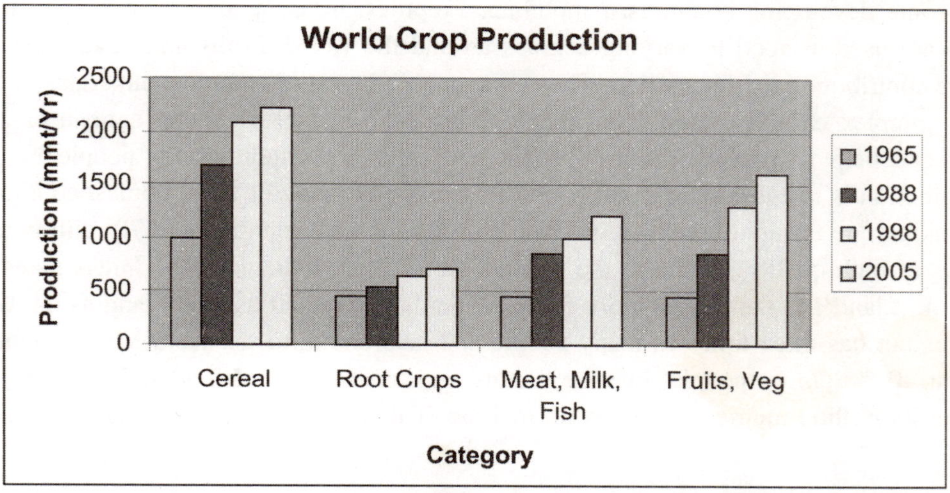

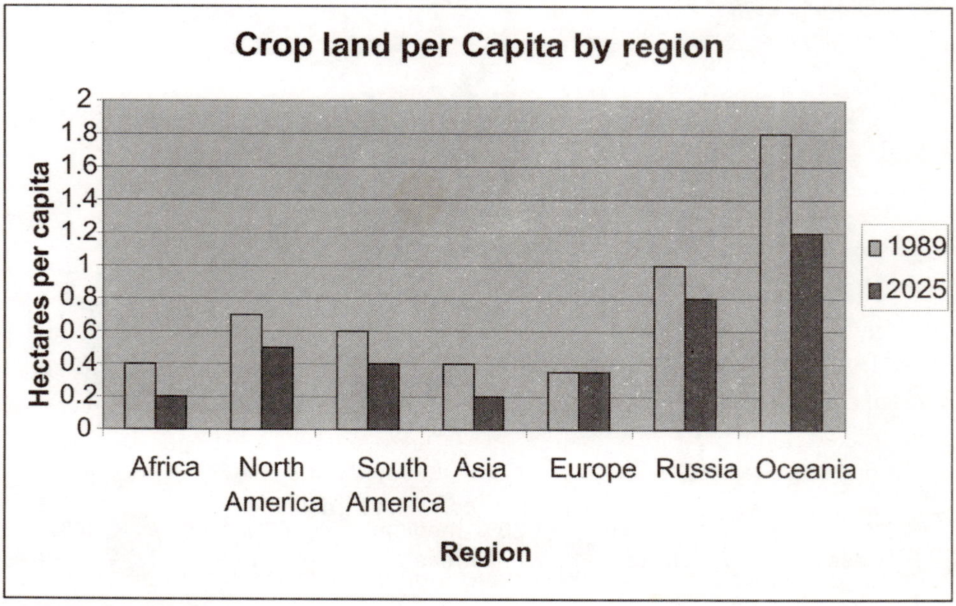

(See colour view also)

from generation to generations. Children in this category suffer from permanently stunted growth and other developmental disorders. One child in every four in the developing world per year dies due to diseases that could be prevented with better diet, clean water, and simple medicines. The most common dietary problem in the richer countries is too many calories. The average daily intake in North America, Europe is about 3500 calories. At least 20% of Americans are seriously overweight which contribute to cardiovascular diseases, cardiac arrest, strikes etc.

All of our food comes from other organisms and plants. Most sea food is harvested from free roaming wild organisms. Sea food provides up to 60 % of the protein supply for some developing countries. Amphibians, reptiles, birds, mammals, and even insects also are used as food to varying degrees around the world. Fruits and vegetables make large contribution to human diets. They contain high levels of vitamins, minerals, and fiber and complex carbohydrates. Even many fruits and vegetables contain anticancer agents that can help to prevent diseases. Meat and milk are important for people but their distribution is unequal. The grazing land that supports many of these domestic animals is shrinking due to rapid urbanization. Fish and sea food contribute about 70 million metric tones of high quality protein to the world's diet. China, India and the United states have the most land and the largest work forces. India has only half as much land as the United States but has three times as many people and gets one third the yield from its land and about 40 % hungry people. In spite of this serious shortage of food India has exported about 24 million metric tones of grain in 1986. This would have been enough to raise the

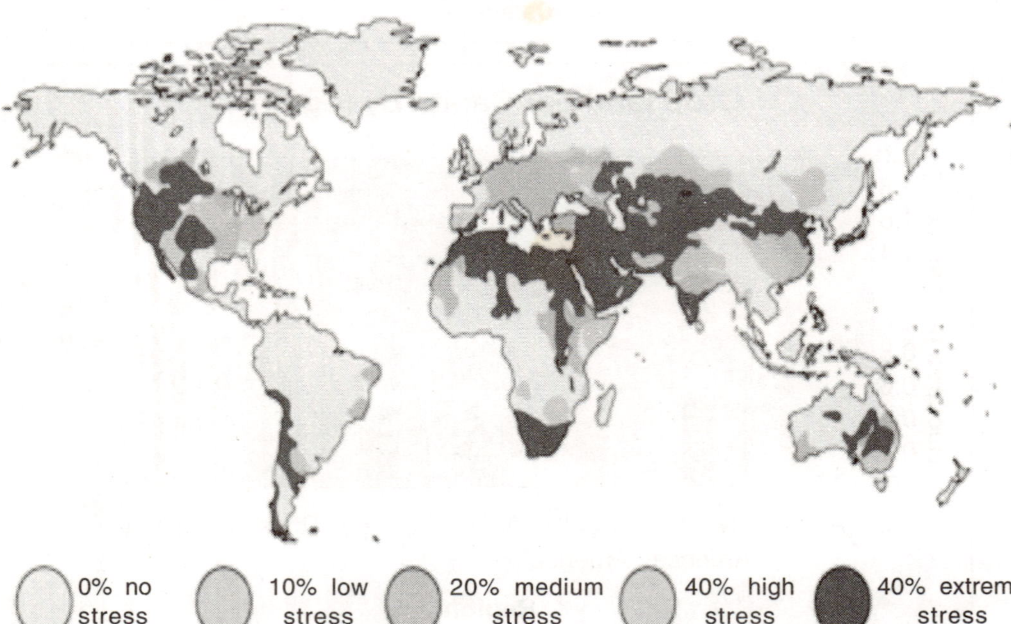

Fig. 3 : World food production and water resources. (See colour view also)

caloric intake of most of India's hungry people to the minimum requirements for normal lives. In between 1959 and 1961 china suffered the greatest famine in the world's history. Now in china, because of Govt policies, farmers cultivate small family plots getting 50% more crops per hector in addition to diversified mixed crops. Farmers are rearing fish, chickens, pigs, ducks and rabbits as a supporting activity. China is now exporting grain to other countries and has enough food to feed every one.

The FAO projects that 64 countries (29 Africa, 21 Asia and 14 in Central America) will be unable to feed there citizens by 2025. Helping these countries to stabilize population, protect their environment and feed the starving should be a world priority for security reasons. Following graph shows cropland/ capita by region which shows that excepting Europe, everywhere else, the quantum is decreasing, thus putting the food security programme to a lot of challenges.

To increase the food production new and improved crops and animal products may be created by biotechnology and genetic engineering. Transgenic plants and animals have tremendous potential for improving agricultural production. Microbes have been developed that consume toxic wastes. Cows could produce more milk, fish could be bigger, and pigs could have leaner meat. Costly drugs including insulin and anti clotting agents can be cheaply produced by bacteria or in the milk of live stock. Genetically engineered bovine growth hormone has already been shown to boost milk production by about 10 %. But the cost of this technology being quite prohibitive can be afforded by only richer farmers. Similar to the green revolution, blue revolution or fish farming has the potential to contribute as much human nutrition as did miracle cereal grains. It involves growing organisms in enclosures such as ponds tanks or net pens anchored in an ocean or estuary. Fish can be grown to high densities under these artificial conditions. About 90 species of fin fish, 13 species of shrimp and numerous shell fishes are grown commercially.

Nutritional deficiencies can affect individuals anywhere in the world who do not get an adequate diet either because food is not available or they do not know how to choose the right things to eat. In recent years agricultural policy makers have shifted their attention from total food availability to food security, the ability to obtain sufficient food on a day today basis. Even when family income is adequate women and children often do not get enough to eat because men spend the money on them selves. We now see that individual food security is really what matters.

SUGGESTED ACTIONS

Following are some steps to improve the food security mainly for rural areas:

- **Improve land use planning :** Due to rapid urbanization, the land under cultivation has come under stress and has been shrinking at a rapid pace. This means that the requirement of food for the growing population has to be met by a lesser area of land. This calls for a lot of innovations on the front of agricultural research and land usage. The scientific community will have to come out with ideas to increase the per acre yield. This can be through better seeds, genetic modifications, improved fertilizers, better water management.

- **Expand educational programs for girls and employment opportunities for women :** Traditionally, the womenfolk are responsible for the overall management of the food in the family. They are also responsible for up-bringing of the children in the family. It is, therefore, very much necessary that they are properly educated and trained on the aspect of the management of nutrition. It is also necessary that they be provided with opportunities to earn through employment of some kind. This will boost their morale and confidence. Cottage industries, self-help groups etc are the best solutions in the Indian context. To improve the family welfare women should allow to manage the income
- **Increase investment and maintenance on rural infrastructure :** If one considers the agricultural and the economy related to it, the same is concentrated in rural areas. Hence, infrastructure plays an important role as it does for any other sphere of the economic activity. Irrigation facilities, seed availability, connectivity, storage facilities, transportation facilities etc are some of the important aspects of the infrastructure related to this activity. Also, availability of easy and affordable credit is a must and can be considered as part of the 'economic' infrastructure that is necessary. A better Irrigation facility will eliminate the dependence of the farmers on the monsoon rains. Storage facility is another very important aspect. Approx 25% of the agricultural and horticultural produce in India is wasted due to deterioration. This can be prevented by establishing a chain of cold storages across the routes. Every kg of the products saved from deterioration is as good as producing the same and, hence, can help a long way in enhancing the availability. Same is the case with Transportation facilities.
- **Preservation of wild species :** There are tribal pockets in India where, the tribal are aware about the medicinal or nutrient value of many plants found in the wild. This information, however, is limited only to a closed group and is not in the public domain. Hence, it is necessary to document the information related to these species. There also needs some research undertaken to extract maximum benefits out of these species. To create awareness about preservation and protection of endangered species among the people is important.

- **New technology :** Rural people should aware of the advantages and disadvantages of the scientific and technological advances and breakthroughs taking place. There should be a mechanism existing to pass on such information to them. Also, they need to be educated about the ones that can be adapted by them under the given climate/ soil condition. The necessary materials need to be made available to the farmers. Also, there need to be held training sessions for them to understand and assimilate the new techniques and put them to the best advantage under given circumstances.

Hunger, poverty, population growth, environmental degradation and social problems form a complex interconnected web. Each is a cause as well as a consequence of the others. One of the most important questions is how many people and at what level of civilization and environmental quality can the world feed?

REFERENCES

1. Environmental Science : a global concern by William P Cunningham and Barbara Woodworth Saigo, WCP Publishers, 3rd Edition, Chapter 10.
2. You and Your Health by Dr. Bhave, Deodhar and Bhave, 6th reprint, published by National Book Trust of India
3. Principles of Nutrition : Nutrition value of common food stuff
4. Website : World Resources Institute

Bioresources for Food Security and Rural Livelihood, Pages 355–365
Edited by : G.K. Kulkarni and B.N. Pandey

ORIGIN AND PHYLOGEOGRAPHY OF TODA BUFFALOES OF NILGIRIS IN SOUTH INDIA

P. Kathiravan, R.S. Kataria, B.P. Mishra, P.K Dubey and B.K Joshi

INTRODUCTION

The domestic river buffalo (*Bubalus bubalis*) stands next only to cattle in its utility to man, mainly in the form of milk, meat, draught and hide. In South Asia, it has traditionally played an important role in satisfying the social and economic needs of many rural and tribal communities. Of late, this species has acquired the status of being an important dairy animal in countries like India and Pakistan. India is having the largest buffalo population in the world (94 million heads) with diverse gene pool ranging from high yielding dairy type (Murrah and Nili-Ravi) to extensively managed semi wild type (Assamese and Toda) animals. Toda buffalo, an endangered breed, derive its name from the Todas, an aboriginal tribe (Figure 1) who inhabit the Nilgiri hills of Tamil Nadu in South India. The Todas have no history, no written character and only a very elementary language. They are supposed to have lived on the plateau of Nilgiris for about 800 years and their livelihood depends mainly in keeping a herd of buffaloes (Gunn, 1909; Littlewood, 1936). The life of Todas is largely centered around these sacred (supposed to be) buffaloes, which form an integral part of their sustenance and cultural heritage. Although Toda buffaloes are predominantly reared by the Toda tribe, a demographic change in the ownership is noticed with communities like Badagas and Kotas, also rearing these animals (Walker, 1986). The Toda buffaloes reared by the Toda tribe are distributed at altitudes more than 2000 metres above mean sea level (MSL) while those reared by non-Todas are found at altitudes around 1000 metres above MSL (Karthikeyan *et al.*, 2002).

DISTRIBUTION AND STATUS

The Toda buffalo herds are distributed throughout the plateaus of the Nilgiri hills. The Todas live with their buffaloes settled in small and tiny hamlet which is popularly called as *munds*. Most of these hamlets with buffaloes are scattered across the Nilgiri hills particularly more on north-western side. Walker (1986) and Karthikeyan *et al.* (2002) observed changing pattern of ownership of these buffaloes between the Todas and other communities. Walker (1986) reviewed the census figures and reported the population of Toda buffaloes in 1848, 1930, 1960 and 1975 as 2171, 1619, 2186 and 2650 respectively. As of May, 1994, the total population of the Toda buffaloes in the breed tract was 3313 and there were 1809 adult females, 70 male young stock, 584 female young stock and 850 calves (Karthikeyan *et al.*, 2002).

Fig. 1 : Men and women of Toda tribe in their traditional attire

Fig. 2 : A herd of Toda buffalo

TODA TRIBE AND THEIR BUFFALOES

Anthropological studies of the Toda tribe reveal their dependence on the buffaloes for their livelihood. The importance of these buffaloes in the life of Todas had been portrayed as early as 17th century by Finicio (1603). Rivers (1906) made a detailed account on the Toda buffaloes, dairy organization and ceremonies accompanied the daily work of the dairy and important events of buffalo life. In spite of the change in the life style of the Todas, still the Toda buffalo forms the basis of the traditional society and hence the religion of the Toda is a highly ritualized buffalo-cult (Walker, 1986). Some of the important cultural practices associated with the buffaloes include naming of female buffaloes, exchange of buffaloes during the marriage ceremony as dowry, ceremonial slaughter of buffalo cows during death of Toda people, salt giving practice to all the buffalo cows, etc. The most sacred of all is the temple located within the same hamlet which also bears the shape of horn and head of the buffalo reflecting the role of buffaloes in their life (Figure 3).

Fig. 3 : Inscription of buffalo head with horns at the entrance of Toda temple

MANAGEMENT PRACTICES

The Toda buffaloes are semi-wild in nature and temperamentally not easily tractable as they become aggressive as a group when approached by strangers. The gregarious nature protects them from the wild predators. The Toda buffalo is the only breed of buffalo in India, maintained still under primitive level of management. Normally, the buffaloes are housed only during night in an open pen (Figure 4). Buffalo calves are kept in pens made of wooden stakes and covered at the top with tiles and dried pine leaves. Breeding of female Toda buffaloes is mostly by the wild stud bulls which roam inside the forest and mingle with the domestic herds while grazing. Toda buffaloes exclusively thrive on grazing alone. The commonly available vegetation in the natural pasture of Nilgiris include kikuyu grass (*Pennesetum clandestinum*) interspersed with white clover (*Trifolium repens*) and hariyali grass (*Cynodon dactylon*).

PHYSICAL CHARACTERISTICS

The Toda buffaloes are medium-sized animals. Body is fairly long with broad and deep chest. Legs are short, strong and sturdy and hooves are black in colour. Usually, the coat colour in buffalo heifers and cows is ash grey. Tail is fairly long extending up to the hock joint and the switch is black. The head is large and heavy, carried to the level of the body and is held uprightly at the sight of a stranger. Generally, the forehead is broad, convex in calves and slightly concave in adults. The head is surmounted by horns which are usually set wide apart with typical crescent shape. Udder is not so prominent and is round in shape. Teats are cylindrical and round at the tip. The milk vein is also not prominent. Two characteristic chevron markings are present, varying from white to light brown in colour; one is present just round the jowl and the other below the neck, just above the brisket region (Figures 5 and 6).

Fig. 4 : Open housing of Toda buffaloes

Fig. 5 : Wild male Toda buffalo Figure

KARYOLOGICAL STUDIES

The Asian buffaloes on the basis of their habitat and phenotype are divided into two types: riverine and swamp type. The karyotype of riverine type of buffalo comprises of 24 pairs of autosomes and a pair of sex chromosomes while the swamp type has only 23 pairs of autosomes apart from sex chromosomes. Of the 24 pairs of autosomes in river buffaloes, 5 pairs were biarmed (metacentric/submetacentric) and 19 pairs were acrocentric. The reduction of a pair of autosomes in swamp buffalo was due to tandem fusion of river buffalo chromosomes 4 (4p) and 9 (DiBerardino and Iannuzzi 1981, Chowdhary *et al.* 1989) resulting in a distinctly larger chromosome than any biarmed pair of river buffaloes. Based on the morphology and behaviour, Toda buffaloes were initially thought to be of swamp type. However, karyological studies, revealed them to be river buffaloes with the diploid chromosome number being 50 typical to that of riverine type (Nair *et al.,* 1986). The first five pairs of autosomes were sub meta centric and the remaining 19 pairs were acrocentric. The third pair was more metacentric than submetacentric. In females both the X-chromosomes were acrocentric and larger than all other acrocentric chromosomes, while in males the X-chromosome was longest among acrocentrics like those of females. The Y-chromosome was among the smaller acrocentrics.

Fig. 6 : Toda buffaloes with chevron markings

GENETIC CHARACTERIZATION OF TODA BUFFALO USING MICROSATELLITE MARKERS

Basic Diversity Indices in Toda Buffalo

Different genetic diversity indices like observed number of alleles, effective number of alleles, observed and expected heterozygosity estimates and polymorphism information content are presented in Table 1. Polymorphism information content varied between 0.272 (ILSTS 031) to 0.739 (ILSTS 052) with a mean value of 0.529 In Toda buffalo population, none of the markers were having PIC values less than 0.25 and 15 out of the 25 loci studied were having values higher than 0.5 indicating their suitability for assessing genetic variation (Botstein *et al.,* 1980). A total of 105 alleles were detected in Toda buffaloes across the 25 loci investigated with an average of 4.2 alleles per locus. The observed allele number per locus varied between 2 (ILSTS 033) to 6 (ILSTS 056, CSSM 066, ILSTS 036, ILSTS 095, ILSTS 052 and ILSTS 061). The mean effective number of alleles per locus was 2.661 and ranged between 1.412 (ILSTS 031) to 4.390 (ILSTS 052) across different loci (Kataria *et al.,* 2009).

Table 1 : Measures of different genetic diversity indices across 25 microsatellite loci in Toda buffaloes

Locus	n_a	n_e	H_o	H_e	Nei's H_e	PIC
CSRM060	4.0	2.281	0.553	0.569	0.562	0.517
ILSTS026	3.0	1.971	0.432	0.498	0.493	0.382
HEL013	3.0	2.123	0.455	0.537	0.529	0.473
ILSTS030	4.0	2.810	1.000	0.652	0.644	0.586
ILSTS033	2.0	1.859	0.475	0.468	0.462	0.355
ILSTS017	4.0	2.316	0.421	0.576	0.568	0.509
ILSTS019	3.0	1.857	0.600	0.467	0.462	0.388
ILSTS045	4.0	2.688	0.756	0.635	0.628	0.551
ILSTS034	5.0	2.846	0.886	0.658	0.649	0.585
ILSTS058	3.0	2.212	0.206	0.556	0.548	0.473
ILSTS056	6.0	3.622	0.500	0.736	0.724	0.676
ILSTS068	5.0	4.042	0.600	0.761	0.753	0.710
CSSM066	6.0	3.627	0.951	0.733	0.724	0.684
ILSTS036	6.0	3.997	1.000	0.760	0.750	0.708
ILSTS095	6.0	2.923	0.349	0.666	0.658	0.605
ILSTS029	3.0	1.578	0.342	0.371	0.366	0.315
ILSTS028	4.0	1.981	0.415	0.501	0.495	0.433
ILSTS025	4.0	2.475	0.700	0.604	0.596	0.513
ILSTS052	6.0	4.390	0.700	0.785	0.772	0.739
ILSTS031	3.0	1.412	0.026	0.295	0.292	0.272
ILSTS073	3.0	2.173	0.913	0.552	0.540	0.434
ILSTS060	3.0	2.922	0.733	0.665	0.658	0.584
BM1818	4.0	2.057	0.381	0.520	0.514	0.476
ILSTS061	6.0	3.504	0.667	0.723	0.715	0.665
ILSTS008	5.0	2.848	0.184	0.658	0.649	0.592
Mean	4.2	2.661	0.570	0.598	0.590	0.529

Heterozygosity Deficit and Hardy Weinberg Equilibrium

The average heterozygosity deficit across 25 microsatellite loci was 0.048 and was significantly positive ($P<0.05$). Fifteen out of the 25 loci investigated were found to have positive F_{IS} values. Although the average heterozygosity deficit was found to be significantly positive (0.048), it was found to be comparatively much lesser than that of other buffalo breeds (Mishra *et al.,* 2009; Kathiravan *et al.,* 2009). It is interesting to note the relatively lower F_{IS} values in Toda buffaloes despite the population being very limited and restricted to a narrow geographical area. Toda buffaloes are mostly found to be sired by the wild males while grazing, thus reducing the possible effect of inbreeding. The test for Hardy-Weinberg equilibrium revealed significant deviations in most of the loci except eight viz. ILSTS 026, HEl013, ILSTS 033, ILSTS 019, ILSTS 029, ILSTS 028, ILSTS 025 and ILSTS 061. The presence of null alleles, small sample size and possible relatedness of the samples (in the absence of parentage records) could be the possible reasons for the deviations from HWE.

Genetic Structure of Toda Buffalo

The genetic relationship between different Indian river buffalo breeds have been reported earlier by Kumar *et al.* (2006) and Vijh *et al.* (2008). The relationship among the maternal lineages of Indian river buffalo breeds and their domestication have been analyzed (Kumar *et al.,* 2007a,b). Toda buffaloes were found to be genetically distinct when compared to seven other river buffalo breeds of the Indian sub-continent when analyzed using microsatellite markers (Kumar *et al.*, 2006). However, the important South Kanara buffalo, the only well defined buffalo breed of South India other than Toda, was not included in these studies. Hence, a total of 3520 genotypes across 20 microsatellite loci were generated to assess the genetic structure of Toda buffalo in relation to three other buffalo breeds viz South Kanara, Marathwada and Murrah. Population differentiation was found to be lowest between South Kanara and Toda indicating close genetic relationship between these breeds. The estimates of Nei's genetic distance varied from 0.191 (South Kanara and Toda) to 0.361 (Marathwada and Murrah) among different breeds analyzed. The UPGMA phylogeny drawn using Nei's distance grouped South Kanara and Toda buffaloes at a single node while Marathwada and Murrah together formed a separate node. Further, the multidimensional scaling display of pair-wise F_{ST} values revealed the closeness of South Kanara and Toda buffaloes while Marathwada and Murrah were found to be distinct entities (Figure 7). The stress value was 0.0108.

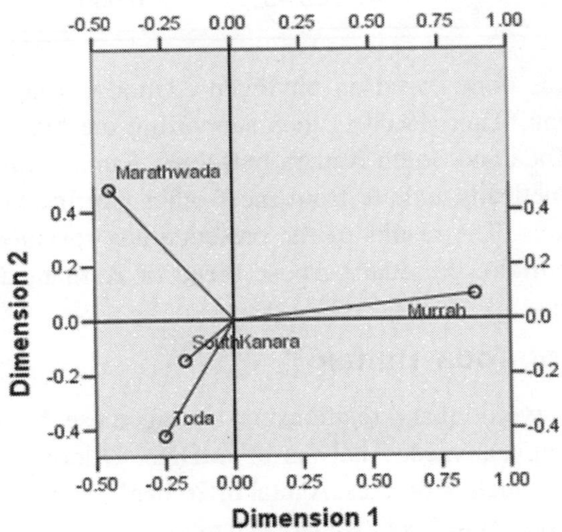

Fig. 7 : Multidimensional scaling plot of pairwise F_{ST} values between different buffalo breeds

Further, to detect the underlying genetic structure among the investigated breeds, analysis of molecular variance was performed. When no grouping was assumed among the breeds, 10.76% of the total variance was explained by between breed differences (Table 2). In order to test the uniqueness of Toda buffalo as reported earlier, AMOVA was performed by grouping Marathwada, South Kanara and Murrah together in one group and Toda in another group. The results revealed 12.57% variation among populations within group while negative variance across different groups. Negative variance components in AMOVA normally indicate that genes from different groups are more related to each other than genes from the same group.

Table 2. Global AMOVA with weighted average over different polymorphic loci

Source of variation	Microsatellite		mtDNA D-loop	
	% variation	P-value	% variation	P-value
No Groups				
Among populations	10.76	0.000	7.70	0.003
Within populations	89.24	0.000	93.30	0.000
Group I				
Among groups	-3.33	0.911	-9.98	0.995
Among populations within group	12.57	0.000	14.04	0.000
Within populations	90.76	0.000	95.94	0.002
Group II				
Among groups	5.44	0.0215	11.64	0.002
Among populations within group	6.13	0.000	-2.06	0.881
Within populations	88.43	0.000	90.41	0.003

When grouping was done based on phylogeny, among group variance was positive (5.44%) and highly significant (P<0.01) thus supporting the hypothesis of close genetic relationship between Toda and South Kanara buffaloes. Kumar *et al.* (2006) reported that Toda buffaloes are genetically unique from most other riverine breeds of buffalo based on microsatellite markers. The results of the present study revealed genetic similarity of Toda with Kanarese buffalo, the dual purpose breed of Karnataka state of South India.

Phylogeography of Toda Buffalo

In order to know the evolutionary relationship among these breeds, D-loop region of mitochondrial DNA was analyzed. Analysis of mtDNA D-loop region of 537bp length revealed 15 polymorphic nucleotide sites. A total of 26 haplotypic sequences were observed with 5, 4, 10 and 7 haplotypes in South Kanara, Marathwada, Murrah and Toda respectively. Among the observed haplotypes only 4 were found to be shared by two or more breeds. Among the five haplotypes of South Kanara buffalo one was found to be shared with Toda while another was shared with all the investigated breeds. The phylogenetic analysis of mt DNA haplotpyes following NJ algorithm with *Bos taurus* as out group revealed three distinct clusters. The major cluster included haplotypes from all the four investigated breeds while the other two minor clusters were formed by South Kanara and Toda haplotypes. Haplotype frequencies were further utilized to estimate pair-wise F_{ST} values which ranged from no differentiation (between South Kanara and Toda) to 0.299 (between

South Kanara and Marathwada). Analysis of molecular variance revealed 7.7% of total variation being explained by differences among breeds. When breeds were grouped, among group variance was negative when Toda was kept in a separate group and it was significantly positive (11.64%; $P<0.01$) when Toda and South Kanara were grouped together. AMOVA results based on mitochondrial DNA variation were broadly similar to the results of microsatellite data, thus pointing to the existence of strong evolutionary relationship among Toda and South Kanara buffaloes.

However, the sequential events that occurred during the evolution and origin of Toda buffalo in a small geographical area are not clear. We performed reduced median network analysis with a total of 46 haplotypes belonging to four investigated breeds while the Chinese swamp haplotype (DQ364186) was kept as outgroup. Most of the singleton haplotypes of Murrah and Marathwada were only few mutational steps away from the nodes formed by the most common haplotypes. However, few haplotypes of South Kanara and Toda were found to be quite distinct from the commonly found haplotypes indicating that these might have been ancestral to all the present day haplotypes. Interestingly, among these sequences, few mutations in two of the haplotypes of South Kanara buffalo were found to have contributed to ancestral haplotypes of Toda buffalo suggesting the possible migration of buffaloes from Kanarese region towards Nilgiris along the Western ghats.

CONCLUSION

Toda and South Kanara buffaloes are the only well defined breeds of buffalo in South India, although both are morphologically distinct from each other. The close genetic relationship between these breeds as revealed by microsatellite and mitochondrial DNA polymorphism is interesting. Although Nilgiri hills, the habitat of Toda buffalo is geographically isolated, it is part of Western ghats (long and continuous mountain range of peninsular India), which forms the eastern boundary of native breed tract of South Kanara buffaloes (Udupi, Mangalore and Shimoga). This contiguity of geographical terrain may give possible explanation for the basis of genetic relationship among these two buffalo breeds. The close genetic relationship between Toda and South Kanara buffaloes supports the view suggested by Grigg (1880) that this fine buffalo breed might be related to Chokatti buffalo (although no literature on this buffalo is available at present) of Mysore region in Karnataka. Interestingly, Griggs's suggestion had been quoted by Kamalapur (1958) to bolster the theory of Mysore origin of Toda tribe although circumstantial and linguistic evidence point to adjoining Wynad region as possible place of origin of Todas (Walker, 1986). In conclusion, there is a strong genetic and evolutionary relationship between Toda and South Kanara buffaloes. The results also support the theory of possible origin and migration of Todas and their buffaloes from Kanarese/old Carnatic region. However, such a proposition needs to be thoroughly corroborated with anthropological, linguistic and archaeological evidence.

REFERENCES

Botstein D, White R L, Skolnick M and Davis R W. 1980. Construction of a genetic linkage map in man using restriction fragment length polymorphisms. American Journal of Human Genetics, **32:** 314–331.

Chowdhary B P, Gustavsson I, Kunnevongkrit A, Lohachit C and Makinen A. 1989. Detailed mitotic description of the tandem fusion translocation differentiating river and swamp buffalo. Buffalo Journal, **1:** 41-49.

DiBerardino D and Iannuzzi L. 1981. Chromosome banding homologies in swamp and Murrah buffalo. Journal of Heredity, **72:** 183-188.

Finicio, Y. 1603. Letter to the Jesuit Vice-provincial in Calicut (in Portuguese) British Museum. Add. Ms. 9853. (Translated by A. de albert) Reprinted in : The Todas (Ed.)

Grigg HB 1880. A manual of the Nilgiri district in the Madras presidency. Madras: E. Keys, Government press.

Gunn WD 1909. Cattle of Southern India. Vol. III, Bulletin No. 60. Department of Agriculture, Madras.

Kamalapur JN 1958. Todas of the Nilgiri hills: Their origin. Quarterly Journal of the Mythic Society (Bangalore) **48:** 113-124.

Karthikeyan MK, Iyue M, Kandasamy N and Panneerselvam S 2002. Characteristics and performance of Toda buffaloes of the Nilgiris, India 1. Habitat, morphology and morphometry. Buffalo Journal 3, 303-313.

Kataria RS, Kathiravan P, Bulandi SS, Yadav NK, Dubey PK and Mishra BP 2009. Assessment of genetic diversity, mutation drift equilibrium and mitochondrial D-loop variation in Toda buffalo-the endangered breed of South India. Journal of Applied Animal Research **35:** 67-72.

Kathiravan P, Mishra BP, Kataria RS and Sadana DK 2009. Evaluation of genetic architecture and mutation drift equilibrium of Marathwada buffalo population in Central India. Livestock Science 121, 288-293.

Kumar S, Gupta J, Niraj K, Dikshit K, Navani N, Jain P and Nagarajan M 2006. Genetic variation and relationships among eight Indian riverine buffalo breeds. Molecular Ecology 15, 593-600.

Kumar S, Nagarajan M, Sandhu S, Kumar N, Behl V and Nishanth G. 2007a. Mitochondrial DNA analyses of Indian water buffalo support a distinct genetic origin of river and swamp buffalo. Animal Genetics **38:** 227-232.

Kumar S, Nagarajan M, Sandhu S, Kumar N and Behl V 2007b. Phylogeography and domestication of Indian river buffalo. BMC Evolutionary Biology **7:** 186-193.

Littlewood R.W. 1936. Livestock of Southern India. Government of Madras, India.

Mishra BP, Kataria RS, Kathiravan P, Bulandi SS, Singh KP and Sadana DK 2009. Evaluation of genetic variability and mutation drift equilibrium of Banni buffalo using multi locus microsatellite markers. Tropical Animal Health and Production **41:** 1203-1211.

Nair PG, Balakrishnan M and Yadav BR 1986. The Toda buffaloes of Niligiris. Buffalo Journal **2:** 167-178.

Rivers, W.H.R. 1906. The Todas. Macmillan and Co. Ltd., England. Reprint: (1986). Rawat Publications, Jaipur.

Selkoe KA and Toonen RJ 2006. Microsatellites for ecologists: a practical guide to using and evaluating microsatellite markers. Ecology Letters **9:** 615-629.

Vijh RK, Tantia MS, Mishra B and Bharani Kumar ST 2008. Genetic relationship and diversity analysis of Indian water buffalo (*Bubalus bubalis*) Journal of Animal Science **86:** 1495-1502.

Walker AR 1986. The Toda of South India. A New Look. Hindustan Publishing Corporation, Delhi. pp. 98-118.

Bioresources for Food Security and Rural Livelihood, Pages 367–372
Edited by : G.K. Kulkarni and B.N. Pandey

QUANTIFICATION OF HISTAMINE CONTENT IN TUNA (THUNNUS SPECIES) DURING PROCESSING

M. Pandian, Nilima Prabhu, Suman S. Pawar and P.D. Raut

ABSTRACT

Fresh Tuna virtually do not have histamine in their muscles however, after decomposition these fish contain substantial amount of histamine. Histamine is potentially hazardous and is the causative agent in fish poisoning. Quantification of histamine content is an imperative necessity during routine inplant monitoring / processing. Many laboratory techniques for isolation and quantification are in practice using different kits available in the market.

In the present investigation HPLC method is used, as it is officially recognized, standardized and widely followed for quantification of histamines.

The histamine content was maximum in the finished product and minimum in raw material. The quantity of histamine showed monthly variations. The histamine content varied from a minimum range of 12 – 27 ppm to a maximum range of 29 – 96 ppm from August 2001 : February 2002. Variations observed in the histamine content may probably be attributed to the variations in the counts of histamine producing bacteria and their ability of producing decarboxylating enzymes. In general microbial counts are highest in the month of September : the post monsoon period, probably due to rich terrigenic organic matter along with microbial biomass brought in by the rainwater. Temperature abuse is also found to be responsible for the increase in histamine producing bacteria.

Keywords : tuna, histamine, HPLC.

INTRODUCTION

Fresh Tuna virtually do not have histamine in their muscles. However, after decomposition these fish contain substantial amount of histamine. The production of histamine in fish can be fairly rapid. Histamine is potentially hazardous, and is the causative agent in fish poisoning (Arnold and Brown 1978). Histamine is produced by bacteria that decarboxylate histidine, a biogenic amine that is abundant in Scombroid fish muscles.

Quantification of histamine content is an imperative necessity during inplant monitoring/ processing. Many laboratory techniques for isolation and quantification of histamine in the sample are developed using chromatographic technique and rapid testing methods approved by AOAC (Association of Official Analytical Chemists,. 1999).

HPLC is one of the methods which is officially recognized, standardized and widely followed for quantification of histamine levels during processing, as described in the AOAC manual (Behling and Taylor,. 1982). Present investigation is a venture to carry out quantification of histamine using HPLC.

Sample Collection & Histamine Quantification

Fish samples belonging to Thunna sp were collected during five different steps of processing viz. raw material, preprocessing, processing and frozen storage. It was carried out by High Performance Liquid Chromatography using the method of Yen and Hsieh, (1991). Histamine dihydrochloride (165.7 mg) standard solution from sigma lab dissolved in 10 ml of deionsed water. From the stock, a concentration of 100 ppm was prepared by diluting 100 μl stock solution in deionized water to an overall final volume of 10 ml. Histamine was extracted using standard method as shown below :

20 gms fish muscle + 10 ml of 0.4 M Perchloric acid Grind in blender for 3 minutes

↓

Add 200 μl of Amine standard solution

↓

Centrifuge homogenate at 12,000 rpm for 10 minutes at 4^0 c

↓

Collect supernatant in 25 ml volumetric flask

↓

Residue re-extracted with equal amount of 0.4 M Perchloric acid by centrifuging

↓

Supernatant collected, both supernatants combined and final volume adjusted to 25 ml with 0.4 M Perchloric acid

The benzoyl derivative of histamine was prepared according to standard method of Redmond and Tseng (1979). Analysis done using Hitachi make HPLC with C18 reverse phase column, consisting of a model of L – 4200 UV-VIS detector set at 254 nm. Gradient isocratic elution systems were used. For the isocratic systems the mobile phase was methanol-water (55 : 45 v/v) at 1.5 ml / min. at room temperature the gradient elution programme was set at 1.1 ml / min. starting with methanol-water mix (55 : 45, v/v) for 2.5 min. The sample volume injected was 20 ìl and the column elution was monitored at 254 nm.

The histamine content present in Tuna samples at different stages of processing were quantified and are shown in Table 1, Table 2 and Table 3.

Table 1 : Histamine content present in the tuna fish at different steps of processing in different months of sampling.

month/year	Detail of samples	No. of samples tested	Histamine (ppm)	
			min.	max.
Aug – 01	Raw material receiving	4	12	33
	Pre-processing	4	17	45
	Processing	4	23	69
	Freezing	4	21	94
	Frozen storage	4	29	104
Sep – 01	Raw material receiving	4	18	46
	Pre-processing	4	23	55
	Processing	4	25	64
	Freezing	4	24	62
	Frozen storage	4	33	111
Oct – 01	Raw material receiving	4	15	31
	Pre-processing	4	17	47
	Processing	4	19	58
	Freezing	4	21	79
	Frozen storage	4	29	96
Nov – 01	Raw material receiving	4	10	41
	Pre-processing	4	14	53
	Processing	4	22	66
	Freezing	4	25	81
	Frozen storage	4	27	95
Dec – 01	Raw material receiving	4	13	23
	Pre-processing	4	18	34
	Processing	4	24	71
	Freezing	4	28	67
	Frozen storage	4	22	108
Jan – 02	Raw material receiving	3	15	26
	Pre-processing	3	19	61
	Processing	3	20	64
	Freezing	3	26	72
	Frozen storage	3	31	89
Feb – 02	Raw material receiving	3	17	37
	Pre-processing	3	17	47
	Processing	3	23	56
	Freezing	3	27	76
	Frozen storage	3	32	106
Mar – 02	Raw material receiving	2	12	29
	Pre-processing	2	16	46
	Processing	2	19	55
	Freezing	2	24	86
	Frozen storage	2	27	96

Table 2 : Increase of Histamine content present in the tuna fish from raw material to finish product.

Month of sampling	Temp.	Histamine content (ppm)		
		Raw material	Finished product	Fold of increase
Aug – 01	Min.	12	29	2.4
	Max.	33	104	3.1
Sep – 01	Min.	18	33	1.8
	Max.	46	111	2.4
Oct – 01	Min.	15	29	1.9
	Max.	31	96	3.1
Nov – 01	Min.	10	27	2.7
	Max.	41	95	2.3
Dec – 01	Min.	13	22	1.7
	Max.	23	108	4.7
Jan – 02	Min.	15	31	2.1
	Max.	26	89	3.4
Feb – 02	Min.	17	32	1.9
	Max.	37	106	2.9
Mar – 02	Min.	12	27	2.3
	Max.	29	96	3.3

Table 3 : ANOVA of HPB and Histamine content as influenced by histamine production and month of sampling.

Fish	Stages of processing	Variable	F – value	Significance at 1% & 5% levels
Tuna	Raw material receiving	Histamine content	53.55	1%
		Month of sampling	1.42	NS
	Pre – Processing	Histamine content	126.23	1%
		Month of sampling	1.36	NS
	Processing	Histamine content	534.86	1%
		Month of sampling	2.24	NS
	Freezing	Histamine content	155.1	1%
		Month of sampling	0.57	NS
	Frozen storage	Histamine content	589.34	1%
		Month of sampling	1.02	NS

NS – not significant

During processing there is an increase in the histamine content and it is more in the finished product. The increase of histamine content is minimum by 2.7 and maximum by 4.7 times during November and December 2001 respectively. The histamine is produced by pathogenic enterobacteria and saprophytes. Both of them grow in fish during varied temperatures and produce the enzyme histidine decarboxylase resulting in the formation of histamine. Therefore the level of histamine increases in the fish during process at different temperature (Olley and Baranowski,. 1985).

The variation in the histamine content is also seen with different temperature abuse. According to Olley and Baranowski (1985) the temperature for enzyme activity in microorganisms that grow at warm temperature is important in histamine formation provided sufficient bacterial numbers have been reached before cold storage. Lehane and Olley (2000) have also strongly suggested that the temperature abuse is responsible for the histamine accumulation in thawed frozen fish.

CONCLUSIONS

The histamine content exhibits monthly variation. This may be attributed to the variation in the ability of producing decarboxylating enzymes (Olley and Baranowski,. 1985). In general, the microbial counts are the highest during the month of September, which coincides with the post monsoon period in the region studied. It may be attributed to the rich terrigenic organic matter along with the microbial biomass brought in by the rainwater.

REFERENCES

Arnold S. H. and Brown W. D., (1978) Histamine toxicity from fish products. Adv. Food Res. **24:** 113-154.

Behling A. R. and Taylor S. L., (1982) Bacterial histamine production as a function of temperature and time of incubation. J.Food Sci. **47:** 1311-1317.

Lehane L. and Olley J. (2000) Histamine fish poisoning revisited. Int. J. Food Microbiol. **58:** 1-37.

Olley J. and Baranowski J. V., (1985) Temperature effects on histamine formation. In histamine marine products: production by bacteria measurement and prediction of formation. FAO Fisheries technical paper No. 252. Food and Agriculture Organisation of the United Nations, Rome. Pp. 14-17.

Redmond J. W. and Tseng A. (1979) High pressure liquid chromatographic of putrescine, cadaverine, spermidine and spermine. J. Chromatogr . 170-479.

Yen G. C. and Hsieh C. L., (1991) Simultaneous analysis of biogenic amines in canned fish by HPLC. J. Food Sci. **56(1):** 158-160.

Bioresources for Food Security and Rural Livelihood, Pages 373–382
Edited by : G.K. Kulkarni and B.N. Pandey

SEASONAL VARIATION IN BIO-CHEMICAL COMPOSITION OF FRESHWATER CRAB: *SARTORIANA SPINIGERA* (WOOD MASON)

R. Tahseen and A. Moitra

INTRODUCTION

Study of the seasonal variation in the bio-chemical composition of an economically important shell fish viz. crab is very much interesting because of its specificity in relation to food values, and for evaluating their physiological needs at different periods of life. The nutrients increase considerably for metabolic needs during premonsoon months for the ensuing spawning season.

A number of workers have obtained varying results, particularly in respect of protein content in different crabs, Govindappa and Rajabai (1976), Sinha and Ahmed (1978), De-Weichen *et.al.* (2006). Similarly significant variations in the bio-chemical composition of different tissues of crab during pre-spawning and post-spawning periods have been observed with special reference to protein, glycogen, cholesterol, fat and water content (Am bore and Venkatachari, 1979; Heath and Barnes,] 970; Sinha and Ahmed, 1978; George and Shah, 1954; George and Patel, 1954-55).

Environmental conditions playa very important role in the physiology of animal (Hoar, 1959). For successful survival it is necessary that crabs must adapt itself properly to changing environment.

Hepatopancreas is usually considered to be the main storage organ in decapods crustaceans. Information relating to cyclic and seasonal changes in growth, moulting cyclic, oxygen consumption and gonadal cycle in crabs are available. But only few reports (Heath and Barnes, 1970; Rosa and Nunes, 2003) are .available on seasonal studies on biochemical composition of crab. It is known that triacylglycerides are synthesized in hepatopancreas and accumulation of labelled acetate in the triacyclglycerol fraction of the hepatopancreas are greater in female than that of male crab (Kallapur, *et. all.*, 1984). A striking relationship between water, protein and lipid content in hepatopancreas (Xugan

Wu *et.al.,* 2007) and relationship between lipid level in ovary and gonadosomatic index (Rosa and Nunes, 2003) of crab have been observed. An increase in the proportion of one leads to the decrease in the content of the other so that their sum is approximately constant.

In addition to the seasonal environmental conditions, certain, other factors such as sex and size (Ambore and Venkatachari, 1973; Sinha and Ahmed, 1978), acclimation (Govindappa and Rajabai, 1976) salinity (Venkatachari & Vasantha, 1973) pond and lake adaptation (Xygan Wu, 2007) causes variation in the biochemical composition of the crab body. Heath and Barnes (1970) have tried to establish the relationship between seasonal biochemical variation and moulting cycle in crab- *Carcinus maenas.*

The water temperature is one of the most important aboitic factor which influence the biochemical composition of crab. Due to change of season, the water temperature also changes and as such they also influence these parameters.

Reports about the variation in protein content in different season in crab is scanty. Ambore and Venkatachari (1973) studied the biochemical parameters of the freshwater crab *Paratelphusa guerini.* They *found* that the amount of protein varies in individual tissues in relation to sex and size, calling *forth* considerable variation in the structural and functional efficiency of the tissues concerned. Govidappa and Rajabai (1976) has investigated some aspect of protein metabolism in the tissue of field crab - *Paratelphusa hydrodromus,* during cold acclimation.

The present work is aimed to study the seasonal variation of biochemical composition of the hepatopancreas in the freshwater crab. *Sartoriana spinigera.* An attempt have been made to study the dynamics of seasonal variation in moisture, lipid, protein, sugar and ash content in the hepatopancreas of *Sartoriana spinigera.*

The intermoult (i.e. animal in normal condition) crabs were brought from river bank of adjoining areas of Bhagalpur to laboratory and acclimatized in large aquarium. The volume of water was adjusted so that the animals remained submerged. They were fed with phytopankton and fish flesh. The water was changed daily. The weight and sex of the crabs were noted prior to anaesthetise the animal. The hepatopancreas mass was taken out in a preweighed petridish and then weight again. *For* different biochemical estimation the following method were employed:

- Protein content was estimated following Lowry *et.al.* (1951).
- Total sugar was estimated using the phenol sulphuric method following Dubois *et.al.* (1986).
- The fat was extracted by soxhlation method using petroleum ether as solvent.
- Moisture was calculated by drying the sample in oven at 100° C till the constant weight achieved.
- Ash content was estimated following Winberg (1971) method.

Protein content, lipid content, total sugar content, moisture content and Ash content of freshwater crab - *Sartoriana spinigera* in hepatopancreas are depicted in Table 1.

Table 1: Seasonal Variation in Bio-chemical Composition of *Sartoriana spinigera*

Parameter	Sex	Body Weight	Season		
			Winter	Summer	Rainy
Protein	Male	58-70	3.97±1.54	8.15±2.82	8.31±2.93
	Female	60-68	3.64±2.13	6.97±3.34	8.62±2.55
Sugar	Male	60-70	1.99±0.57	3.61±2.00	6.29±2.09
	Female	60-70	1.57±0.36	3.19±1.74	6.73±0.61
Lipid	Male	62-70	8.40±1.24	9.77±1.3 7	10.35±1.38
	Female	65-70	8.27±1.22	8.55±1.00	10.75±1.32
Moisture	Male	60-68	68.23±1.79	62.91±3.27	65.03±2.80
	Female	58-65	67.11±2.38	60.84±2.73	64.16±2.21
Ash	Male	60-68	19.77±1.86	23.99±2.30	15.02±1.23
	Female	65-70	18.83±1.63	19.94±2.06	14.39±1.02

Amount of each constituent expressed in mg/l00mg of tissues

The protein content in the hepatopancreas varied between 3.97% to 8.31 % in male and 3.64% to 8.62% in female. The highest record of hepatopancreas protein was recorded in hepatopancreas 8.31% in male 8.62% in female during rainy season followed by 8.15% in male and 6.97% in female during and summer months and 3.97% in male and 3.64% in female in winter season.

The present lipid content in the hepatopancreas of *Sartoriana spinigera* was found to be between 10.35% to 8.40% in male and 10.75% to 8.27% in female. The lipid content also records maximum in rainy season, medium in summer season and least in winter season.

In *Sartoriana spinigera,* the total sugar exhibited highest percentage in different season i.e. 6.29% in male, 6.73% in female followed by 3.61 % in male and 3.19% in female followed by 1.99% in male and 1.57% in female during rainy, summer and winter season respectively.

The percentage moisture content in the hepatopancreas of crab- *Sartoriana spinigera* varied from 62.91 % to 68.23% in male and 60.16% to 67.11 % in female over the period of observation. The highest in winter (68.23% in male and 67.11 % in female) followed by rainy (65.03% in male and 64.16% in female) and lowest in summer season (62.91% in male and 60.84% in female). The Ash content in crab - *Sartoriana spinigera* seem to vary from 15.02% to 23.99% in male and 14.39% to 18.83% in female throughout different season.

The seasonal dynamics of the bio-chemical composition is determined to a great extent by seasonal changes in water temperature and to some extent environmental condition. Present study revealed that the hepatopancreatic protein in freshwater crab- *Sartoriana spinigera* varied from 3.97% to 8.31 % in male, 3.64% to 8.62% in female. It was reported that the large amount of protein found in *Cancer pagarus* 6.22% fresh weight (Renaud, 1949) and almost the high percentage (8.21%) observed in river crab *Sesarma boulgengeri* (Sinha and Ahmed, 1978) indicates that the hepatopancreas is not the seat of carbohydrate and fat metabolism but also protein metabolism. The result obtained from the present observations (Table 1) have clearly established that there is a gradual rise in the protein content from March to November when the crab approaches sexual maturity. The highest percentage of protein in peak summer month. May is marked with peak period of maturation and showing depletion in this content at the end of spawning. Percentage of hepatopancreatic protein content in freshwater crab - *Sartoriana spinigera* varied according to season and maturity of animal. The highest protein content was recorded in hepatopancreas 8.68% followed by 6.55% and 3.55% in different season. Storage of protein in hepatopancreas increases in both sexes, after winter hibernation. Highest value of protein in the month of April and May in both sexes are indicative of gonadal maturation period. In pre-breeding season (June and July) percentage of protein in female decreases in hepatopancreas and it may be utilize in ova formation. In rainy season animal feed voraciously as confirmed by their Gastro somatic index and the percentage of abundance in gorged stomach. It seems that this may be due to the compensation of energy lost in breeding season and also stored the material before next-hibernation. Abundance of food material in environment causes increased feeding activity in rainy season and also protein stored up in hepatopancreas before the onset of winter.

Govindappa and Rajabai (1976) has reported the protein nitrogen and non-protein - nitrogen content in hepatopancreas of field crab - *Paratelphusa hydrodromus.* They reported the rise in protein nitrogen was very low in hepatopancreas during cold acclimation, suggesting a low rate of turnover of the nitrogen metabolism in this tissue. These observations clearly suggested that the tissues were actively involved in the synthesis of protein with a simultaneous suppression of this degradation, thus the tissues of cold adapted crab were involved in stepping up the anabolic phase and suppressing the catabolic phase of protein metabolism. This observation support the lower percentage of protein storage during winter season.

A comparative study shows that hepatopancreatic and muscle protein are more concentrated in males than in females except for the muscle protein in the large size females (Orton, 1936, Ramamurthi, 1966).

Protein is found to vary inversely with the moisture and directly with the fat content. The high protein values in the hepatopancreas of *Sartori ana spinigera* during pre-spawning period, may be due to the fundamental nitrogen demands for the maturation energy requirements of the crab. The low value of protein observed in the post-spawning period (i.e. August). Health and Barnes (1970) observed the fluctuations in proteins nitrogen values in shore crab-*Carcinus maenas* shows higher during August to November, which almost is similar to the present observations in *Sartoriana spinigera.*

Seasonal variation in the sugar content was observed in *Sartoriana spinigera.* It was minimum from 1.99% in male and 1.57% female and moderate during the rest period of the year except July when the sugar content was low. Sugar content in hepatopancreatic mass is almost same in both the sexes of *Sartoriana spinigera.*

Heath and Barnes (1970) also observed little variation in glycogen content varied in shore crab - *Carcinas maenas* 1.27% to 7.12% in male where as 1.65% to 8.76% in female. This value is very close to value of *Sartoriana Spinigera* in which male had 1.99% to 6.29% where as in female it is varied between 1.57% to 6.73%. In *Sartoriana spinigera* the total sugar exhibited highest percentage in rainy season in both sexes, 6.29% in male and 6.73% in female followed by 3.61% in male and 3.19% in female in summer season. It is very low in winter season.

Glycogen content increases in rainy season as varieties of plankton are available in the surroundings. During winter season storage of glycogen in hepatopancreas decreases because animal remains in hibernating condition. In pre-breeding season glycogen content increases and gastrosomatic index value observed maximum in this season. It is also observed that the sugar content in female crab is comparatively higher than male. This observation corroborates the finding of Dean and Vernberg (1965), who observed a rise in the glucose level in *Callinectes* just before spawning, but decreased in animals carrying the egg mass.

Raymout and Krishnaswamy (1960) observed in marine planktonic crustaceans a high carbohydrate content of 20% are found in mixed planktonic population.

In comparision to other crustacean the carbohydrate content are maximum in fairy shrimps *Streptocephalus dichotomus* (7.43%) and *Branchinella kugenumaensis* (8.93%) reported by Bernice (1972), whereas in Lobster *Panulirus polyphagus* it is 3.39% and in Prawn *Palaemon carcinus* 4.16% (George & Shah, 1954).

Animal utilize glycogen for the synthesis of chitin which is essential for the formation of exoskeleton and the production of chitinous membrane. Vorne (1924) examined the storage of glycogen and its conversion into chitin in both Carcinus maenas and Cancer pagurus.

The lipid content in different species of crab varies within wide limits. Slim *et.al.* (2007) reported in green crab *Carcinus mediterraneus* total saturated fatty acicts was higher in hepatopancreas (25.15% to 26.24% of total FAS). Mehmet, *et.al.* (2003) observed the percentage of total saturated fatty acids in Blue crab *(Callinecites sapidus)* was higher in the hepatopancreas than in the claw or breast meats. The total no fatty acids were 8.61 % in hepatopancreas. Rosa and Nunes (2003) has observed in *Aristeus antematus* (Crustacea, decapoda) the higher proportion of neutral lipids, mainly triacylglycrols was found in both ovary and hepatopancreas. Eric *et.al.* (2000) Studies in

Meganyctiphanes norvegica reveal that it has triacylglycerols (TAG) as the major neutral lipid class with high levels in the digestive gland or hepatopancreas.

In *Streptocebhalus dichotomus* and in *Branchinella Kugenumaensis* the lipid content are 11.01 % and 17.14% respectively (Bernice, 1970).

Studies on cyclic changes in lipid content of *Sartoriana spinigera* reveal that it has limited quantity of lipid content in its hepatopancreas. In this crab lipid content was recorded minimum in winter season (8.40% in male and 8.27% in female) and maximum in rainy season (10.35% in male and 10.75 in female).

Fat values are found to be low during the spawning period and high during the pre and post spawning periods. This indicates that the fat reserve is more in fully mature crab, which has to go the spawning exhaustion. This high values of fat coincide with the low values of water content and *vice versa* (June to August).

From the present investigation, it can be mentioned that the period from September to November might be regarded as the fat storage period in case of *Sartoriana spinigera.* Fat value are found to be low during the spawning period (July to August) and high during the pre and post spawning period (May and October).

The conclusion drawn from the table available is that the stored fat is utilized for development of egg only in female crab but as reserve material in both the sexes. Heath and Barnes (1970) observed in *Carcinus* the storage of fat content in hepatopancreas vary little in the month and April and June; at this time animal have already under gone their first moult after winter.

The period extending from March to May is an active period for crab. During this period the fat content of the hepatopancreas varies. It is 9.95% male and 8.98% in female in March and 9.28% in male and 7.69% in female in June. As a matter of fact the fat content in the hepatopancreas of male crab is slightly reduced. While in the female, fat content in hepatopancreas is greatly reduced (8.26% in summer). This observation is supported by many authors. Rosa and Nunes (2003) in *Aristeus antennatus* (Decapoda) because stored fatsare utilized for egg formation in female but in male the gonad does not undergo marked fluctuation from month to month.

In *Carcinus meanas* phospholipid content varies from 1.00 to 1.87%, during entire season (Heath and Barnes, 1970). While lipid content varies in *Sartoriana spinigera* ranged from 8.40% to 10.35% in male and 8.27% to 10.75% in female. To overcome, the period of starvation the crab *(Paratelphusa)* accumulate maximum quantity of fat in their body before the onset of winter which are gradually utilized as a source of energy through out the entire period of hibernation (November to February). During summer months crabs remains very active. Most of their lipid becomes oxidized to provide extra energy to the crab required for the purpose of reproduction. Rosa and Nunes (2003) observed in *Aristeus anennatus* the gonadosomati index increased significantly in June and July and during the maturation process. Hepatosomatic index also increased throughout the ovarian maturation. They found a positive correlation between lipid levels in the ovary and gonado somatic index and hepatopancreatic index. In crustaceans, the hepatopancreas is the principal energy supplier, and a large part of this energy is stored as lipid (Chapell, 1977; Mourents and Rodriguez, 1991; Albesserd *et.al.* 2001; Ravid *et. al.,* 1999). The main functions of these stored lipids are reserves for vitellogenesis and moulting (Dall, 1981).

From the present investigation it is revealed that moisture percent in the hepatopancreas of *Sartoriana spinigera* ranges between 62.91 % to 68.23% in male and 60.84% to 67.11% in female from different season (Table 1). This values are very close to the finding of Ambore and Venkatachari (1979) in *Barytelphula spinigera* (61.2% to 77.17% in male and 56.31% to 85.0% in female) and observation, by Heath and Barnes (1970) in *Carcinus maenas* it is 64.6% to 79.00% in male, 63.9% to 74.00% in female.

Previous work also reveal that the water content varies within limited range in different species of crustaceans. In fairy shrimps *Streptocephalus dichotomus,* and *Branchinella kugenumaensis,* the water content value is very high, 90.1% to 91.20% respectively. In Scorpio *Carpaurus fuscus* the percentage of water content of the hepatopancreas ranged between 47.8% and 68.4% and varied seasonally as well as between the genders. In *Caenestheriella* it is 70.2% (Michael and Chandran, 1967). Ambore and Venkatachari (1973)

has shown the water content differs in different tissues and it decreases in order of gill, muscle, heart and hepatopancreas in respective of sex and age. A view of the present study in *Sartoriana spinigera* reflects that this crab undergoes a regular cycle of hydration and dehydration.

The water content of *Munida subrugosa* present two maxima, in August and December. These changes can be attributed to the moulting (Nunes, 2003). Tapella (2002 b) showed that M. subrogosa > 10 mm CL has two periods of moult duting the year in spring and summer. At moulting, decapods absorb a great quantity of water, which is later replaced by the organic matter that constitutes tissues (Vemet and Charmantier-Daures, 1994). Nevertheless, the hepatopancreatic energetic increase during February in probably associated with the energy gained for summer moulting and ovarian growth. Normally the water content of the living tissues decrease throughout life and the most marked variations of water content takes place in the early stages of embryonic development (Bertolini, 1969). In most mammals, the tissue water content decrease with age (Bertolini, 1969).

As regards ash present in this animal it was minimum in rainy season and maximum in summer season and it is inversely proportion to water content. The value of Ash content in hepatopancreas ranges between 15.02% to 23.99% in male while 14.39% to 19.94% in female. George and Shah (1954) reported ash contents in *Panulirus polyphagus* 1.09% and in *Palaemon carcinus* it is 0.75%.

From the present findings, it is very much clear that there is seasonal variation in different bio-chemical composition in freshwater crab- *Sartoriana splmgera.* The variation in bio-chemical composition in hepatopancreatic tissues in crab are affected *not* only by different season, it is also related with moulting cycle. In summer months the crabs are characterised by higher protein, lipid and ash content but lower values of water content whereas in winter months the water content are higher and other bio-chemical composition are lower.

As recorded the gastro somatic index value in pre-breeding season (April to May) the full stomach were maximum and this results to storage of protein, lipid and glycogen. It might be due to the reason that the animal in this season require *more* nutrients for the proliferation of gonads. Similar observation in the pre-breeding season in the case of fishes were also recorded by Thakur (1992). In rainy season, these composition (Protein, Lipid, Glycogen) increases due to avaibility of food in environment and also observed the comparatively high percentage of occurrence of food filled stomach. It seems that this may be due to the compensation of energy lost in the breeding season as well as due to recovery of gonads (Thakur, 1992).

It is possible that large amount of lipids and proteins are also used *for* the development of egg, Lipids and proteins are both used for the production of energy and the development of eggs, both the constituents should show paralled trends in quantitative increase or decrease (Barnes *et. al.,* 1963).

Variation in composition in the crab organ itself which is likely to change depending on the physiological state as well as availability of sufficient quantity of food in the habitat.

REFERENCES

Albessard, E.; Mayzaud, P. and Roudy, J.C. (2001) : Variation of lipid classes among organs of the Northern Krill *Meganyctiphanes norveglca* with respect to reproduction. Compo Biochem. Physiol., **129 A:** 373-390.

Ambore, N.E. and Venkatachari, S.A.T. (1979) : Age and sex related studies on the chemical architecture of the tissue of the freshwater crab, *Barytelphusa guerini,* 1. Water Content, 1 Anim. Morphol. Physiol., Vol. **26.** Nos. 1 and 2, pp. 126 to 131.

Barnes, H., M. Barnes and D.M. Filayson, (1963 a) : The seasonal changes in the body weight, biochemical composition, and oxygen uptake of two common borea-arctic cirripedes, *Balanus balanoides* (L.) and B. *balanus* (L.) da costa. J. mar. biol. Ass. U.K., Vol. **43,** pp. 185-211.

Bertolini, A.M. (1969) : Gerontologic metabolism. Charles C. Thomas. U.S.A. Bernice R. (1972) : Biochemical composition of *streptocephalus dichotomus* Baired Branchinella kugenumaeasis (Ishikawa) Hydrobiologia, Vol. **39, 2**, pp. 155-164.

Celik M.; Tureli, C., Celik, M.; Yanar, Y.; Erdem, U.; and Kucukgulmez, A. *et.al.* (2004) : Fatty acid composition of the blue crab *(Callinectes Sapidus* Rathbun, 1896) in the north eastern Mediterranean. Food Chemistry, Volume 88, Issue 2, November 2004, pp. 271-273.

Chapelle, S. (1977) : Lipid composition of tissues of marine crustaceans, Biochem. Syst. Ecol. **5:** 241-249.

Cherif, S.; Frikha P.; Gargouri. Y.; and Miled, N. (2008) : Fatty acid composition of green crab *(Carcinus mediterraneus)* from the Tunisian mediterranean coasts, Food chemistry Volume III, Issue- 4, 15 December 2008, pp. 930-933.

Dall, W. (1981) : Lipid absorption in the Norwegian Lobster, *Nephrops norvegicus* (L.) J. Exp. Mar. Biol. Ecol., **50:** 33-45.

Dean, J.M. and F.J. Vernberg (1965): Variation in the blood glucose level of crustacea (Callinectes, Hepatus, Libinia, Parapeus, Uca). Compo Biochem. Physiol., Vol. 14, pp. 29-34. Dubois, M., K. Gilles, J.K. Hamilton, P.A. Robers and F. Smith (1986) : Colorimetric method for determination of sugar and related substances. Anal. Chern., **28:** 350-356.

George, J.C. and Patel, B.S. (1954-55) : The seasonal variation in the fat content of the liver and gonads in a marine and freshwater Decapod.

George, J.C.. and Shah V.C. (1954): A comparative study of the chemical composition of the muscle of the lobster *Punulirus polyphagus* (Herb st) and the prawn *Palaemon cacinus* (Henderson and Mathai).

Govindappa, S. and Rajabai, B.S. (1976) : Some aspects of Protein metabolism in crab, *Paratelphitsa hydrodromus* (Herbst) during the acclimation, J. Anim. Morphol. Physiol., Vol. 23, No. 1, pp. 76 to 84.

Heath, J.R and Barnes H. (1970) : Some changes in biochemical composition with season and during the moulting cycle of the common shore crab, *Carcinus maenas* (L.), J. expo mar. Biol. Ecol., 1970, Vo1. 5, pp. 199-233; North-Holland Publishing Company, Amsterdam.

Hoar, W.S.H., D.H. Randall and J.R Brett (1979): Fish Physiology. Vol. III. Bioenergetics and growth. Academic Press, INC. London Ltd.

Kallapur, V.L.; Ramamohanrao, Y.; Nasasubhai, A.V. (1984) : Triacylglycerol synthesis in the premolt field crab *Paratelphusa hydrodromus* Milne Edwards (Crustacea), Arch Int. Physiol. Biochem. Aug., **92 (2):** 119-24 Links.

Lawry, O.H.N.J. Rossenbrugh, A.L. Fan and RJ. Randol (1951): Protein measured with the Folin ciocaltece reagent J. BioI, Chern., **193:** 265-275.

Michael, RG. and Chandran, M.R (1967): On the biochemical composition of the conchostracan, caenestheriella Biol. Zentrabl., **86:** 495-499.

Mourente, G. and A. Rodriquez (1991) : Variation in the lipid content of wild- caught females of the marine shrimps *Penaeus kerahturus* during sexual maturation. Mar. BioI. **110:** 21-28.

Orton, lH. (1936): Experiments in sea on the rate of growth of some crustacean decapods. l Mar. BioI. Assoc. U.K., **20:** 673-689.

Ramamurthi, R (1966) : Succinic dehydrogenase activity in a freshwater crab in relation to salinity stress. Compo Biochem. Physiol. **19:** 645-648.

Ravid, T.; Tietz, A.; Khayat, M.; Boehm, E.; Michells, Rand Lubzens, E. (1999): Lipid accumulation in the ovaries of a marine shrimp *Penaeus semisulcatus* (De Haen). J. Exp. BioI. **202:** 1819-1829.

Renaud, L: (1949): Le cycle des reserve urganiques chezles crustace's decapodes. Annls. Inst. Ocenagr. Monaco, N.S., T. 24, pp. 259-287.

Rosa R, Nunes, M.L., (2003) : Tissue biochemical composition in relation to the reproductive cycle of deep sea decapod *Aristeus antennatus* in the Portuguse South Coast; Journal of the Marine Biological Association of the UK (2003), **83 :** 5: 963-970. Cambridge University Press.

Sasamorajski, T. and Roslester, E. (1973) Neuribiological aspects of maturation and ageing. ed. D.H. Ford. Elsevier, Amsterodam. 253.

Sinha, RC and Ahmed, H.K. (1978) : Some changes in Biochemical composition with sex and size of the crab, *Sesarma boulengeri* Claman, Hydrobiogia Vol. **61, 1,** pp. 15-19.

Tapelley, F., Lovrich, G.A.; Romero, M. C. and Thatije (2002 b) : Reproductive biology of the crab *Munida subrogosa* (Decapoda : Anomura : Galatheidae) in Beagle Channel, Argentina. J. Mar. BioI. Ass. U.K. **82 :** 589-595.

Thakur, S.S. (1992) : Ecological study of a freshwater teleost fish *Channa punctatus* (Bloch.) in respect of its introductive activity and general study of water pollutants and their effects on reproduction. Ph.D. Thesis, D.H.G. Vishwadhalaya, Sagar, M.P. 363 pp.

Vankatachari, S.A.T. and Ambore, N.E. (1973): Effect of sex and size on the protein composition in the tissues of the freshwater crab, *Paratelphusa* (Berytelphusa) guerini Hydrobiologia, **43:** 157-169.

Venkatachari, S.A.T. and Ambore, N.E., (1973) : Effect of sex and size on the protein composition in the tissue of the freshwater crab, *Paratelphusa* (Barytelphusa) *guerini* Milne Edwards, Hydrobiologia, Vol. **43,** 1-2 pp. 157-169.

Verne, J. (1924) : Note histochemique sur l.e. metabolism du glycogene, pendant la mue chez les crustaces. C.r. Seane. Soc. Biol., T. 90. pp. 186-188.

Vernet, G. and Charmantier - Daures, M. (1994): Mue autotomic et regeneration. in. J. Forest, editor. Tralte de Zoo logie, Crustaces. 7. pp. l07-194.

Winberg, G.E. (1971) : Methods of the estimation of production of Aquatic Animals. 179. pp. London and New York; Academic Press.

Xugan Wu; Cheng, Y.; Sui, L.; Yang, X.; Nan, T. and Wang, J.; (2007): Biochemical composition of pond-reared and lake stocked Chinese mitten crab *Eriocheir Sinensis* (H. Milne-Edwards) broodstock Aquaculture Research Vol. 38, Issue 14, pp. 1459-1467. (Published Online).

Bioresources for Food Security and Rural Livelihood, Pages 383–394
Edited by : G.K. Kulkarni and B.N. Pandey

DIETARY FIBERS: IN DAIRY PRODUCTS

D.M. Choudhari and J.N. Khedkar

ABSTRACT

Dietary fiber is the group of food components that pass through the stomach and small intestine in undigested state and reach the large intestine virtually unchanged. Most of other nutrients are digested and are being used in other parts of the body by this stage. During its passage through the large intestine some components of dietary fiber are broken down by varying degrees and absorbed by the body. The remaining components are excreted in the fasces; however, fiber does play a very important role in human health. All plant based foods that we eat are full of fiber. Whole foods containing two essential forms of fiber viz., soluble and insoluble fiber. Both types of fiber are needed to maintain a healthy intestinal tract and promote healthy and regular bowl movements. Most of the people do not eat the dietary fiber as a major food component. But it available in daily diet in one or other source through food gradients i.e. cereals, fruits, vegetables and nuts etc. It is tradition of vegetarian people to incorporate the cereals, fruits, vegetables etc., in the milk and/ or milk products. This leads to the easy consumption and beneficial to the human health. Incorporation of plant origin materials in milk or milk products directly or indirectly adds dietary fiber in human food. So far the research on incorporation of plant origin material focused on value addition, taste and flavour development, thickening agents etc. This aspect may be viewed and think from the point of its significance in nutrition and therapeutic value. The Literature reviewed on milk and milk products prepare and are the incorporation of plant original materials replete with dietary fiber which is not thought directly as a source of dietary fiber. Therefore, the consumption of milk and milk products prepared by incorporation of plant origin edible materials is not the out of the place of advise of consumption of fiber containing foods viz., fruits, vegetable, legumes cereals etc. Because, nutritional and therapeutic benefits by the consumption of milk in isolation can be accelerated by the incorporation of plant original material which is not thought of. There is need to educate the consumers with the aim to consume the milk and milk products containing plant origin material in general and dietary fiber in particular. Through this, the consumers may attract towards such products thereby the one can increase the market to the above referred dairy products, accelerate the dairy industry physically monitory benefits also.

Keywords : Diet, Fiber, Milk, Products.

INTRODUCTION

Since the 1970s the importance of dietary fiber for human health has been ever increasingly acknowledged and investigated. While the term dietary fiber is familiar to most of the people, most do not clearly understand the nature of dietary fiber, its role in the diet even though and it is normal constituent of healthy food.

Dietary fiber is the group of food components that pass through the stomach and small intestine in undigested state and reach the large intestine virtually unchanged. Most of other nutrients are digested and are being used in other parts of the body by this stage. During its passage through the large intestine some components of dietary fiber are broken down by varying degrees and absorbed by the body. The remaining components are excreted in the fasces; however, fiber does play a very important role in human health. All plant based foods that we eat are full of fiber. Whole foods containing two essential forms of fiber viz., soluble and insoluble fiber. Both types of fiber are needed to maintain a healthy intestinal tract and promote healthy and regular bowl movements.

Table 1 : Role of Fiber in Human diet

Type of fiber	Source	Role in human diet
Insoluble fiber	Wheat bran, wheat bran products, whole grain products and vegetables viz., carrots, peas and the stain of fruits	• To help move the waste through the intestine • Helps to promote regularity in body • Prevents constipation
Soluble fiber	Dried peas, beans and some vegetables and fruits	• Helps to reduced cholesterol level • May helps to prevent the heart disease • Prevents the formations of abnormal protrusion in the bowels • May also protect against many types of cancer Stevens (2007)

Table 2 : Components of Dietary fibers

Component	Source
Cellulose	All food plants
Hemicelluloses	All food plants especially cereal bran
Pectin	Mainly fruits
Lignin	Mainly cereals and woody vegetables
Gums and some foods thickeners	Woody plants

DAILY REQUIREMENT OF DIETARY FIBER

According to WHO, the requirement of dietary fiber is 23-27 g/day/capita, while the National Institute of Nutrition it is 40-50 g/day/capita.

Most of the dietary fiber is consumed through the cereals, fruits and vegetables. Incorporation of plant origin materials in milk or milk products directly or indirectly adds dietary fiber to human food. So far the research on incorporation of plant origin material has been focused on value addition, taste and flavour development, thickening agents etc. This aspect is being reviewed in this paper from the point of its significance in nutrition and therapeutic value.

The considerable information in available on the use of cereals, fruits, vegetables and nuts in milk and milk products as documented in Table 3 and milk product considered containing of dietary fiber but studied with different purpose are documented in Table 4.

Table 3 : Availability of dietary fibers in some of the plants products.

Sr. No.	Name of plants products	Dietary fiber contain g/100 g of product
A.	**Cereals**	
	1. Rice	0.2
	2. Maida	0.3
	3. Wheat	71.2
B.	**Pulses**	
	1. Soybean	3
	2. Gram	3
C.	**Fruits**	
	1. Mango (Ripened)	16.2
	2. Coconut (dry)	3.6
	3. Cashew nut	6.6
	4. Guava	5.2
	5. Apple	1.0
	6. Banana	0.4
	7. Orange	0.5
	8. Papaya	0.8
	9. Fig (Ripened)	6.4
	10. Chiku	2.6
	11. Custard apple	3.1
	12. Wood apple	5.0
	13. Grape	2.9
	14. Strawberry	-
	15. Pomegranate	5.0
D.	**Vegetable**	
	1. Tomato	0.8
	2. Carrot	1.2
E.	Nuts	
	1. Almonds	1.7
	2. Cashew nut	1.3
	3. Pista	2.1

(Waghmare *et al.*, 2007)

Table 4 : The milk product considered containing of dietary fiber but consumed with a different purpose.

Sr. No.	Name of milk and milk product	Use of cereals, fruit, vegetable and nuts
A.	**Flavour milk**	Following flavour or fruit syrup used Strawberry, orange, lemon, pineapple, blackberry and Raspberry (De et al., 1996)
B.	**Milk shake**	1. Custard apple (Taware, 2002") 2. Apple 3. Carrot
C.	**Heat co-agulated**	
	1. Burfi	1. Fig 2. Wood apple (Sakate *et al.*, 2000) 3. Papaya 4. Sapota 5. Orange 6. Coconut (De et al., 1996)
	2. Gulabjamun	Maida, Sugi, Cardamum (Aneja et al., 2000)
	3. Gajar Ka Halwa	Carrot (Aneja et al., 2000)
	4. Kheer	Rice (De et al., 1976)Dry fruits
D.	**Acid co-agulated**	
	1. Panner	Blending with soya milk (Jadhawar, 2002
E.	**Fermented**	
	1. Youghurt	1. Guava 2. Pomegranate (Avatade, 2007)
	2. Lassi	1. Papaya 2. Pine apple 3. Pomegranate
	3. Shrikhand	1. Sapota (Jagtap, 1998) 2. Mango 3. Strawberry 4. Cocapowder
	4. Shriknad wadi	1. Coconut (Bagal et al., 2007)
	5. Kadhi	1. Basan 2. Karhi leaves (Aneja et al., 2002)
F.	**Ice Cream**	1. Mango (Gaikwad et al., 2007) 2. Custard apple 3. Strawberry 4. Dry fruits
G	**Whey bevreage**	
	Type of	Fruits flavour/pulp/juice used
	1. Cheese whey	Mango, Chiku, citrus, Guava, Pomegranate, pineapple etc.
	2. Chakka whey	
	3. Channa/paneer whey	

Table 5 : Estimated dietary fiber content in certain dairy products.

Sr. No.	Product	Dietary fiber content (g/kg) of product
1.	Milk shake (Custard apple)	3.1
2.	Burfi - Wood apple (20 %)	10.00
	- Fig (3 %)	0.9
3.	- Kaju burfi (Aneja *et al.*, 2000)	2.1
4.	Kheer - Rice (5 %)	0.1
5.	Payasam - Rice suji (50 %)	1.0
	- Wheat suji (40 %)	284.8
6.	Paneer - Soya milk (25 %)	7.5
7.	Shrikhand - Sapota (15 %)	3.9
8.	Shrikhand wadi - Coconut gratting (6 %)	3.6
9.	Whey beverage - Mango (1.8 %)	2.94
10.	Ice cream - Mango (10 %)	16.2
11.	Yoghurt - Papaya (10 %)	0.8
	- Mango (20 %)	16.2
12.	Kadhi – Besan (8 %)	2.4

Process have been developed for the incorporation of dietary fiber in traditional Indian dairy products and illustrated in the flow diagram given below:

(a) Custard apple milk shake (Taware, 2002)

Milk
↓
Filtration
↓
Pasteurization (63^0C for 30 min)
↓
Cooling to 7 ^{0}C
↓
Additional of fruit pulp (@ 10 andcane sugar (@ 7 % level)
↓
Mixing (90-120 sec.)
↓
Filling of bottles
↓
Cooling and storage ($7^0 \pm 0.5\ ^0$C)
↓
Custard apple milk shake

(b) Rice milk kheer/puddings (Choudhari *et al.*, 2007)

Milk
↓
Filtration
↓
Heating
↓
Concentration (25 % TS) reduction of original volume
↓
Addition of rice which is fried in ghee
↓
Boiling
↓
Addition of sugar @ 7 % of milk
↓
Cooking over flame in container till rice cooked in milk
↓
Ready to serve

(c) Payasam

Wheat suji (400 g)/ Rice suji (500 g)
↓
Roasted in ghee
↓
Soaked in water (5-10 min)
↓
Add coconut paste i.e. 200 g
↓
Add in milk i.e. 10 lit.
↓
Cooked
↓
Sugar added (600 g)
↓
Add cardamom saffron according taste
↓
Payasam

(d) Wood apple burfi (Sakate, 2000)

Milk
↓
Filtration
↓
Standardization (4 % fat)
↓
Continuous heating and stirring in Karahi
↓
Addition of sugar (@ 45 % for wood apple burfi)
↓
Concentrating the product to dough stage
↓
Broken into bits
↓
Addition of wood apple pulp(@ 30 per cent)
↓
Gentle heating to desired consistency
↓
Spreading the mixture in greased tray
↓
Cooling (10-12 hrs)
↓
Cutting into pieces
↓
Wood apple Burfi

(e) Cow milk + Soya milk paneer (Jadhavar, 2002)

Cow milk+ soy milk (75 % + 25 %)
↓
Heat the milk (82^0 C for 5 min)
↓
Cool (75^0 C)
↓
Add coagulant citric acid @ 2 %
↓
Hoop the curd
↓
Drain the whey by pressing the hoopedcurd for 15 min @ 3 kg/cm^2
↓
Dip the curd block (chilled water 1-2 hrs)
↓
Cut
↓
Packaging
↓
Storage (5)

(f) Fruit Yoghurt : Jha *et al.*. (2006)

Milk
↓
Standardization 3 % fat
↓
Add 4 % SMP, 7 % sugar and stabilizer
(0.4 % sodium alginate)
↓
Pasteurization at 90^0C for 5 min
↓
Cooled (38^0C)
↓
Innoculation
↓
Added pulp (10% Papaya or 20% mango pulp)
↓
Incubation $42^0 \pm 1^0$C 4-5 hrs
↓
Product is ready to serve

(g) Sapota shrikhand (Jagtap, 1998)

Dahi
↓
Drain out whey by hanging in
cloth bag
↓
Chakka
↓
Add sapota pulp 15
% of chakka
↓
Add sugar 40 %
↓
Sieve and knead to a
smooth paste
↓
Cool and store at 4-6^0C

(h) Coconut grating Shrikhand wadi (Bagal *et al*.., 2007)

Dahi

↓

Drainage of whey

↓

Chakka

↓

Addition grinded sugar 50 %,Cardamom
and coconut grating (6 %)

↓

Heating on open pan

↓

Shrikhand wadi

(i) Fruit Whey beverage

Whey

↓

Neutralization with $CaCO_3$ (0.75 %)

↓

Addition of sugar (14 %)

↓

Heating of whey 60^0C for 30 min

↓

Cooling at room temperature

↓

Addition of fruit juice/pulp (Mango, Orange,
Guava, Pomegranate etc.)

↓

Mixing

↓

Filtration

↓

Filling in pouches

↓

Packaging and sealing

↓

Cooling and storage at room temperature

(j) Kadhi : Aneja ***et al.*****. (2002)**

Butter milk

↓

Add 5-8 % besan

↓

Equal quantity of water added

↓

Stirring

↓

Mix is heating (10-15 min)

↓

Add appropriate quantity of turmeric powder,
spice, karhi leaves and salts

↓

Boil it upto total solids increase to 14-16%

↓

Serve it

(k) Fruit Ice cream : De ***et al.*****. (1994)**

Selection of ingredients

↓

Figuring the mix

↓

Making the mix

↓

Pasteurizing the mix (63^{0}C for 30 min)

↓

Homogenizing the mix (First stage 2500 psi)

↓

Coaling and aging the mix (0-4^{0}C)

↓

Freezing the mix (-4 to -5^{0}C)

↓

Packaging of ice cream

↓

Hardening and storage of ice cream (-23 to -29 ^{0}C)

CONCLUSIONS

It is evident from the literature reviewed the incorporation of materials of plant replete with dietary fiber which is not thought direct source of dietary fiber. Therefore, the consumption of milk and milk products prepared by incorporation of plant origin edible materials is not the out of the place of advice of consumption of fiber containing foods viz., fruits, vegetable, legumes cereals etc. because, nutritional and therapeutic benefits by the consumption of milk in isolation can be enhanced by the incorporation of materials of plant origin this is not thought of. There is need to educate the consumers with the aim to consume the milk and milk products containing plant origin material in general and dietary fiber in particular. Through this, the consumers may attract towards such products thereby the increasing the market potential of the above referred dairy products, which benefit the dairy industry also.

REFERENCES

Aneja, R.P., B.N. Mathur, R.C. Chandan and A.K. Banerjee. 2002. Technology of Indian Milk products. A Dairy India Publication Delhi.

Anonymous, 2008. www.google.com/dietary fiber.com.

Avtade, N.B. 2007. Utilization of lassi by using pomegranate juice. M.Sc. (Agri.) Thesis submitted to P.D.K.V., Akola.

Bagal, V.R. 2007. Preparation of shrikhand wadi blending with coconut grating from cow skim milk. M.Sc. (Agri.) Thesis submitted to P.D.K.V., Akola.

Choudhari, D.M. 2007. Suitability of Aromatic varieties of rice far preparation of kheer. M.Sc. (Agri.) Thesis submitted to P.D.K.V., Akola.

De, S. 1994. Outline of Dairy Technology. 12th Ed. Oxford University Press, New Delhi.

Gaikwad, N.B. 2007. Preparation of ice cream by using mango pulp. M.Sc. (Agri.) Thesis submitted to P.D.K.V., Akola.

Hapse, S.S. 2004. Preparation of chakka whey beverage using mango pulp. M.Sc. (Agri.) Thesis submitted to M.P.K.V., Rahuri.

Jadhavar, V.V. 2002. Studies on chemical and sensory quality of paneer prepared from cow milk and Soya mix milk. M.Sc. (Agri.) Thesis submitted to M.P.K.V., Rahuri.

Jagtap, M.M. 1998. Use of sapota pulp in shrikhand preparation. M.Sc. (Agri.) Thesis submitted to M.P.K.V., Rahuri.

Jha, Y.K. and C.S. Chopra. 2006. Studies on preparation of fruit yoghurt. Beverage and of Food World. 6: 67-69.

Sakate, R.J. 2000. Studies on preparation of wood apple burfi: M.Sc. (Agri.) Thesis submitted to M.P.K.V., Rahuri.

Stevens, 2007. How to better understand fiber. www.google.com/ dietary fibre.

Taware, M.G.2002. Studies on preparation of custard apple milk shake. M.Sc. (Agri.) Thesis submitted to M.P.K.V., Rahuri.

Waghmare. 2008. A text book of Poshan and Aahar, pp. 25-27

Bioresources for Food Security and Rural Livelihood, Pages 395–404
Edited by : G.K. Kulkarni and B.N. Pandey

ECOLOGICAL TRENDS AND ENVIRONMENTAL REQUIREMENTS OF SOME FISH SPECIES IN LOWER ZONE OF THE RIVER GANGES

Debabrata Das

ABSTRACT

Intensive water samples are taken from the four sampling centers during the period of four years (2002-2006) to develop a water quality database of the lower part of the river Ganges for fisheries applications. In this communication ranges of water quality parameters and ecological trends are derived from such water quality database. Certain common fish species are also identified for all the centers in same river stretch. The requirements of suitable ranges of physicochemical parameters are measured for such common fish species available in lower part of the holy river. The suitable species identified for the measured ranges of water qualities are namely *M rosenbergii, A. aor R. rita, E. vacha, R. corsula ,G. guris, W attu, L. rohita, L. bata, C. cat/a, C. mrigala, A. coila, C. garua, N (Chitala) chitala, N notopterus, P. paradis ius, M vittatus, P. saphore, P pama.* The cited species are need to be enhanced naturally or through establishment of hatcheries.

Keywords: Environmental requirement, fish database, river Ganges, ecological trends.

INTRODUCTION

The river Ganges has many fisheries habitats. Fish species available in the certain river stretch may not survive in entire river or other rivers due to requirement of their specific physico-chemical reaction and physiological needs. Different fish species have different physico-chemical needs. In certain stretch of river Ganges, a unique range of physico-chemical requirements can be measured. In present paper forty-nine water quality parameters including microbial loads and pesticide dissolved in river water and their range are measured. A common of nineteen fish species are found in river stretch (Berhampur to Uluberia) covering four sampling centres (Berhampur, Palta, Dakshineswar and Uluberia) which is derived from the four point databases of fisheries of individual sampling centres. Water quality parameter takes an important role in inland fishery. It has been observed that physicochemical factors affecting fish densities (Sabo et. al, 1991) and the abundance and species richness of the fishes (Reash and Jimmie, 1990) since physiological and physicochemical factors associated with fish maturation and spawning (Donaldson, 1975). It has been reported that analyses of pond water and mud samples show that nitrifying bacteria (including ammonifying bacteria, nitrite bacteria, nitrobacteria and denitrifying bacteria) are in general closely correlated with various physico-chemical factors (Jun et. al, 2000). In case of inland fishery, all the fish species survive under a certain range of physico-chemical parameters. Beyond the certain ranges of those water quality parameters fishes may survive under stress. Under extreme values of physico-chemical parameters fish species may not survive thus species become endanger. Disturbance, stressful environments, and the ability of fish to cope with physicochemical challenges all affect local fish assemblage composition or dynamics (Matthews, 1998) and hence it is important to know the suitable physico- chemical values and their range under which species survive without stress. Number of fish species identified within an entire river stretch is critical and this can be easily observed maintaining a good database as databases are used for inland fisheries enhancement (Anon., 1999). On the other hand range of physico-chemical parameters and their values can be obtained using any spread sheet or a RDBMS package over the same period of time. Usually the time period of three to four years may be ideal.

Monthly sampling of forty-nine water quality parameters from four sampling centres namely Berhampur, Palta, Dakshineswar and Uluberia over the period of four years (2002 to 2006) are kept in database. A total of 382 records are available and statistically the range of their values are identified for each forty-nine water quality parameters including microbial loads and pesticides dissolved in the river water. Notification of fish species each of four sampling canters are kept in separate databases, these are point databases of four locations. The common species of four point databases can be obtained using query command like intersect. Development database based upon the catch observation of fish species over the four years of study. Nineteen common fish species i.e. species available in all the four centres are identified. Database of water quality can be normalized in three different databases like physicochemical parameters, microbial parameters, heavy metals and pesticide parameters and their range and mean values are obtained from the database files during the course of study.

ECOLOGICAL TRENDS

Ecological trends (Fig 1 to Fig. 11) plotting the monthly mean values during the four years of study are identified and presented in this research communication. In all the figures month 1 indicates January, month 2 denotes February and so on.

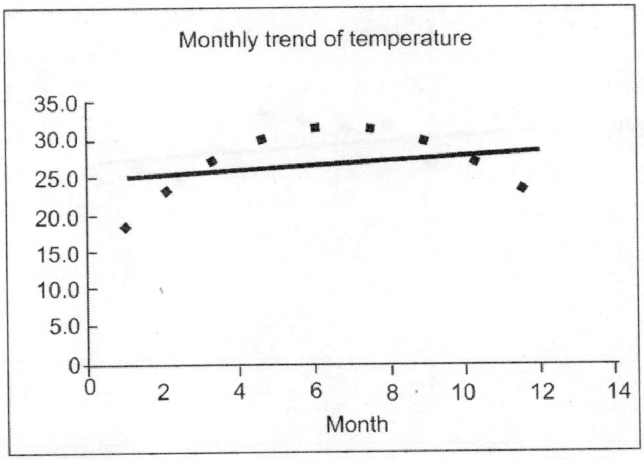

Fig. 1 : Water temperature

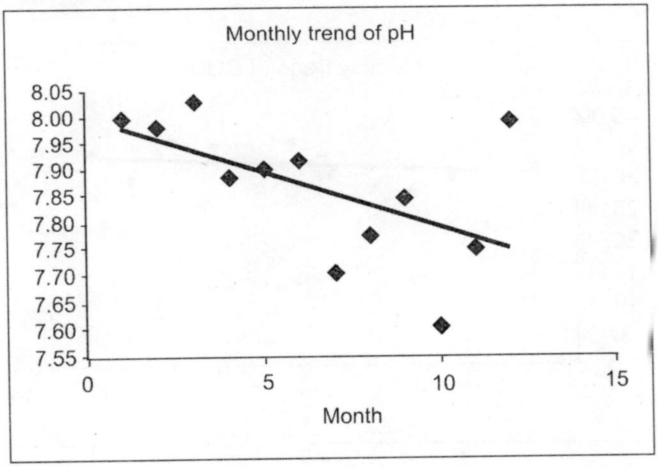

Fig. 2 : Water pH

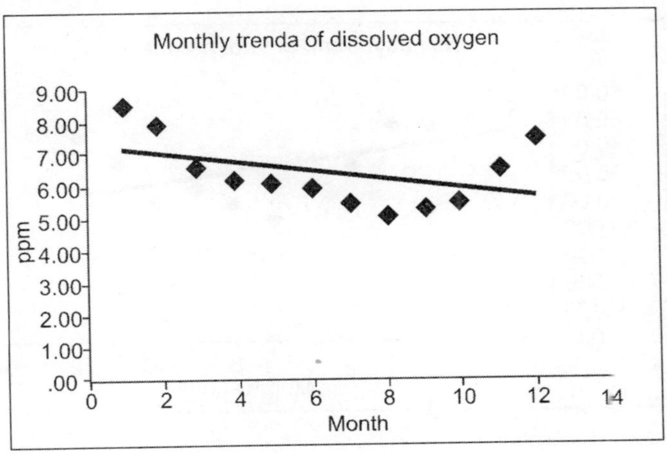

Fig. 3 : Dissolved oxygen

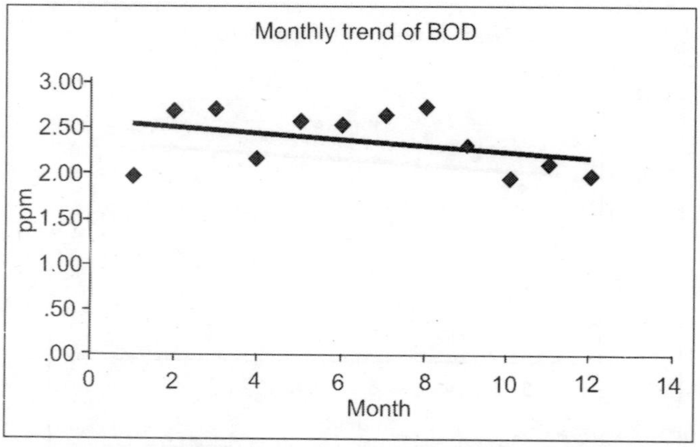

Fig. 4 : BOD

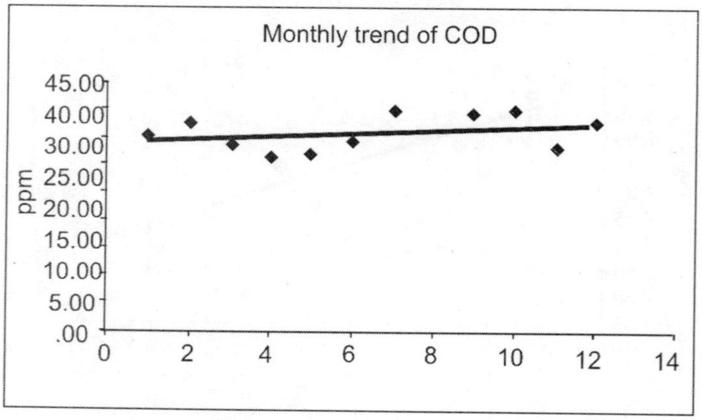

Fig. 5 : COD

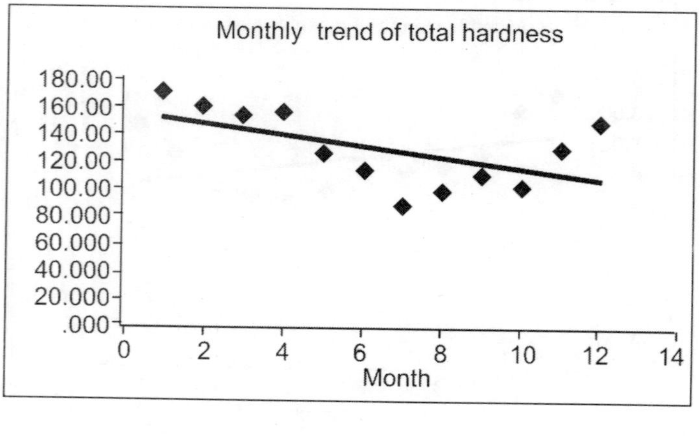

Fig. 6 : Total hardness

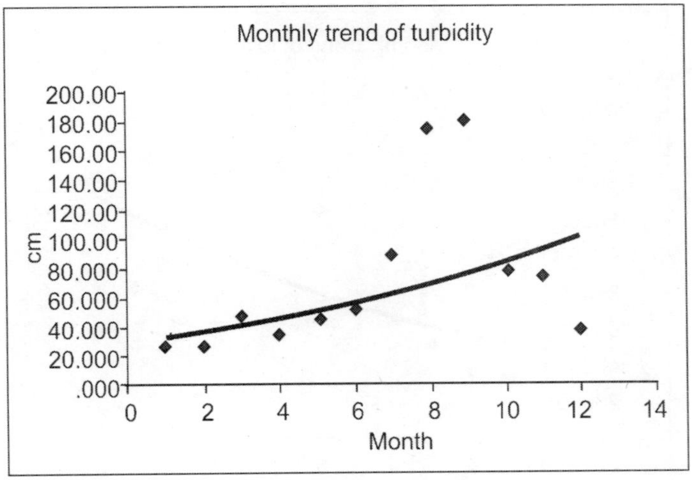

Fig. 7 : Turbidity

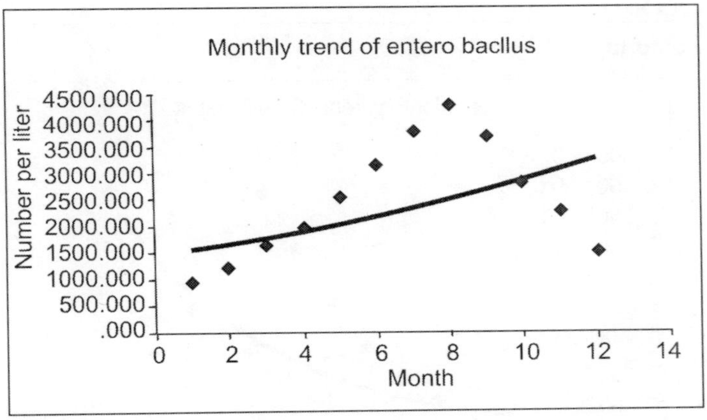

Fig. 8 : Entero bacillus

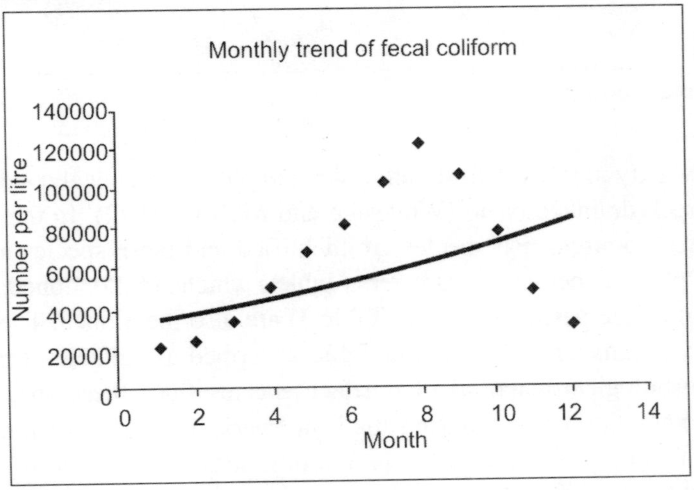

Fig. 9 : Fecal coliform

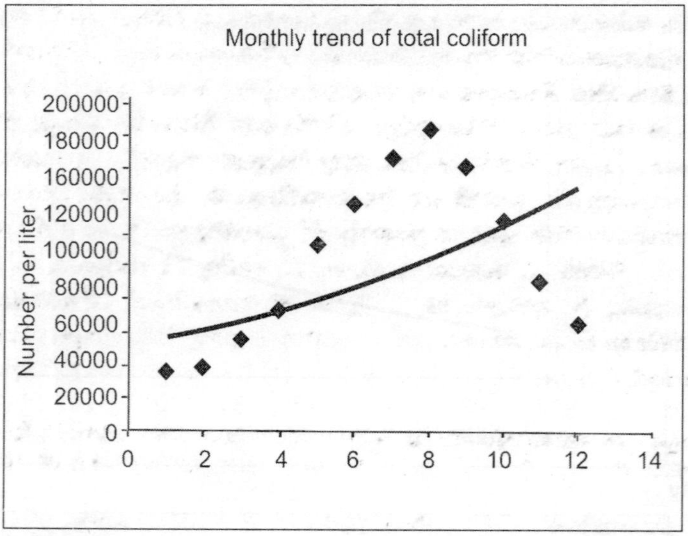

Fig. 10 : Total coliform

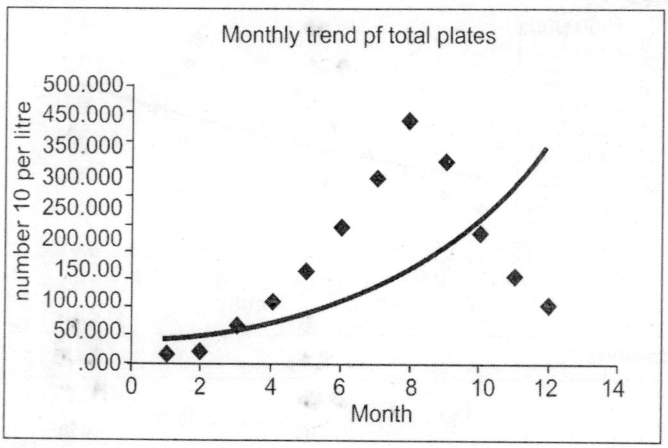

Fig. 11 : Total plate count

Fish species diversity in relation to stream order and physicochemical condition is observed in the Plum Creek drainage basin (Whiteside and McNatt, 1972). In the present research article a nineteen common fish species are identified and these species are ideally suited in the range of physicochemical parameters (Tablet) which are mentioned. Microbial loads (Table 2) and tolerable pesticide ranges (Table 3) are also mentioned. The value of ranges of water quality parameters (Table 1 and Table 2) termed as suitable, if not optimum, for growth and physiological reactions of nineteen species since these species are available in all the four sampling centres during the four years. The value of range of pesticides and heavy metals (Table 3) may be the permissible limit for survival of nineteen species since no mortality of these fish species are being reported during the course of study. In an earlier works relationship between physicochemical factors and abundance of fishes in tributary confluences of the lower Channelized Missouri River (Braaten and Guy, 1999) was attempted. It is also observed that maintaining the water quality in a steady, optimum state for fishes is important (Woiwodge, 1996) and when the range of physiochemical parameters become more water bodies may become stressful to some fishes and less number of species ideally suited for the condition in the entire stretch. Present paper describes the nineteen fish species namely *M rosenbergii A. aor R. rita, E. vacha, R. corsula,* G. *guris, Wattu, 1. rohita, 1. bata,* C. *catla,* C. *mrigala, A. coila,* C. *garua, N. (Chitala) chitala, N. notopterus, T ilisha, P. paradisius, M vittatus, P. saphore, P pama* are available in all the four sampling centres namely Berhampur, Palta, Dakshineswar and Uluberiea and their suitable water quality ranges are derived (Table 1 and Table 2).

Table 1 : Ranges of water quality parameters in the river stretch from Berhampur to Uluberia of river Ganga during the four year (2002-2006) of study.

Water quality parameters	Min (ppm)	Max (ppm)	Average (ppm)
Water Temperature(°C)	16	33.5	27.167
Ph	7.06	8.7	7.865
DO	2.75	10.5	6.472
BOD	0.6	6.6	2.389
COD	12	76	36.316
Turbidity	4.3	654	72.991
Sulphate	6.45	78.9	18.799
Conductivity	0.15	0.66	0.339
Chloride	1.25	51.23	13.830
Na	0	112.97	14.473
Ca	2.4	96.06	26.345
Mg	0.98	46.83	20.220
Alkalinity	50	196	126.587
Total hardness	60	254	131.872
Phosphate	0.02	2.86	0.331
Ammonium-nitrogen	0.004	2.49	0.312
Nitrate-nitrogen	0.005	2.22	0.425
Total-nitrogen	0.27	20.12	3.103
Fe	0	6.20	1.361
Mn	0	2.72	0.288
Zn	0	0.37	0.071

Table 2: Ranges of microbial loads in the lower parts of river Ganges (in number per litre of water

Microbial loads	Minimum	Maximum	Average
Total coliform	1200	400000	104711
Fecal coliform	500	300000	64808
Total plates	$0.2x10^9$	800x10:1	$185x10^9$
Entero bacillus	200	9000	2479

Table 3 : Ranges of heavy metals and pesticides in the river stretch Berhampur to Uluberia of river Ganga during the four year (2002-2006)

Heavy metals/Pesticides	Min (ppm)	Max (ppm)	Average (ppm)
Cu	0	0.11	0.011
Cd	0	0.04	0.004
Cr	0	1.71	0.053
Ni	0	0.375	0.049
Pb	0	1.35	0.098
HCH1	0	1.14	0.019
HCH2	0	18.65	0.959
HCH3	0	1.86	0.045
HCH4	0	1.69	0.056
Total-HCH	0	18.65	1.079
pp-DDE	0	2.09	0.045
Op-DDD	0	3.69	0.046
Op-DDT	0	0.48	0.008
pp-DDT	0	0.45	0.006
Total DDT	0	3.69	0.105
Endo-1	0	2.23	0.027
Endo-2	0	1.52	0.014
Endo-S	0	0.40	0.006
Endo-T	0	2.23	0.047
Total Organo-chlorine	0	18.78	1.231
Dimethoid	0	3.132	0.082
Methyl parathion	0	3.05	0.219
Malathion	0	4.83	0.369
Total Organo-phosphorus	0	6.28	0.671

CONCLUSIONS

Ecological trends of temperature, dissolved oxygen, biochemical oxygen demand (BOD), chemical oxygen demand (COD), pH, turbidity, alkalinity, microbial loads like fecal coliform, total coliform, entero bacillus, total plates are identified. These are having some definite monthly trends year the round. Fish species which are most suitable in the entire river stretch of Berhampur to Uluberia of the river Ganga are mentioned in this communication. Requirement of the values of physico-chemical and physiological parameters for those species in terms of ranges are measured. The value of ranges of water quality parameters shown in the Table I and Table 2 are concluded as suitable for those species mentioned in the article. The reason is that river is open water system and the value of ranges of physico-chemical parameters in certain river stretch if not suitable for those fish species, they will migrate to upstream or downstream to find their more suitable physicochemical condition. Another aspect of this communication is to development of fish species database in certain river stretch like Berhampur to Uluberia. Database on water qualities of the river Ganga in the entire stretch helps to find out the seasonalities and extremities of those parameters. The practical importance of the communication may be to set up culture based capture fisheries development programme like hatcheries in the adjoining areas for the mentioned species or to conserve them.

ACKNOWLEDGEMENT

The author is grateful to Dr. K K Vass, the Director, Central Inland Fisheries Research Institute Barrrckpore for encouragement and Dr. R K Kolev. BCKV. Mohanpur.

REFERENCES

Anon. *FAO Fisheries Circulars,* FIRI/C945(945), 1, 1999.

Braaten, P.J and Guy, C.S. *American Fisheries Society, Transactions,* **128(6):** 1213, 1999.

Donaldson, E.M. *Food and Agriculture Organization of the United Nations, European Inland Fisheries Advisory Commission, Technical Papers,* Issue No. 25, 53, 1975.

Jun, X. and Xiuzheng, F. *NAG A: I C LA R M Quarterly,* **23(4):** 16, 2000.

Matthews, W. J. *Freshwater Fish Ecology,* 318, 1998.

Reash, Rob J. and Jimmie Pigg. *Oklahoma Academy of Science, Proceedings,* 70, 23, 1990.

Sabo, Matthew J., William E. Kelso, C. Frederick Bryan and D. Allen Rutherford. *Regulated Rivers: Research and Management.* **6(2):** 109. 1991.

Whiteside, B. G. and McNatt, R. M. *American Midland Naturalist,* **88(1):** 90,1972

Woiwode, lG. *The Role of Aquaculture in World Fisheries.* **6:** 214, 1996.